Mechanisms and Machine Science

Volume 37

Series editor

Marco Ceccarelli, Cassino, Italy

More information about this series at http://www.springer.com/series/8779

Saïd Zeghloul · Med Amine Laribi
Jean-Pierre Gazeau
Editors

Robotics and Mechatronics

Proceedings of the 4th IFToMM International Symposium on Robotics and Mechatronics

Springer

Editors
Saïd Zeghloul
Institut PPRIME, UPR 3346
University of Poitiers
Poitiers
France

Jean-Pierre Gazeau
Institut PPRIME, UPR 3346
University of Poitiers
Poitiers
France

Med Amine Laribi
Institut PPRIME, UPR 3346
University of Poitiers
Poitiers
France

ISSN 2211-0984 ISSN 2211-0992 (electronic)
Mechanisms and Machine Science
ISBN 978-3-319-22367-4 ISBN 978-3-319-22368-1 (eBook)
DOI 10.1007/978-3-319-22368-1

Library of Congress Control Number: 2015946758

Springer Cham Heidelberg New York Dordrecht London

Printed on acid-free paper

Springer International Publishing AG Switzerland is part of Springer Science+Business Media
(www.springer.com)

Preface

ISRM 2015, IFToMM International Symposium on Robotics and Mechatronics, is the fourth event of a series that started in 2009 as a specific conference activity on robotics and mechatronics. The first event was held at the Hanoi University of Science and Technology, Vietnam in September 2009, the second was held at Shanghai Jiao Tong University, Shanghai, China in November 2011, and the third was held at the Nanyang Technological University, Singapore in October 2013.

The aim of the ISRM symposium is to bring together researchers, industrialists, and students involved in a broad range of disciplines related to Robotics and Mechatronics, in an intimate and stimulating environment in order to disseminate their results and to exchange about future works, trends, and challenges.

ISRM 2015 received more than 40 papers, and after careful review with at least two reviews for each paper, 31 papers were considered suitable for publication in this book and were presented in the conference. The oral presentations were organized into a 2-day conference with seven technical sessions held from 24 to 25 June in University of Poitiers, France.

The ISRM 2015 proceeding presents state-of-art research findings in robotics and authored mainly from the IFToMM community from China, France, Greece, Italy, Kazakhstan, Mexico, Morocco, Russia, Singapore, Spain, Taiwan, Tunisia, and United States of America. Major topics of the papers are related with robotics and mechatronics, including but not limited to: mechanism design, modeling and simulation, kinematics and dynamics of multibody systems, control methods, navigation and motion planning, sensors and actuators, bio-robotics, micro/nano-robotics, complex robotic systems, walking machines, humanoids, parallel kinematic structures: analysis and synthesis, smart devices, new design, application, and prototypes.

In conjunction with ISRM 2015, a technical day "Open and Collaborative Robotics" was held. The aim was dedicated to highlight developments and research advances in industrial robotics and automation, covered by industry professionals, professors, and researchers working on collaborative robotics issues. This technical day was co-organized by B&R Automation and Pprime institute of the University of Poitiers on Tuesday 23 June at the Futuroscope campus in Poitiers

We would like to express grateful thanks to the members of the current International Scientific Committee for ISRM Symposium for cooperating enthusiastically for the success of the 2015 event:

I-Ming Chen (Singapore) as Chair of the IFToMM Technical Committee on Robotics and Mechatronics

Marco Ceccarelli (Italy)
Feng Gao (China-Beijing)
Manfred Hiller (Germany)
Qiang Huang (China-Beijing)
Shuo-Hung Chang (China-Taipei)
Nguyen Phong Dien (Vietnam)
Lotfi Romdhane (Tunisia)
Yukio Takeda (Japan)
Min-June Tsai (China-Taipei)
Nguyen Van Khang (Vietnam)
Saïd Zeghloul (France)
Teresa Zielinska (Poland)

We thank the authors who have contributed with very interesting papers on several subjects, covering many fields of Robotics and Mechatronics and additionally for their cooperation in revising papers in a short time in agreement with reviewers' comments. We are grateful to the reviewers for the time and efforts they spent in evaluating the papers with a tight schedule that has permitted the publication of this proceedings volume in time for the symposium.

We thank University of Poitiers, in particular, the Fundamental and Applied Science Faculty, for having hosted the ISRM 2015 event.

We also thank the support of International Federation for the Promotion of Mechanism and Machine Science (IFToMM). The symposium received generous support from local sponsors, namely the University of Poitiers, The Grand Poitiers, The Poitou-Charente region, and the industrial partner B&R Automation, which were critical to make this symposium of low registration cost possible.

We thank the publisher Springer and its Editorial staff for accepting and helping in the publication of this Proceedings volume within the book series on Mechanism and Machine Science (MMS).

Poitiers
June 2015

Saïd Zeghloul
Med Amine Laribi
Jean-Pierre Gazeau

Contents

Part VII Advanced Robotics

Part I
Mechanism and Advanced Mechanical Design

A Study of Structural Stress Analysis of Reducers for Supporting Reliability Design

Yuo-Tern Tsai, Kuan-Hong Lin and Kuo-Shong Wang

Abstract Reducer is extensively used in many devices for speed reduction of rotational motions. It must possess the properties including high rigidness, large reduction ratio and structure compactness when it is applied in robots. This paper presents an innovative design for reducers based on differential displacements of deceleration gears for satisfying the above properties. The structural stresses including vibration frequencies of the reducer are analyzed using Finite Element Methods (**FEM**) to observe the designed weaknesses of the related components. The kinematic characteristics of the reducer are simulated by SolidWorks motion analysis to identify the accuracy of the mechanisms. The geometric models are leaded into ANSYS to analyze the structural stresses and the vibrations. The analyzed results are further integrated with probabilistic theories to perform reliability design for the reducer. The studied results can provide useful information including the allowed loading and the reliabilities for the high speed-reduction reducer in using.

Keywords Reducer · Reliability · Stress and vibration

Y.-T. Tsai (✉)
Department of Mechanical Engineering, De-Lin Institute of Technology,
New Taipei City, 236, Taiwan, R.O.C
e-mail: yttsai@dlit.edu.tw

K.-H. Lin
Department of Mechanical Engineering, Tungnan University, New Taipei City, 222,
Taiwan, R.O.C
e-mail: khlin@mail.tnu.edu.tw

K.-S. Wang
Department of Mechanical Engineering, National Central University,
Chungli, Taiwan, R.O.C
e-mail: kswang@cc.ncu.edu.tw

S. Zeghloul et al. (eds.), *Robotics and Mechatronics*,
Mechanisms and Machine Science 37, DOI 10.1007/978-3-319-22368-1_1

1 Introduction

Robots play a critical role on automatic production and are more and more common in industrial applications. The movements of robots usually require positioning accuracy, geometric agility and wide ranges of speed and torque changing to meet the needs of different working conditions [1]. The reducer (speed-reduced mechanism) is a key component of robots which is put on the joint axis for reducing the rotational speeds of motors. It must own the properties of high reduction ratio and small volume due to the function requirements and the geometric constraints of robots. The harmonic gear decelerators were commonly used in many robots in the past because it owned the properties required by the robots. The website introduces the information of the harmonic drivers [2]. The existed problems of harmonic gear decelerator are the transmission torques limited due to the flexible design of the transmitted mechanisms [3] although it possessed high reduction ratio and structural compactness. To satisfactory the properties of high rigidness, large reduction ratio and structure compactness, an innovated design named as Differential Reducer (**DR**) is presented in this paper.

The kinematic characters and the stresses of the **DR** in using are analyzed for obtaining high reliability products. The business software, Computer Aided Design/Engineering (**CAD/CAE**), is applied to perform the activities of designing and analyzing for the **DR**. The similar studies integrating **CAD/CAE** to develop high reliability products had been presented by the authors [4]. It integrated probabilistic design methods with **CAD/CAE** to perform reliability design optimization for the arms of robots. Reliability design was developed mainly based on Stress-Strength Interference (**SSI**) theory. The failed positions of a design can be established by failure mode effects analysis and/or fault tree analysis. No sooner are the critical points identified, than the design variables related to the critical points (such as forces, materials, geometric dimensions, etc.) can be decided subjected to the expected reliabilities [5].

ANSYS is a Finite Element Analysis (**FEA**) tool which is extensively used in supporting engineering design. The **CAD** system generates a solid model of a design. The geometry is transferred or the component is meshed within the **CAD** system and the finite element model is transferred to **ANSYS** [6]. It is sometimes slightly infeasible to obtain an **ANSYS** solution due to sometimes slightly violating the constraints. To improve the efficiency of optimization, the response surface method can be used to replace the time-consuming finite element simulations. A practical application of the **ANSYS** simulations is to evaluating the fatigue life of dental implants for which there are only limited test data [7]. The frequency response of system is one of the most important features in forced vibration. It is very harmful if resonance occurs in a mechanical system. The purpose of vibration analysis is to predict the frequency of resonance occurrence and then to determine what steps would be taken to prevent it from occurring [8].

In this paper, a Differential Reducer (**DR**) based on differential displacements is presented to obtain the properties of structural rigidness, high reduction ratios and compact design. The geometric structures are designed by SolidWorks software including motion analysis to ensure the accuracy of designing as well as avoiding interference occurred. The **ANSYS** software is used to analyze the structural stresses and the frequency responses of the **DR** for obtaining the allowed loading and to prevent resonance from occurrence. The analyzed results are further combined with reliability theory to perform reliability design for the reducer subjected to the given reliability needs.

2 Reliability Calculation

Reliability calculated is formulated based on **SSI** theory. The reliability index can be calculated based on the second-order moment method if the strength random variable X is normally distributed with a mean value of μ_X and standard deviation of σ_X, and the stress random variable Y is also normally distributed with parameters μ_Y and σ_Y. The geometric distance for reliability measurement can be rated by

$$z = \frac{\mu_X - \mu_Y}{\sqrt{\sigma_X^2 + \sigma_Y^2}} \tag{1}$$

The value of z would be the reliability index and the reliability can be computed as

$$R = 1 - \Phi(-z) = \Phi(z) \tag{2}$$

A typical target value for reliability index z = 3 is the reliability R = 0.99865.

In general conditions, the reliability for mechanical devices is evaluated from the strength and the stress. The induced stress (μ_Y, σ_Y) is the designed variants if the strength information of materials (μ_X, σ_X) is known. When the coefficients of variation of the designed variants γ_Y are known, the standard deviation of the induced stress can be evaluated by

$$\sigma_Y = \gamma_Y \mu_Y. \tag{3}$$

Substituting Eq. (3) into Eq. (1), we can derive the following equation when the reliability target is set to R ≥ Φ(z):

$$\left[1 - (\gamma_Y z)^2\right] \mu_Y^2 - 2\mu_X \mu_Y + \mu_X^2 - (\sigma_X z)^2 \geq 0. \tag{4}$$

Solving the above equation, the design solution for the induced stress must satisfy the condition

$$\mu_Y \leq \frac{\mu_X - \sqrt{\mu_X^2 - (1 - \gamma_Y^2 z^2)(\mu_X^2 - \sigma_X^2 z^2)}}{1 - \gamma_Y^2 z^2}. \tag{5}$$

The variant coefficients γ_Y can be decided from the means and the variations of the variables in the stress equation. For example, for (F, A), representing the force and the area of a design, respectively, the mean of the stress will be $\mu_Y = \mu_F/\mu_A$. The variation coefficient will be $\gamma_Y^2 = \gamma_F^2 + \gamma_A^2$. No sooner has the γ_Y been decided, than the allowed value of the induced stress can be computed by Eq. (5), when the strength information and the needed reliability are given. This form is commonly adopted when the performance index belongs to the-smaller-the-better.

On the other hand, if the known information is the expected performance given directly by the customers (the requirement marked as Y), the designed variant will be the real performance of the design (the supply marked as X). Likely, the following equation is derived from Eq. (1):

$$\left[1 - (\gamma_X z)^2\right]\mu_X^2 - 2\mu_Y\mu_X + \mu_Y^2 - (\sigma_Y z)^2 \geq 0. \tag{6}$$

The design solution for the strength must satisfy

$$\mu_X \geq \frac{\mu_Y + \sqrt{\mu_Y^2 - (1 - \gamma_X^2 z^2)(\mu_Y^2 - \sigma_Y^2 z^2)}}{1 - \gamma_X^2 z^2}. \tag{7}$$

Here, γ_X represents the same behavior as γ_Y. The performance of design supply μ_X must satisfy the above equation. This form is usually used when the performance index is the-bigger-the-better.

3 Frequency Response Function (FRF)

The harmonic response is analyzed for the **DR** to find the natural frequency and the frequency response. The analyzed results can show the range of the frequency that is suitable for the **DR** as well as to prevent resonance from occurrence.

The vibration of mechanisms can be described by a single coordinate, the natural frequency depends on two system properties; mass and stiffness. The circular natural frequency, ω_n, can be found using the following equation:

$$\omega_n^2 = \frac{k}{m} \tag{8}$$

where:

k = stiffness of the beam (Newton/Meters or N/m)
m = mass of weight (kg)
ω_n = circular natural frequency (radians per second)

From the circular frequency, the natural frequency, f_n, can be found by simply dividing ω_n by 2π. The natural frequency can be formulated as:

$$f_n = \frac{1}{2\pi}\sqrt{\frac{k}{m}} \tag{9}$$

where: f_n = natural frequency in hertz (1/seconds)

Aiming to the forced vibration, the behavior of the spring mass damper model needs to add a harmonic force. A force of this type could be generated by a rotating imbalance.

$$F = F_0 \cos(\omega t) \tag{10}$$

For a single degree of freedom oscillator, the sum the forces on the mass are calculate using following ordinary differential equation:

$$m\ddot{x} + c\dot{x} + kx = F_0 \cos(\omega t) \tag{11}$$

The result states that the mass will oscillate at the same frequency, f, of the applied force, but with a phase shift φ. The scale of the vibration amplitude to the harmonic force is defined by the following formula.

$$\frac{X}{F_0} = \frac{1}{k}\left[\frac{\omega^2}{\sqrt{(\omega_n^2 - \omega^2)^2 + (2\zeta\omega\omega_n)^2}}\right] \tag{12}$$

where ζ is the damping factor which is defined as $c/m = 2\zeta\omega_n$, The phase shift, φ, would be

$$\phi(\omega) = \tan^{-1}\left(\frac{2\zeta\omega_n}{\omega_n^2 - \omega^2}\right) \tag{13}$$

An efficient method to reduce the magnitude of the vibration is by adding the damping. On other hand, the magnitude can also be reduced if the natural frequency can be shifted away from the forcing frequency by changing the stiffness or mass of the system. The other approach is to change the frequency of the harmonic force.

4 Design of Differential Reducer (DR)

The **DR** is designed using the principles of differential displacements of the reduction gears which are similar to the moving and aligning of the fixed and the slid meters in the traditional caliper. The reduction gears are designed as gear rings which is driven by an off-centre cam. The geometric structure of the **DR** is shown in Fig. 1. The outer

Fig. 1 Geometric structure of the differential mechanism

gear rings have two the fixed and the slid ones which are different in tooth numbers but having the same module of tooth shape. The fixed ring is fixed on the support box for engaging and leading the intermediary ring generating the planetary motions. The slid ring is forced to generate differential movements while the intermediary ring is rotating circling the fixed ring. The differential movements are produced when the teeth of the slid ring is to align to the teeth of the fixed ring. The intermediary ring is driven by an off-centre cam which is connecting to the power shaft. The power output is connected to the slid ring. The scales of reduction ratios are decided depending on the difference of the tooth numbers of the two outer gear rings.

The mechanism is designed using SolidWorks software. The gear rings are design with module m = 1, pressure angle θ = 20° for engaging transmission. The tooth numbers and widths for the gear rings are set to {100, 12}, {101, 12} and {80, 24} mm for the fixed, slid and intermediary rings, respectively. The off-centre cam was designed to deviate 10 mm from the rotational center. These components were assembled as a whole to simulate the kinematic behaviors. The slid ring would be shifted to generate small displacements when the intermediary ring rotates circling the fixed ring. Figure 2 shows the kinematic simulation for the outer gear and

Fig. 2 Kinematic simulation for the gear rings

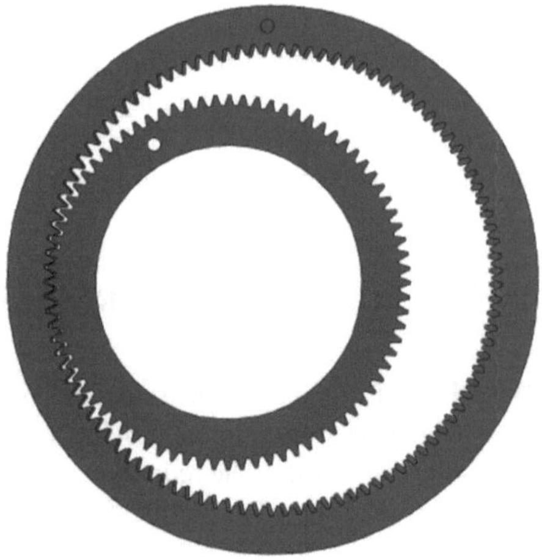

intermediary rings. The simulated results showed that the **DR** can run smoothly and no interference occurred.

5 Analysis of Differential Reducer

The structural stresses of the **DR** are analyzed to perform reliability design. Reviewing the motion transmission of the mechanism, the designed weaknesses may be at the teeth of the gear rings destroyed either wear-out or broken-down when the transmitted loading over the designed strength.

(1) Structural stress

The structural design was leaded to ANSYS to analyze the induced stresses. The connection of the paired gear rings are set to no separation. The outer ring is set to fixed support as well as the inner ring to frictionless support. The loading is supposed to transmitting 200 N-m by the intermediary ring.

The analyzed results for structural stresses and deformation are shown in Fig. 3. The results show that the maximum stress is 103 MPa occurring at the position of the teeth engaged and the maximum deformation occurs at the other side of the teeth no being contacted (see Fig. 3b). The contact pressures of the engaged teeth can also be obtained by contact tool. The maximum contact pressure for the engaged teeth is 1409 MPa.

The analyzed results of structural stress can be applied to set the allowed loading of the **DR** as well as to determine the design variables for the given transmitted torques. For example, the allowed torque of transmission for the **DR** would be 485 N-m if the material strength is 250 MPa.

(2) Vibration Analysis

The frequency response of the **DR** are further investigated for determining the suitable speeds of the **DR** in running. The geometric shapes of the support box of

(a) (b)

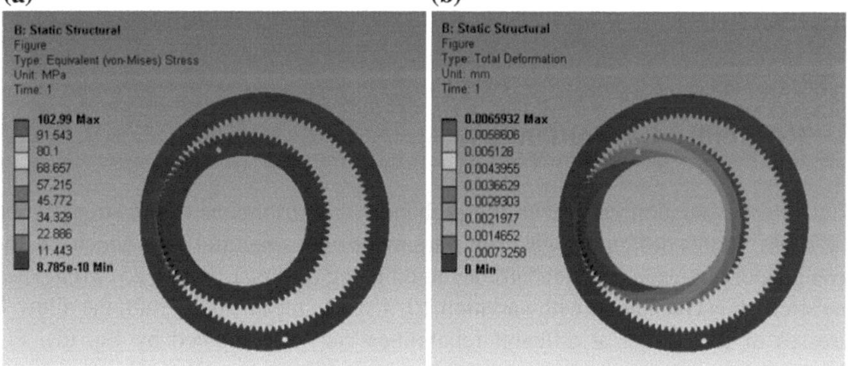

Fig. 3 The structural stresses and the total deformations, **a** Equivalent stress, **b** Total deformations

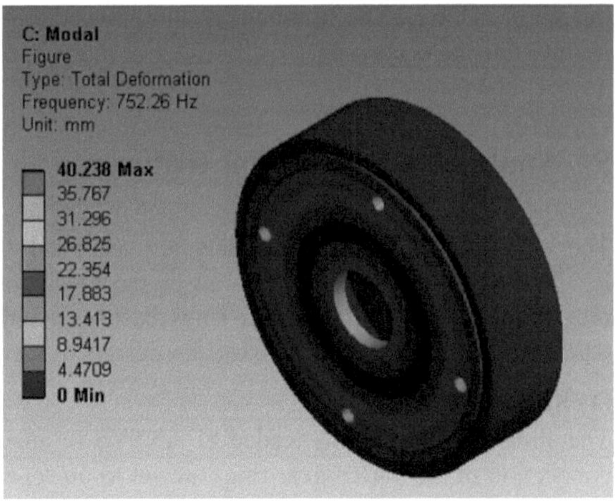

Fig. 4 The vibration shape for mode 1

the **DR** are constructed according to the practical design. The gear rings are enclosed inside since it is fixed on the support box. Here, the gear rings are replaced with the round rings which have the same sizes for simplifying the analyzed models and shorting the calculated time of computer simulation. The modal analysis is first done. The analyzed results for the front 6 modes are {752, 4395, 4418, 6125, 7105, 7408} Hz.

The vibration shape of the first order mode is at the rotational direction as shown in Fig. 4. The natural frequency of the structure is 752 Hz. The frequency implies that the **DR** occurs resonance being low because the rotational speed of the power shaft usually is far smaller than the frequency.

The frequency response of the design is further analyzed to observe the stresses of vibration. The harmonic forces of the support box are induced by the cam shaft when it drives the intermediary ring generating planetary motion circling the fixed ring. The harmonic forces can be obtained by dividing the transmitted torque to the rotational radius. Here, it is set to 2000 N as bearing loads. The analyzed results are shown in Fig. 5 where the maximum stress is 32.4 MP occurring at the natural frequency.

6 Reliability Evaluation

Reliability evaluation deals with the probabilistic distributions of the strength and the stress. In this design, the gear rings are made with structural steel which has the strengths of the mean and standard deviation being $(\mu_X, \sigma_X) = (250, 25)$ MPa when the strength is considered with variation 10 % of the mean. As a result, the allowed stresses of the design at different reliabilities can be computed by Eq. (5). For

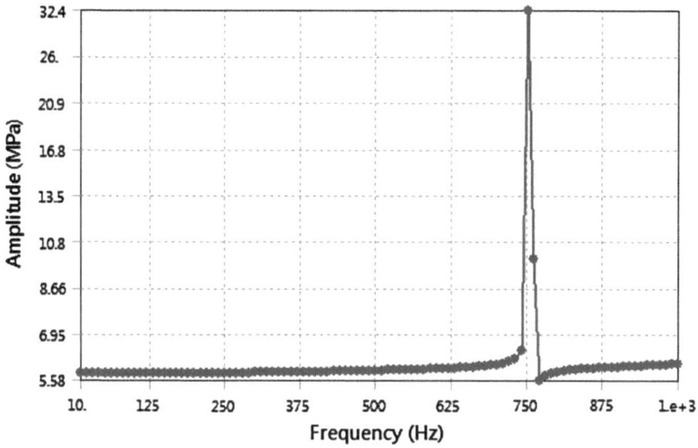

Fig. 5 Frequency response of x-axis for normal stress

Fig. 6 Changing of the maximum stress and the contact pressure

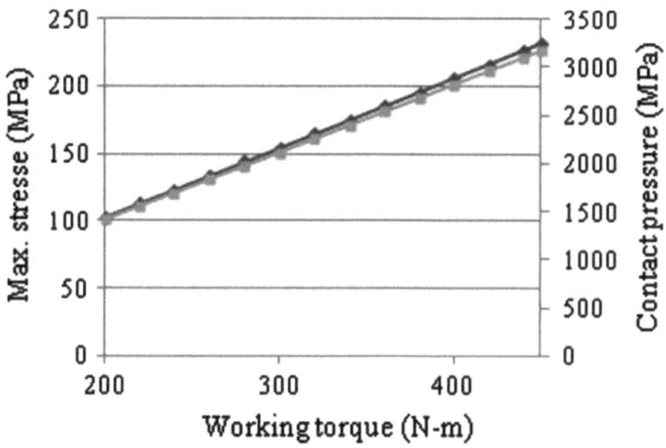

example, the allowed stresses of the design would be $\mu_Y = 209$ Mpa if the reliability need is set to R = 0.95. The allowed stresses would be 198 MPa if the stress variation is considered as the same as the strength. According to the analyzed results, the induced maximum stresses and contact pressures for the gear rings in terms of transmitted torques are shown in Fig. 6, The changing of stresses and contact pressures depending on the torques is linear.

Furthermore, the reliabilities of the design can be evaluated by integrating the analyzed results and the **SSI** theory. Figure 7 shows the reliability changing of the design depending on the torque loadings. The results show that the reliabilities have faster degradation when considering the variations of the stresses.

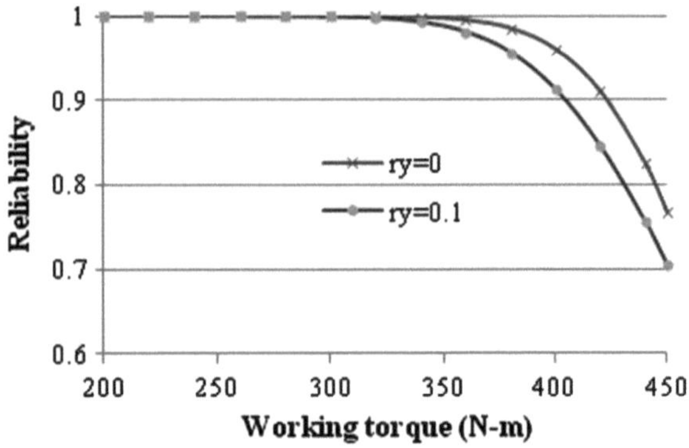

Fig. 7 Reliability changing of the design depending on the torques

7 Conclusion

This paper presents a creative design for **DR** possessing the properties, high rigidness, large reduction ratio and structure simple. The **DR** can satisfy the requirements of design space limited as well as large speed reduction ratio. The analyzed results in **FEM** are used to identify the designed weaknesses and to provide the information of design modification. The first maximum amplitude for normal stress and deformation would happen at 724 **Hz** according to the modal analyzes results. The results indicate that the **DR** occurs resonance being low because the input shaft usually is far lower the frequency. The allowed transmission loadings of the **DR** would be 380 N-m according to the analyzed results subjected to the needed reliability being 0.95 and the strength variation being 0.1. The results are useful for considering the safety as well as the reliability of the **DR** in using.

Acknowledgment The work was supported by a grant from the National Science Council under contract No: MOST 103-2221-E-237-002. The authors would like to appreciate the reviewers for their valuable suggestions.

References

1. Ono, K., Hayashi, T., Fujii, M., Shibasaki, N., Sonehara, M.: Development for industrial robotics applications. IHI Eng. Rev. **42**(2), 103–107 (2009)
2. http://harmonicdrive.net
3. Ghorbel, F.H., Gandi, P.S., Alpeter, F.: On the kinematics error in harmonic drive gears. J. Mech. Des. **123**(1), 90–97 (2001)
4. Tsai, Y.T., Lin, K. H., Hsu, Y.Y.: Reliability optimization design for mechanical devices based on modeling processes. J. Eng. Des. **24**(12), 849–863 (2013)
5. Tsai, Y.T., Chang, H.C.: Reliability-based optimum design for mechanical problems using genetic algorithms. Proc. IMechE, Part C: J. Mech. Eng. Sci. **222**(C9), 1791–1799 (2008)
6. Robert, D.: Cook, Concepts and Applications of Finite Element Analysis. Wiley, New York (2001)

7. Tsai, Y.T., Wang, K.S., Woo, J.C.: Fatigue-life and reliability evaluations of dental implants based on computer simulation and limited test data. Proc. IMechE, Part C: J. Mech. Eng. Sci. **227**(3), 554–564 (2013)
8. Sofian, M., Hazry, D., Saifullah, K., Tasyrif, M., Salleh, K., Ishak, I.: A study of vibration analysis for gearbox casing using finite element analysis. Proceedings of International Conference on Applications and Design in Mechanical Engineering (ICADME), Batu Ferringhi, Penang, Malaysia, 11–13 Oct 2009

Structural and Dimensional Synthesis of Parallel Manipulator with Two End—Effectors

Zh. Baigunchekov, M. Kalimoldaev, M. Utenov, B. Arymbekov and T. Baigunchekov

Abstract In this paper the methods of structural and dimensional synthesis of a parallel manipulator with two end-effectors are presented. Parallel manipulator with two end-effectors is formed from two serial manipulator by connecting of their links by binary links with two revolute kinematic pairs. Geometrical parameters of the binary link have been determined at three, four and five finitely separated position of the end-effectors.

Keywords Parallel manipulator · End–effector · Dimensional synthesis · Finitely separated positions

1 Introduction

In technological processes with stereotyped motion of the manipulating object it is advisable to use a parallel manipulator (PM) with fewer DOF or with one DOF instead of robot with many DOF. PM is synthesized by given laws of motion of the end-effectors and input links. Therefore PM unlike the serial manipulator operates at certain kinematic diagram and geometrical parameters of its links. There are

---

Zh. Baigunchekov (✉) · B. Arymbekov · T. Baigunchekov
Laboratory of Mechatronics and Robotics, Kazakh-British Technical University, Almaty, Republic of Kazakhstan
e-mail: bzh47@mail.ru

M. Kalimoldaev
Laboratory of Mathematical Modeling and Cybernetics, Institute of Information and Computing Technologies, Almaty, Republic of Kazakhstan
e-mail: mnk@ipic.kz

M. Utenov
Department of Mechanics, Kazakh National University, Almaty, Republic of Kazakhstan
e-mail: umu53@mail.ru

S. Zeghloul et al. (eds.), *Robotics and Mechatronics*,
Mechanisms and Machine Science 37, DOI 10.1007/978-3-319-22368-1_2

many works on the structural and dimensional synthesis of mechanisms [1–4]. In these works the structural synthesis of mechanisms is carried out separately, and the dimensional synthesis of mechanisms is carried out at the certain kinematic diagram. We propose a method of structural-parametric synthesis of mechanisms and manipulators [5–7]. According to this method any mechanism irrespective of functional purpose (function generation, path generation, body guidance) is formed from the actuating kinematic chain (AKC) and the closing kinematic chain (CKC). A kinematic chain with many DOF which implements the given laws of motions of the end-effectors and the input link is called AKC. A kinematic chain with negative DOF which connects the links of the AKC and forms a mechanism (manipulator) is called CKC. AKC and CKC are the structural modules. In this paper the methods of structural and dimensional synthesis of PM with two end-effectors on the basis of the method of structural-parametric synthesis are presented. Two serial manipulators (AKC) and the binary link with two revolute kinematic pairs (CKC) are the structural modules of the PM with two end-effectors.

2 Structural Synthesis of PM with Two End-Effectors

Consider the problem of manipulating of two objects. Depending on the type of technological operation work the manipulation of two objects may be implemented in two modes: a simultaneous manipulation and a consecutive manipulation.

In the simultaneous manipulation in the initial position of the serial manipulators *ABC* and *DEF* (Fig. 1a) a griping operation of two objects P_1 and P_2 is carried out by the grippers *C* and *F* (Fig. 1a). Then a transference of the objects to the positional point $P_{1(2),N}$ is carried out (Fig. 1b). After positioning of the manipulating objects the grippers back to their initial positions (Fig. 1a).

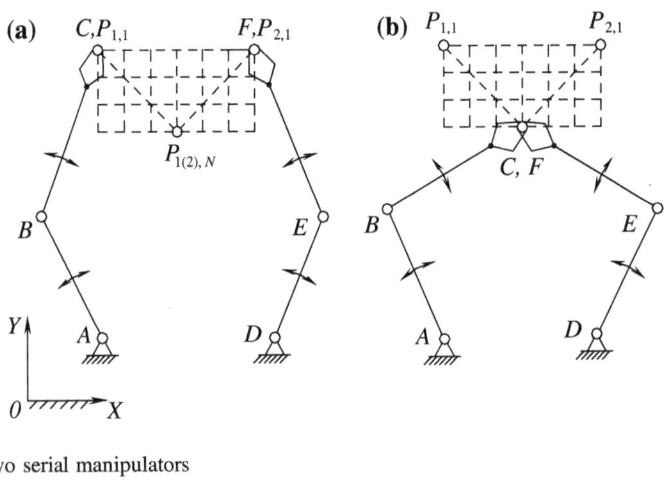

Fig. 1 Two serial manipulators

In the consecutive manipulation in the initial positions of the serial manipulators *ABC* and *DEF* a griping operation of the object P_1 is carried out by the gripper *C* of the first serial manipulator *ABC* (Fig. 1a). Then a transference of the object to an intermediate position $P_{1(2),N}$ is carried out (Fig. 1b). At the same time the gripper *F* of the second manipulator *DEF* simultaneously positioned in the intermediate point. In this intermediate point $P_{1(2),N}$ a re-grasping operation of the object occurs by the second manipulator *DEF*. Then both manipulators back to their initial positions (Fig. 1b).

A novel PM (parallel manipulator) with two end-effectors and with two DOF to manipulate of two objects has been developed (Fig. 2). This PM provides a set of laws of motion of two objects in two modes. Closed kinematic chain of the novel PM increases its load capacity and positioning accuracy. Furthermore, the number of drives is reduced from four to two, and two drives in mobile joints *B* and *E* are excluded which is also an advantage of the novel PM.

Structural synthesis of the novel PM with two end-effectors is made as follows. Two links *BC* and *EF* of the two serial manipulators *ABC* and *DEF* are connected by binary link *GH* with two revolute kinematic pairs. As a binary link with two revolute kinematic pairs has one negative DOF and it imposes one geometrical constraint to the relative motions of the links we obtain a kinematic chain *ABGHED* with three DOF. If we connect the link *GH* of this kinematic chain with a frame by a binary link *IK* with two revolute kinematic pairs then we obtain a kinematic diagram of the PM with two end-effectors and two DOF.

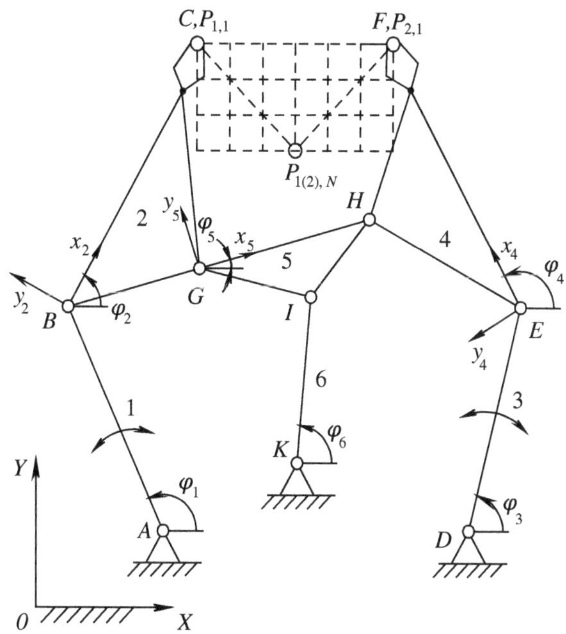

Fig. 2 PM with two end-effectors

Number of DOF of PM is determined by Chebyshev's formula [8]

$$W = 3n - 2p_5, \tag{1}$$

where n is number of mobile links, p_5 is number of kinematic pairs of the fifth class. For the considered PM $n = 6$ and $p_5 = 8$ then $W = 3 \cdot 6 - 2 \cdot 8 = 2$. For the binary link with two revolute kinematic pairs $n = 1$ and $p_5 = 2$ then $W = 3 \cdot 1 - 2 \cdot 2 = -1$.

3 Dimensional Synthesis of PM with Two End-Effectors

The end-effectors C and F have been moved from their initial positions $P_{1,1}$ and $P_{2,1}$ to the ultimate position $P_{1(2),N}$ along the segments $P_{1,1}$ $P_{1(2),N}$ and $P_{2,1}$ $P_{1(2)N}$ (Fig. 2). Set the coordinates $X_{P_{1,i}}$, $Y_{P_{1,i}}$ and $X_{P_{2,i}}$, $Y_{P_{2,i}}$ of the points P_1 and P_2 along these segments in the absolute coordinate system OXY. Also set the coordinates X_A, Y_A and X_D, Y_D of the fixed joints A and D, the lengths of links $l_{AB,}$ l_{BC} and $l_{DE,}$ l_{EF} of two serial manipulators ABC and DEF. Determine the angles φ_1, φ_2 and φ_3, φ_4

$$\begin{aligned} \varphi_{1i} &= \varphi_{ACi} + \cos^{-1}\frac{l_{AB}^2 + l_{ACi}^2 - l_{BC}^2}{2l_{AB}l_{ACi}}, \\ \varphi_{3i} &= \varphi_{DFi} - \cos^{-1}\frac{l_{DE}^2 + l_{DFi}^2 - l_{EF}^2}{2l_{DE}l_{DFi}}, \end{aligned} \tag{2}$$

$$\varphi_{2i} = \mathrm{tg}^{-1}\frac{Y_{P_{1,i}} - Y_{B_i}}{X_{P_{1,i}} - X_{B_i}}, \quad \varphi_{4i} = \mathrm{tg}^{-1}\frac{Y_{P_{2,i}} - Y_{E_i}}{X_{P_{2,i}} - X_{E_i}}, \tag{3}$$

where

$$\begin{bmatrix} X_{Bi} \\ Y_{Bi} \end{bmatrix} = \begin{bmatrix} X_A \\ Y_A \end{bmatrix} + l_{AB} \cdot \begin{bmatrix} \cos\varphi_{1i} \\ \sin\varphi_{1i} \end{bmatrix}, \begin{bmatrix} X_{Ei} \\ Y_{Ei} \end{bmatrix} = \begin{bmatrix} X_D \\ Y_D \end{bmatrix} + l_{DE} \cdot \begin{bmatrix} \cos\varphi_{3i} \\ \sin\varphi_{3i} \end{bmatrix}. \tag{4}$$

Determine the geometrical parameters $x_G^{(2)}, y_G^{(2)}, x_H^{(4)}, y_H^{(4)}, l_{GH}$ of the binary link GH, where $x_G^{(2)}, y_G^{(2)}$ and $x_H^{(4)}, y_H^{(4)}$ are the coordinates of the joints G and H in the local systems of coordinates Bx_2y_2 and Ex_4y_4 of the links 2 and 4 respectively, l_{GH} is the length of the link GH. Consider the inverse motion of the system of coordinate Ex_4y_4 relative to the system of coordinate Bx_2y_2. Then the joint H moves along the circular arc with center at the joint G and with radius l_{GH}. Write the equation of the circle

$$(x_{Hi}^{(2)} - x_G^{(2)})^2 + (y_{Hi}^{(2)} - y_G^{(2)}) = l_{GH}^2, \quad i = 1, 2, \ldots, N, \tag{5}$$

where $x_{Hi}^{(2)}$ and $y_{Hi}^{(2)}$ are the coordinates of the joint H in the system of of coordinates Bx_2y_2 of the link 2; N is number of given positions of the systems of coordinates Bx_2y_2 and Ex_4y_4.

Coordinates $x_{Hi}^{(2)}$ and $y_{Hi}^{(2)}$ of the joint H in the system of coordinates Bx_2y_2 of the link 2 are defined by the equation

$$\begin{bmatrix} x_{Hi}^{(2)} \\ y_{Hi}^{(2)} \end{bmatrix} = \begin{bmatrix} \cos\phi_{2i} & \sin\phi_{2i} \\ -\sin\phi_{2i} & \cos\phi_{2i} \end{bmatrix} \cdot \begin{bmatrix} X_{Hi} - X_{Bi} \\ Y_{Hi} - Y_{Bi} \end{bmatrix}, \tag{6}$$

where X_{Hi}, Y_{Hi} and X_{Bi}, Y_{Bi} are the coordinates of the joints H and B in the absolute system of coordinates OXY which are defined by the Eq. (4) and the equation

$$\begin{bmatrix} X_{Hi} \\ Y_{Hi} \end{bmatrix} = \begin{bmatrix} X_{Ei} \\ Y_{Ei} \end{bmatrix} + \begin{bmatrix} \cos\phi_{4i} & -\sin\phi_{4i} \\ \sin\phi_{4i} & \cos\phi_{4i} \end{bmatrix} \cdot \begin{bmatrix} x_H^{(4)} \\ y_H^{(4)} \end{bmatrix}. \tag{7}$$

Substituting the expressions (6) and (7) into the Eq. (5) we obtain

$$(x_{Hi}^{(2)} \cdot x_G^{(2)} + y_{Hi}^{(2)} \cdot y_G^{(2)}) + \frac{1}{2}(l_{GH}^2 - x_G^{(2)^2} - y_G^{(2)^2}) = \frac{1}{2}(x_{Hi}^{(2)^2} + y_{Hi}^{(2)^2}). \tag{8}$$

Denote

$$h = \frac{1}{2}(l_{GH}^2 - x_{Gi}^{(2)^2} - y_{Gi}^{(2)^2}), \quad r_{Hi}^{(2)^2} = x_{Hi}^{(2)^2} + y_{Hi}^{(2)^2}. \tag{9}$$

Then the Eq. (8) takes the form

$$x_{Hi}^{(2)} \cdot x_G^{(2)} + y_{Hi}^{(2)} \cdot y_G^{(2)} + h = \frac{1}{2} r_{Hi}^{(2)^2}, \quad i = 1, 2, \ldots, N. \tag{10}$$

The resulting system of Eq. (10) is linear of the coordinates $x_G^{(2)}, y_G^{(2)}$ of the center G of the circle and its radius l_{GH} expressed through the parameter h.

Let the number of given positions $N = 3$. Then we obtain a system of three linear equations with three unknowns $x_G^{(2)}$, $y_G^{(2)}$, h the matrix form which has a form

$$\begin{bmatrix} x_{H_1}^{(2)} & y_{H_1}^{(2)} & 1 \\ x_{H_2}^{(2)} & y_{H_2}^{(2)} & 1 \\ x_{H_3}^{(2)} & y_{H_3}^{(2)} & 1 \end{bmatrix} \cdot \begin{bmatrix} x_G^{(2)} \\ y_G^{(2)} \\ h \end{bmatrix} = \frac{1}{2} \begin{bmatrix} r_{H_1}^{(2)^2} \\ r_{H_2}^{(2)^2} \\ r_{H_3}^{(2)^2} \end{bmatrix} \tag{11}$$

or

$$\boldsymbol{A}\boldsymbol{x} = \boldsymbol{b}, \tag{12}$$

where

$$\boldsymbol{A} = \begin{bmatrix} x_{H_1}^{(2)} & y_{H_1}^{(2)} & 1 \\ x_{H_2}^{(2)} & y_{H_2}^{(2)} & 1 \\ x_{H_3}^{(2)} & y_{H_3}^{(2)} & 1 \end{bmatrix}, \ \boldsymbol{x} = \left[x_G^{(2)} y_G^{(2)} h\right]^T, \ \boldsymbol{b} = \frac{1}{2}\left[r_{H_1}^{(2)} r_{H_2}^{(2)} r_{H_3}^{(2)}\right]^T.$$

Vector $\boldsymbol{x}$ of the binary link GH parameters is determined by the equation

$$\boldsymbol{x} = \mathbf{A}^{-1}\boldsymbol{b} \tag{13}$$

when $\det(A) \neq 0$. If $\det(A) = 0$ then the radius l_{GH} of the circle is equal to ∞, i.e. the point G will move in a straight line. Consequently, the joint G should be transformed into slider.

Thus, when $N = 3$ we can find a point (joint) G on a plane of the link BC and length l_{GH} of the link GH for all points lying on a plane of the link EF in condition $\det(A) \neq 0$.

Let number of given positions $N = 4$. Then we obtain a system of four linear equations from (10). This system of equations has a solution when the determinant of the expanded matrix is zero

$$w_{1,2,3,4} = \begin{vmatrix} x_{H_1}^{(2)} & y_{H_1}^{(2)} & 1 & \frac{1}{2} r_{H_1}^{(2)^2} \\ x_{H_2}^{(2)} & y_{H_2}^{(2)} & 1 & \frac{1}{2} r_{H_2}^{(2)^2} \\ x_{H_3}^{(2)} & y_{H_3}^{(2)} & 1 & \frac{1}{2} r_{H_3}^{(2)^2} \\ x_{H_4}^{(2)} & y_{H_4}^{(2)} & 1 & \frac{1}{2} r_{H_4}^{(2)^2} \end{vmatrix} = 0. \tag{14}$$

If substitute into the determinant (14) the values of $x_{Hi}^{(2)}$, $y_{Hi}^{(2)}$, $r_{Hi}^{(2)}$, (i = 1, 2, 3, 4) defined by the Eqs. (6), (7) and (9) then we obtain the equation of the fourth order which is a curve of circular points. The points on this curve describe the arc of the circle. If we select a point on the curve of circular points it is sufficiently to determine the center of a circle and its radius from any three equations of the system (10). Then this circle passes through the fourth position.

Let the number of given position $N = 5$. Then we obtain a system of five linear equations from (10). This system of five linear equations has a solution if the rank of expanded matrix of this system is three. Consequently, all the minors of the fourth order of this matrix is equal to zero

$$w_{1,2,3,4} = 0, w_{1,2,3,5} = 0, w_{1,2,4,5} = 0, w_{2,3,4,5} = 0, w_{1,3,4,5} = 0. \tag{15}$$

Each of determinants (15) has a view similar to (143). Substituting to the determinants the values of $x_{Hi}^{(2)}$, $y_{Hi}^{(2)}$, $r_{Hi}^{(2)}$, $(i = 1, 2,\ldots, 5)$ defined by the Eqs. (6) and (9) we obtain the curves equations of the fourth order. Solutions of the system (15) is determined by the points of intersection of any two curves of this system, for example by the points of intersections of the curves $w_{1,2,3,4}$ and $w_{1,2,3,5}$. These points are circular points each of which has five positions on one circle. Other three curves also pass through the circular points.

As a result of synthesis of the binary link *GH* we obtain a mechanism *ABGHED* with three DOF. If we connect the link *GH* with a frame by a binary link *IK* then we obtain a kinematic diagram of the PM with two DOF.

For synthesis of the binary link *IK* we write the equation of circle in absolute motion

$$(X_{I_i} - X_K)^2 + (Y_{I_i} - Y_K)^2 = l_{IK}^2, \tag{16}$$

where

$$\begin{bmatrix} X_{I_i} \\ Y_{I_i} \end{bmatrix} = \begin{bmatrix} X_{G_i} \\ Y_{G_i} \end{bmatrix} + \begin{bmatrix} \cos \varphi_{5_i} & -\sin \varphi_{5_i} \\ \sin \varphi_{5_i} & \cos \varphi_{5_i} \end{bmatrix} \cdot \begin{bmatrix} x_I^{(5)} \\ y_I^{(5)} \end{bmatrix}. \tag{17}$$

Substituting the expression (17) into the Eq. (16) we obtain

$$(X_{I_i} \cdot X_K + Y_{I_i} \cdot Y_K) + \frac{1}{2}(l_{IK}^2 - X_K^2 - Y_K^2) = \frac{1}{2}(X_{I_i}^2 + Y_{I_i}^2). \tag{18}$$

Denote

$$H = \frac{1}{2}(l_{IK}^2 - X_K^2 - Y_K^2), \;\; R_{H_i}^2 = X_{I_i}^2 + Y_{I_i}^2. \tag{19}$$

Then the Eq. (18) takes the form

$$X_{I_i} \cdot X_K + Y_{I_i} \cdot Y_K + H = \frac{1}{2} R_{H_i}^2, \quad i = 1, 2, \ldots, N. \tag{20}$$

The resulting system of Eq. (20) is a linear system with respect to the coordinates $X_{K,}$ Y_K of the center *K* of the circle and its radius l_{IK} expressed through the parameter *H*. Further we determine the coordinates $x_I^{(5)}$, $y_I^{(5)}$ of circular point *I*, the coordinates $X_{K,}$ Y_K of the center *K* and the radius l_{IK} of the circle similarly the synthesis of the binary link *GH*.

Table 1 The coordinate values of the points P_1 and P_2

i	1	2	3	4
X_{P_1}	20.0	24.0	28.0	32.0
Y_{P_1}, Y_{P_2}	68.0	64.0	60.0	56.0
X_{P_2}	44.0	40.0	36.0	32.0

4 Numerical Example

Let the grippers C and F of PM move from their initial positions $P_{1,1}$ and $P_{2,1}$ to the positional position $P_{1(2),4}$ on straight segments $P_{1,1}$ $P_{1(2),4}$ and $P_{2,1}$ $P_{1(2),4}$ in the absolute system of coordinates OXY. The coordinate values of the points are given in Table 1. It is necessary to synthesize PM with two end-effectors.

Set the coordinates $X_A = 16.0;$ $Y_A = Y_D = 9.5;$ $X_D = 54.5$ of the fixed joints A, D and the lengths of the links l_{AB}, $l_{DE} = 30.0;$ l_{BC}, $l_{EF} = 35.0$ of two serial manipulators ABC and DEF and calculate the positions of links and coordinates by the Eqs. (2)–(4). On the base of the above developed method the geometrical parameters of the binary links GH and IK are calculated: $x_G^{(2)} = 12.348;$ $y_G^{(2)} = -12.512;$ $x_H^{(4)} = 18.107$, $y_H^{(4)} = 12.014;$ $l_{GH} = 22.170;$ $x_I^{(5)} = 12.503;$ $y_I^{(5)} = -7.134;$ $X_K = 32.701;$ $Y_K = 17.502;$ $l_{IK} = 20.062$. Results of kinematic analysis have been presented in [9].

5 Conclusions

The methods of structural and dimensional synthesis of the PM with two end-effectors are developed on the base of structural-parametric synthesis. According to these methods PM is formed from the actuating and closing kinematic chains (AKC) and (CKC) which are the structural modules. Two serial manipulators (AKC) and the binary link with two revolute kinematic pairs (CKC) are the structural modules of PM. Geometrical parameters of the serial manipulator's links are given or varied and geometrical parameters of the binary link are determined from the equation of geometrical constraint imposed to the relative motions of the AKC links. The geometrical parameters of the binary link are determined at three, four and five finitely separated positions of the end-effectors.

References

1. Erdman, A.G., Sandor, G.N., Kota, S.: Mechanism Design: Analysis and Synthesis, vol. 1, 4th edn. Prentice Hall, Englewood Cliffs, NJ (2011)
2. Hunt, K.H.: Kinematic Geometry of Mechanism. Oxford University Press, Oxford (1978)
3. Soni, A.H.: Mechanism Synthesis and Analysis. McGraw-Hill, New York (1974)

4. Hartenberg, R.S., Denavit, J.: Kinematic Synthesis of Mechanisms. McGraw-Hill, New York (1964)
5. Baigunchekov, Z.Z., et al.: Modular approach for synthesizing of planar one-DOF and adjustable mechanisms of high classes. Proceedings of the Ninth IFToMM World Congress on TMM. Milan, Italy, 29 Aug–4 Sept 1995
6. Baigunchekov, Z.Z., et al.: Modular synthesis of spatial manipulating devices of high classes. Proceeding of the Twelth International Conference on CAD/CAM Robotics and Factories of the Future. Middlesex University, London, UK, pp. 685–690, 14–16 Aug 1996
7. Baigunchekov, Z.Z., et al.: The basis of structural and parametric synthesis of the parallel manipulators with functionally independent drives. Part I&II. Proceedings of the 16th International Conference on Gearing, Transmissions and Mechanical Systems, The Nottingham Trent University, UK, pp. 1–19, 3–6 July 2000
8. Artobolevskii, I.I.: Theory of Mechanisms and Machines. Nauka, Moscow (1975)
9. Baigunchekov, Z.Z., et al.: Kinematics of parallel manipulators with two end- effectors. 2015 IFToMM World Congress, Taipei, Taiwan, presented, 25–30 Oct 2015

Parametric Design Optimization of Two Link Robotic Manipulator

F.Z. Baghli, L. El Bakkali and O. Hamdoun

Abstract In this work we present an optimal design for a type of serial link manipulators based on multi objective optimization. The objective functions considered are Global Conditioning index (GCI) and workspace volume, the first indice is based on condition number of robot manipulator Jacobian matrix and the objective functions are optimized simultaneously to improve the dexterity as well as the workspace volume which represents the working capacity.

Keywords Robot manipulator · Global conditioning index · Workspace · Design optimization

1 Introduction

Robot arms or manipulators which are mechanical devices driven by electrical motors or actuators have been used in many applications in industry such as painting, drilling, material handling and pick and place. Since the main aim is to increase the productivity in these fields.

Kinematic optimal is a crucial key in designing serial manipulators and it was received considerable attention by researchers during the post decade. An optimal design method was proposed by many researchers, which considers the global workspace and Global Conditioning index (GCI) which based on condition number of robot manipulator Jacobian matrix.

Many methods exist in literatures which are based on analytical or numerical methods for determination of manipulator's workspace.

F.Z. Baghli (✉) · L. El Bakkali · O. Hamdoun
Modeling and Simulation of Mechanical Systems Laboratory, Faculty of Sciences, University Abdelmalek Essaadi, BP.2121, M'hannech, 93002 Tetouan, Morocco
e-mail: baghli.fatimazahra@gmail.com

S. Zeghloul et al. (eds.), *Robotics and Mechatronics*,
Mechanisms and Machine Science 37, DOI 10.1007/978-3-319-22368-1_3

The present work proposes a optimization method based on semi-infinitely constrained programming for workspace evaluation with the definition of GCI as the objective function, a formulation of link length optimization of manipulator is presented her. An algorithm is presented for numerical solution of the optimization, with evaluation of GCI in each solution iteration.

This paper is organized in serval sections. Section 2 presents the kinematic and dynamic modeling, to introduce the design parameters. In Sect. 3, the global performance indices are presented: Condition number and Global Conditioning Index. In Sect. 4 the optimization problem of robot manipulator is formulated. Section 5 presents the results and discussions. Finally Sect. 6 covers the conclusion.

2 Kinematic and Dynamic Models of Robot

The kinematic and dynamic models of the manipulator are developed considering the manipulator shown in Fig. 1. The two links have lengths l_1, l_2 respectively.

2.1 Kinematic Model

The Denavit-Hartenberg (D-H) parameters for the arm manipulator are defined in Table 1:

Let M(x,y) be the target position of end-effector in the Cartesian workspace of the manipulator, the point of interest, for which the performance in terms of

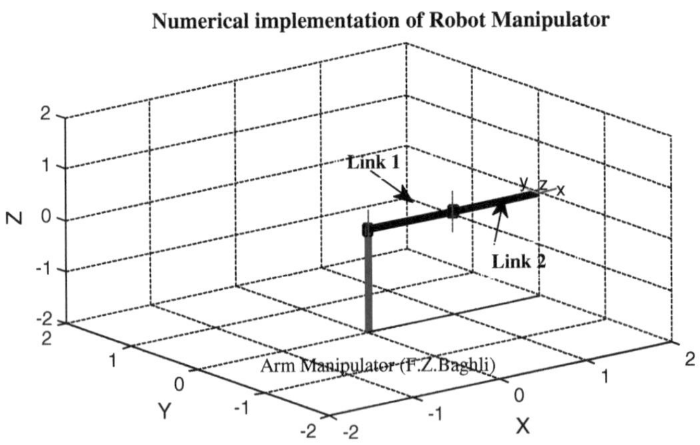

Fig. 1 Numerical implementation of robot manipulator

Table 1 DH parameters of robot

Link	a_i	α_i	d_i	q_i
1	l_1	0	0	q_1
2	l_2	0	0	q_2

positional accuracy has to be modeled and optimized. The coordinates of **M** in Cartesian coordinate for joint angles q_1 and q_2 are given by:

$$M_X = l_1c_1 + l_2c_{12} \tag{1}$$

$$M_y = l_1s_1 + l_2s_{12} \tag{2}$$

where:

$$c_1 = \cos(q_1);\ s_1 = \sin(q_1);\ c_{12} = \cos(q_1 + q_2);\ s_{12} = \sin(q_1 + q_2)$$

From Eqs. (1) and (2), joint variables q_1 and q_2 are obtained as:

$$q_1 = ATAN2(S_1, C_2) \tag{3}$$

$$q_2 = ATAN2(S_2, C_2) \tag{4}$$

where

$$\begin{aligned} s_1 &= \frac{(l_1 + l_2c_2)M_y - l_2s_2M_x}{M_X^2 + M_y^2} \\ c_1 &= \frac{(l_1 + l_2c_2)M_x + l_2s_2M_y}{M_X^2 + M_y^2} \\ c_2 &= \frac{1}{2l_1l_2}(M_X^2 + M_y^2 - l_1^2 - l_2^2) \\ s_2 &= \pm\sqrt{1 - c_2^2} \end{aligned}$$

Differentiating Eqs. (1) and (2) and solving for joint velocities $\dot{q}_1$ and $\dot{q}_2$ gives:

$$\dot{q}_1 = \frac{\dot{M}_xc_{12} + \dot{M}_ys_{12}}{l_1s_2} \tag{5}$$

$$\dot{q}_2 = -\frac{\dot{M}_xM_x + \dot{M}_yM_y}{l_1l_2s_2} \tag{6}$$

3 Dexterity of Robot

3.1 Jacobian and Inverse Jacobian Matrix

The dexterity index is based on the condition number of the jacobian matrix. This quantity which is a mesure of the local dexterity.

The jacobian matrix (J) of the robot manipulator is defined as the matrix representing the transformation mapping the joint rates into the Cartesian velocities transformation is written as:

$$\begin{bmatrix} \dot{M}_{x_1} \\ \dot{M}_{y_1} \end{bmatrix} = J \begin{bmatrix} \dot{q}_1 \\ \dot{q}_2 \end{bmatrix} \tag{9}$$

where $\dot{q}$ is the vector of joint rates and $\dot{M}_x, \dot{M}_y$ are the vector of Cartesian velocities.

However, the inverse jacobian matrix is defined as:

$$\begin{bmatrix} \dot{q}_1 \\ \dot{q}_2 \end{bmatrix} = J^{-1} \begin{bmatrix} \dot{M}_x \\ \dot{M}_y \end{bmatrix} \tag{10}$$

where J of the robot manipulator can be expressed as:

$$J = \begin{pmatrix} -l_2 s_2 & -l_2 s_2 \\ l_1 + l_2 c_2 & l_2 c_2 \end{pmatrix} \tag{11}$$

Therfore, we have:

$$J^{-1} = \frac{1}{\Delta} \begin{pmatrix} l_2 c_2 & l_2 s_2 \\ -(l_1 + l_2 c_2) & -l_2 s_2 \end{pmatrix} \tag{12}$$

where:

$$\Delta = l_1 l_2 s_2 \tag{13}$$

4 Condition Number

The accuracy of the control of the manipulator is dependent on the condition number of the Jacobian matrix. This is so because it represents an indication of the amplification of the error on the position or force at the gripper for a given accuracy of the actuators. This number is to be kept as small as possible, the smallest values that can be attend being ‘1’.

According to definition, the conditioning number of the jacobian matrix is defined by:

$$\kappa(J) = \left\|J^{-1}\right\| \cdot \|J\| \tag{14}$$

where $\kappa, \sigma_{\max}$ and $\sigma_{\min}$ are condition number, and the maximum and minimum singular values of the Jacobian respectively.

The condition number of Jacobian matrix for manipulator can be computed and is expressed as following:

$$\kappa = \frac{(l_1^2 + 2l_2^2 + 2l_1 l_2 c_2)}{2l_1 l_2 s_2} \tag{15}$$

5 Optimization Design

The mathematical formulation of design optimization problem can be written as following:

$$\begin{aligned} &\text{minimizing} \quad f(l) \\ &\text{subject to} \quad g_i(l) \leq 0;\ i = 1, 2, 3, \ldots, k \\ &\qquad\qquad\quad l^- \leq l \leq l^+ \end{aligned} \tag{16}$$

where $l \in R^{2\times1}$ denotes the vector of the design variables, l^- and l^+ are the lower and upper bounds of the design variables, respectively. The objective function is denoted by $f(l)$ and $g_i(l)$ is the constraint function.

The design objective is to maximize the workspace area covered by the robot manipulator.

l_1, l_2 are the two design variables, l^{-1}, l^{+1} are the lower and the upper bounds on the ratio l_1/l_2. The objective function to be optimized is used to maximize the workspace covered by the manipulator.

5.1 Workspace Optimization

Workspace is a significant factor for describing kinematic performance of robot manipulator. Besides deciding the limits of the attainable workspace of system, choosing the feasible workspace is also a difficult question. In order to obtain optimal kinematic performance, it is necessary to determine the position and style of the feasible workspace. In this work, we used the global conditioning index (GCI) as an objective function for maximization to determine the feasible workspace, which is defined as:

$$\eta = \int_{w} \frac{1}{\kappa} dw \bigg/ \int_{w} dw \tag{17}$$

where η is the global conditioning index, κ is the condition number of the Jacobian, dw is a differential workspace of the manipulator.

In the context of the optimum design of robotic manipulator, the GCI is to be maximized over the space of robot manipulator parameters. Thus, the closer to unity the index is the better the overall behavior of the condition number and hence of the robot. The normality condition necessary for a stationary value of η is given by:

$$\frac{\partial \eta}{\partial v} = 0 \tag{18}$$

where v is the vector containing the parameters defining the architecture of the robot manipulator (for example parameters of DH).

To compute the robot manipulator's global conditioning index we have to integrate the reciprocal of condition number over the workspace. Therfore we have,

$$\eta = \frac{1}{4\pi l_1 l_2} \int \int \left(\frac{2s_2}{1/v + 2v + 2c_2}\right) l_1 l_2 s_2 dq_1 dq_2 \, ; \quad v = l_2/l_1 \tag{19}$$

6 Results and Discussions

The optimal design manipulator parameters are obtained to be $l_1 = 0.5$ and $l_2 = 0.3608$. The corresponding workspace 2.2671. In Fig. 2, we have plotted the condition number κ as function of q for each iteration to show the improvement in the condition number from its initial range to its optimum range. In this optimum range of the condition number, when $q = 2.351$ we see the condition number is almost unity.

This angle is the most desirable configuration for the robot manipulator in terms of its dexterity.

In order, the limits of the global workspace can be described as follows:

The global conditioning index of the robot manipulator as a function of v is shown in Fig. 3. So for $v = \frac{l_2}{l_1} = 0.7$ the maximum value of GCI is $\eta_{\max} = 0.7374$.

The workspace area of the robot arm is shown below (Fig. 4):

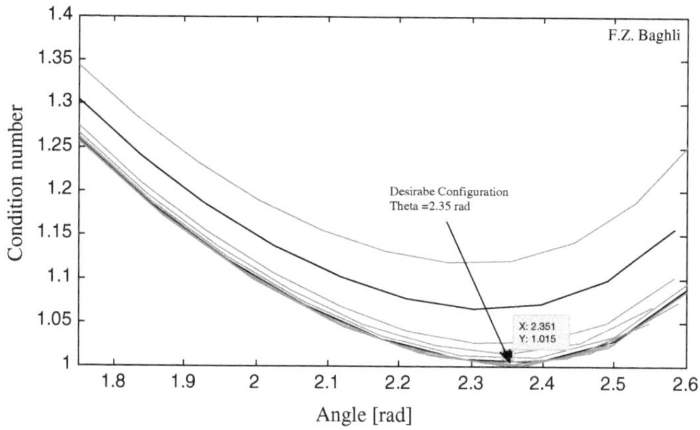

Fig. 2 Condition number of the Robot Manipulator as a function of q

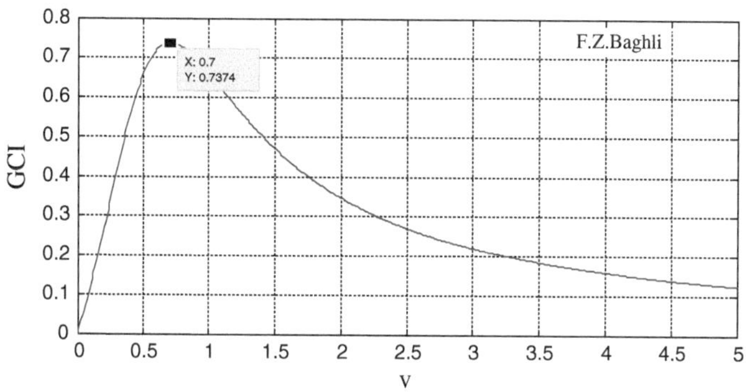

Fig. 3 Global condition index of the robot manipulator

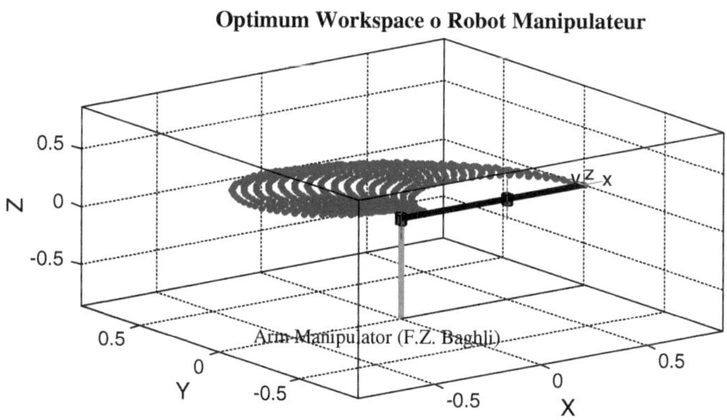

Fig. 4 Workspace of the two link manipulator

7 Conclusions

In this work, the mathematical formulation of complete kinematics and dynamics of two link manipulator is presented.

The optimal configuration is obtained by the principale of the condition number of Jacobian matrix and the GCI. Finally the dimension optimization design is performed according to the objective function.

Investigation of the Behaviour of a New Miniature Carbon-Paraffin Phase-Change Actuator

P. Lazarou and C. Rotinat-Libersa

Abstract This paper presents a new millimeter-sized membrane actuator utilizing the volume expansion, resulting from a phase-change composite material, induced by self-heating through the Joule effect. The composite consists of carbon particles as the conductive elements and of paraffin wax as the polymer matrix. The behavior of the actuator is experimentally characterised by measuring its external temperature, its electrical resistance and the membrane deflection. Several factors influencing these measurements are identified and briefly discussed. A detailed analysis of the experimental findings follows, so as to provide a deeper physical understanding of the observed actuator behavior, thus allowing future design and performance improvements.

Keywords Miniature membrane actuator · Paraffin-carbon composite · Phase-change · Characterisation

1 Introduction

In the last years many low voltage electromechanical implementations based on various actuation principles have been presented, such as electrostatic, piezoelectric, magnetic, thermal, as well as phase-change ones. The last category includes shape-memory actuators utilizing solid-to-solid transformation and solid-to-liquid phase-change materials, where the phase transition incurs a change in geometry, thus providing the required force and movement. Phase-change materials (such as paraffin, hydrogels) have the property of volumetric expansion when changing from solid-to-liquid (or liquid-to-gas) phase. Paraffin wax in particular provides a volumetric expansion of around 10–15 % and it is a very suitable material for generating large hydraulic forces when phase-change takes place.

P. Lazarou · C. Rotinat-Libersa (✉)
Interactive Robotics Laboratory, CEA, LIST, Gif-sur-Yvette F-91191, France
e-mail: christine.rotinat-libersa@cea.fr

S. Zeghloul et al. (eds.), *Robotics and Mechatronics*,
Mechanisms and Machine Science 37, DOI 10.1007/978-3-319-22368-1_4

The force, stroke and response time of paraffin actuators are highly dependent on their geometry and size and their heating/cooling system. A large phase-change, paraffin membrane actuator that is heated externally by thermal conduction has been reported in [1]. A Braille cell filled with paraffin wax and sealed off with a silicone rubber membrane is heated by embedded microheaters [2]. A different approach is introduced in [3], by mixing graphite particles with paraffin wax; the same principle is also used in [4]. This mixing creates an electrically conductive phase-change material, making it possible to heat the paraffin directly with the application of an electrical current, through Joule heating.

This paper introduces a new millimeter-sized membrane actuator that is inspired by the concepts of [3, 4], utilizing the same actuation principle but with a different geometry and composite (carbon particles mixed with paraffin wax). A detailed discussion of this principle and of the multiple physics involved has been previously presented by the authors in [5]. Here, the emphasis is given on the experimental characterisation of the actuator prototype by measurements of its external electrode temperature, the composite's electrical resistance and the membrane deflection when electrically powered by a constant current. Several factors affecting these measurements, such as electrical contact conditions between electrodes and composite, are then identified and discussed. Finally, an analysis of the experimental results ensues, providing a deeper understanding of the observed physical actuator behavior and allowing future design and performance optimisation.

2 Presentation of the New Actuator

Due to fabrication and integration constraints, a small, millimeter-sized cylindrical concentric "pin-cylinder" electrodes concept is adopted for the construction of the actuator (Fig. 1). In the center of the actuator chamber lies the vertical pin-type electrode, surrounded by the phase-change composite material and the cylinder, acting as the outer electrode. The composite presents both conductive (due to the carbon particles and aggregates forming conductive paths) and insulating properties (due to the polymer matrix of paraffin), therefore allowing self-heating and phase-change through the Joule effect, as explained in [3–5]. The configuration is sealed on top with a flexible silicon rubber membrane. With the application of voltage between the two electrodes, a current passes through the bulk of the carbon-paraffin composite, resulting in the volume dilatation, the phase-change of the composite and thus in the deflection of the membrane. The cylindrical symmetry of the configuration induces a radial, non-homogeneous electric field.

The base and the electrodes of the actuator are made of a nickel berillium alloy covered by a gold spatter layer (Fig. 1 Right), with the following internal chamber dimensions: cylinder Ø3.2 mm, pin Ø0.96 mm, height 4.26 mm and chamber volume 31.2 μl. Four support pillars aid in the placement of other elements, such as temperature sensors and electrical connections to the cylinder's outer surface. For the preparation of the composite, the Sigma Aldrich 76228 paraffin wax (melting

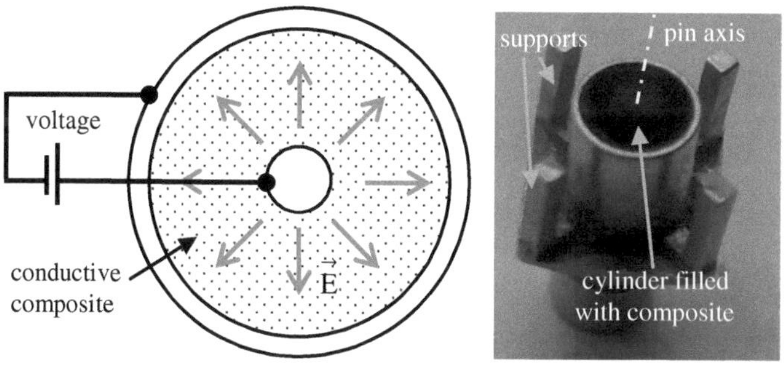

Fig. 1 *Left* top view of "pin-cylinder" actuator concept geometry and resulting radial electric field, *Right* the prototype of a gold layered "pin-cylinder" actuator cell without membrane (the pin is not visible due to a small top layer of composite material that covers it)

point at 44–46 °C according to the supplier's data) and the Degussa PRINTEX XE2-B carbon particles are used. While the wax is melted on a hotplate, the carbon particles are dispersed and thoroughly mixed with the hotplate's magnetic stirrer for 1 h. The mixture is left to cool down for a day and is afterwards used to fill the actuator. The concentration of carbon in the mixture is 5 wt% and lies well above the experimentally found electrical percolation threshold of 1.5 wt%; in other words, it lies well above the minimum required carbon concentration that guarantees electrically conductive paths in the bulk of the composite, and thus allows the flow of current between the electrodes [6]. Finally, the silicon rubber Dow Corning 734 is applied on the top of the actuator by brushing, and is cured at room temperature for 24 h, thus forming the flexible membrane. Its average thickness is around 100 μm.

3 Experimental Characterisation of the Actuator

This section describes the setup used for the characterisation tests and shows representative results of the actuator's electric resistance and membrane deflection measurements, without external mechanical load applied.

3.1 Experimental Setup

A complete setup with the actuator upon its base, the two electrical connections to its electrodes, a laser sensor for the membrane deflection measurement, and a temperature sensor, is shown in Fig. 2. Measurements are taken through suitable electronic circuits and a computer. For a constant electrical current applied on the actuator, its behaviour can be characterised by its membrane deflection D and its

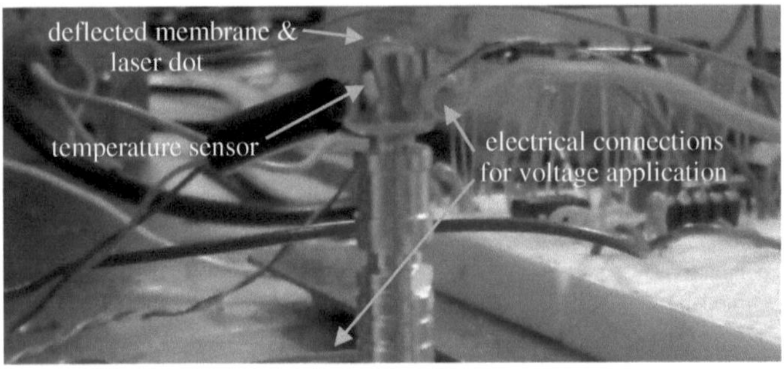

Fig. 2 Actuator cell on a base, the temperature sensor, the membrane with the laser dot for deflection measurement, and electrical connections to the pin and cylinder electrodes

bulk electric resistance R_{load} that is temperature dependent (the composite's electric resistivity is a non-linear function of temperature [5, 7]). It should be emphasized that this work's goal is the identification and characterization of the actuator's behavior, not a performance evaluation; therefore actuation times are not taken into consideration.

3.2 Resistance and Deflection Measurements

For the experiments, currents equal or superior to 50 mA are used in order to ensure the composite's phase-change. Smaller currents will not produce sufficient heat to allow the increase of its temperature up to the phase-change, especially in the case of the paraffin wax which has a high latent heat. Many measurements without external mechanical load on the membrane are made. After a series of around 30 cycles of actuation, the actuator's behavior remains stable, yielding reproducible results. Representative R_{load} and D measurements versus the external electrode temperature are shown in Fig. 3 for an applied heating current of 50 mA. The plot of R_{load} presents two peaks, while the plot of D is increasing until reaching a maximum.

Specifically, as the actuator heats up from room temperature, R_{load} increases almost linearly. The first peak appears around 35 °C, then R_{load} briefly falls and again rises to the second peak around 42 °C. Further heating of the actuator results in a rapid drop of R_{load}, until its value remains almost constant after 44 °C (indicating fully liquid state of the composite). Similarly, D begins to mildly increase as the actuator heats up from room temperature. A big change starts to appear at the area of 40–42 °C (second R_{load} peak), after which the deflection very rapidly reaches its peak value of 257 μm in fully liquid composite state.

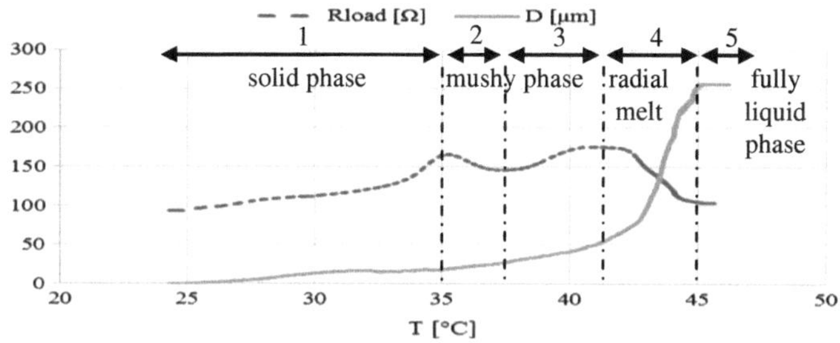

Fig. 3 Load resistance and membrane deflection measurements over temperature for a heating current of 50 mA and definition of regions characterising the actuator's behaviour (Sect. 4.2)

4 Argumented Analysis of the Experimental Results

In this section, a physical comprehension is made, considering the factors influencing the measurements. This analysis is important in order to understand what is happening exactly inside the actuator and how these factors affect the measured quantities, in order to lead us to design and performance improvements.

4.1 Factors Influencing the Measurements

In previous experiments presented in [8], while the general form of R_{load} over T is similar for application of various applied currents, its value range slightly changes, potentially leading to the conclusion that it would be also a function of the current applied. However, reference tests made on constant resistors and composite samples covered with conductive paint, concluded that electrical contact quality severely influences the resistance measurement; all the above exhibited very good electrical contact and the measurements were very accurate. In fact, these preliminary tests verified that there are important electrical contact issues between the actuator's electrodes and the composite, explaining the current dependence's previous assumption. Possible reasons include electrical contact resistance [9], surface roughness, surface tension effects, air gaps or even thin paraffin insulating films preventing carbon particles from directly touching the electrodes according to [10, 11].

Additional factors affecting the behavior and thus the measurements are: (a) the cylindrical configuration of the actuator's Joule powering (Fig. 1 Left), which results in a non-homogeneous heat production, (b) the break-down of conductive carbon paths in the composite during heating [6], due to the non-homogeneous paraffin's phase-change to liquid, (c) sedimentation of the heavier carbon particles and aggregates that have been observed experimentally after several heating/cooling cycles, increasing the non-homogenity of the composite, (d) the local irregularities and the non-constant thickness of the membrane at the microscopic level, (e) the composite volume inside the dilated "domed" membrane that only partially contributes to the resistance measurement (weak electric field) and (f) the selected paraffin's melting beginning at 37 °C and ending at 48 °C (spanning over a range of 11 °C as mentioned in [2]), instead of the 44–46 ° C supplier's data.

4.2 Behavioural Analysis of the Actuator

In light of the above-mentioned measurement issues, the behaviour of the actuator can be explained by taking into consideration the plots of R_{load} and D of Fig. 3. Five different regions are thus identified and analyzed individually. Beginning with region 1, a PTC (positive temperature coefficient) effect is observed for the resistance, arriving to the first peak at 35 °C. With the increase of temperature, the composite matrix (solid paraffin) thermally expands, thereby forcing the conductive pathways to be broken up and the carbon particles to be separated and distanced (increase of the tunneling barrier [12]). The deflection plot shows a mild increase, with the initially flat membrane starting to resemble a small dome.

The beginning of region 2 corresponds to an intermediate point of the solid to liquid phase-change, which has locally begun at and around the area of the pin electrode (for a 35 °C temperature at the external electrode, the temperature at the internal pin is 4–5 °C higher as we verified by thermal FEM simulations [8]). Thus the previously crystalline structure of the solid paraffin wax transitions towards a more relaxed state, effectively turning the solid to a "mushy"-intermediate phase between solid and liquid. This phenomenon facilitates the electrical contact between composite and pin, therefore decreasing the bulk resistance of the actuator. As only a small volume portion of the composite has been transformed, the deflection is not affected much and continues to increase.

In region 3 a new incline of R_{load} begins, leading to the second and highest peak at around 42 °C. With the increase of the temperature, a radially expanding dilatation of the mushy phase towards the cylindrical electrode takes place. The increase of the deflection rate is related to the augmentation of the "domed" membrane volume; this volume removes conductive carbon pathways from between the electrodes, therefore effectively increasing the bulk electric resistivity.

In region 4 and at 42 °C (max internal temperature estimated at 48 °C using FEM simulations—Fig. 4 left), the second R_{load} peak appears. Afterwards a rapid decline follows, as the volume around the pin transitions to fully liquid phase, further facilitating the electrical contact with the electrode. This constitutes a typical NTC (negative temperature coefficient) behavior for R_{load}. Electrical conductivity is increased at the liquid phase, as the wax goes into an amorphous state, thereby enabling carbon particles, aggregations and pathways to orient, align along the applied electric field lines, reform and/or create new conductive pathways; similar findings are presented in [7]. With increasing temperature, the rest of the composite volume transitions fully into liquid phase. The deflection also presents a very

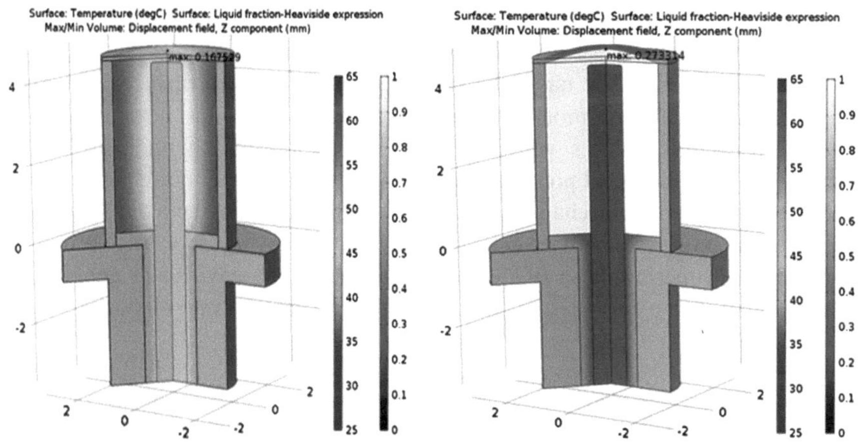

Fig. 4 Two FEM simulation result steps, showing the actuator temperature, the composite's non-homogeneous phase change (ranging from the value '0'-fully solid to the value '1'-fully liquid) and the membrane deflection (indicative, not necessarily representative of the actual experimental one) [8]

significant increase in this region, relative to the volume expansion due to the complete transition of the composite into liquid phase. It should be emphasized that the phase-change that propagates from the central pin electrode towards the external cylindrical electrode is due to: (a) cylindrical geometry inducing a non-homogeneous electric field, (b) very low paraffin wax thermal conductivity (0.21 W m^{-1} K^{-1}), meaning that heat propagation is very slow throughout the composite and (c) a latent heat of 250–260 kJ kg^{-1}, meaning it absorbs a large amount of heat in order to change phase.

In region 5 and at around 45 °C and afterwards, the composite is in fully liquid phase, as seen in Fig. 4 right (reducing the electrical contact resistance at the cylindrical electrode). Both the resistance and the deflection remain constant.

5 Conclusions

A new millimeter-sized conductive paraffin phase-change actuator is presented in this work, made of concentric cylindrical electrodes filled with a carbon-paraffin composite material and sealed with a silicon rubber membrane. The behavior of the actuator without external mechanical load is experimentally characterised through measurements of its external electrode temperature, the bulk electrical resistance of the composite and the membrane deflection, providing repeatable and reproducible results. However, the latter are subject to the influence of identified factors, such as electrical contact conditions between electrodes and composite, non-homogeneous heating due to the geometry etc. An analysis of the measured resistance and deflection versus temperature follows, providing a deeper understanding of the observed actuator behavior and its phase-change procedure.

Future work will firstly target the measurement of the performance capabilities of the existing actuator prototype, as its characterisation has been completed. Specifically, target metrics will include stroke performance with loads on the membrane, input electrical power and output mechanical power correlation, as well as time-related dynamic characterisation. In addition, the issues and challenges identified in this paper will be taken into consideration in order to improve the design and the performance of the actuator. One of the envisioned goals is the integration of such a costless and high power density actuator into miniature robotic devices.

Acknowledgments The authors would like to warmly thank Mr J. Catherine and P. Allain, who have worked respectively on the initial concept that inspired this miniature actuator, and the first experimental test bench and measurements. The authors are also very grateful to the French Ministry of economy, industry and employment (Direction Générale de la Compétitivité, de l'Industrie et des Services—DGCIS) who financially supported part of this work under the grant 102930231 (Mediate, ITEA2 labeled project, EUREKA cluster program).

References

1. Klintberg, L., et al.: A large stroke, high force paraffin phase transition actuator. Sens. Actuators A **96**, 189–195 (2002)
2. Lee, J.S., Lucyszyn, S.: A micromachined refreshable Braille cell. J. Microelectromech. Syst. **14**, 673–682 (2005)
3. Goldschmidtboing, F., Katus, P., Geipel, A., Woias, P.: A novel self-heating paraffin membrane micro-actuator. MEMS IEEE 21st International Conference, pp. 531–534 (2008)
4. Sant, H.J., Ho, T., Gale, B.K.: An in situ heater for a phase-change-material-based actuation system. J. Micromech. Microeng. **20**(8) (2010)
5. Lazarou, P., Rotinat, C.: Multiphysics simulation of a self-heating paraffin membrane microactuator. COMSOL Conference 2013, Rotterdam, Netherlands (2013)
6. Medalia, A.I.: Effective degree of immobilization of rubber occluded within carbon black aggregates. Rubber Chem. Technol. **45**(5), 1171–1194 (1972)
7. Zhang, K., Han, B., Yu, X.: Electrically conductive carbon nanofiber/paraffin wax composites for electric thermal storage. Energy Convers. Manag. **64**, 62–67 (2012)
8. Lazarou, P.: Multiphysic modelling and experimental validation of the thermoelectromechanical behavior of a phase-change conductive paraffin microactuator. CEA internal technical report. Ref. CEA DIASI/14-344/Rev. 0. 102 p (2014)
9. Roy, D., Zandt, M., Wolters, R.: Bias dependent specific contact resistance of phase-change material to metal contacts. Proceedings of STW.ICT Conference 2010. Veldhoven, The Netherlands, pp. 147–149 (2010)
10. Kim, S., Drzal, L.T.: High latent heat storage and high thermal conductive phase-change materials using exfoliated graphite nanoplatelets. Sol. Energy Mater. Sol. Cells **93**(1), 136–142 (2009)
11. Ruschau, G.R., Yoshikawa, S., Newnham, R.E.: Resistivities of conductive composites. J. Appl. Phys. **72**(3), 953–959 (1992)
12. Zois, H., Apekis, L., Omastová, M.: Electrical properties of carbon black-filled polymer composites. Macromol. Symp. 249–256 (2001)

Enumeration of Driving Mechanisms in Robotics by Combinatorial Analysis Method

P. Mitrouchev, J. Chen, F. Mafray and Y. Zheng

Abstract This research deals with the development of a new method for restricted morphological number enumeration for planar pin-jointed driving mechanisms of mobility 1 and 2. It is based on combinatorial analysis by using the *symmetries* of kinematic chains. Since the restricting criterion is the position of the robot's frame, different cases of symmetries are addressed. New expressions for calculating the number of positions of the actuators in a mechanism are presented. These help to reduce the number of chains by avoiding those that are isomorphic with regard to the position of the robot frame. The expressions offer a corrected formulation for counting the number of actuators for planar mechanisms in robotics. Different examples of applications for various planar pin-jointed driving mechanisms of mobility 1 and 2 are presented.

Keywords Kinematics chains · Driving mechanisms · Combinatorial analysis · Actuators · Symmetry · Conceptual design

1 Introduction

Since the 1960s, a great deal of scientific contributions in mechanism design has addressed the type (morphological) synthesis of kinematic chains. Numerous design methods have been proposed, allowing techniques for development in conceptual design [1–3]. Such methods for number synthesis of planar "pin-jointed" chains (linkages) are: *intuition* and *inspection* [4–6] or *transformation of binary chains* [7].

P. Mitrouchev (✉) · J. Chen · F. Mafray
Université Grenoble Alpes, G-SCOP, 38000 Grenoble, France
e-mail: Peter.Mitrouchev@g-scop.inpg.fr

Y. Zheng
College of Mechanical Engineering and Automation, Huaqiao University, Quanzhou 362021, PR China
e-mail: yq_zheng@hqu.edu.cn

S. Zeghloul et al. (eds.), *Robotics and Mechatronics*,
Mechanisms and Machine Science 37, DOI 10.1007/978-3-319-22368-1_5

Others methods are based on the concept *of Assur's groups* [8, 9] or *Franke's notation* [10]. It can be noted that little work has dealt with the subject of the position of the actuators on the robot in the past decades. Crossley was among the first [4], who stated that it was necessary to position a number of actuators equal to the number of degrees of mobility of the mechanism. In 1985, Lee and Tortorelli [1] proposed to retain as variables of design the position of the actuators. However this did not produce a solution, allowing knowing the number of the possibilities for positioning the actuators in a mechanism. Two years later, Manolescu et al. [8] proposed a method for classifying the articulated planar motor-mechanism with two degrees of mobility (DOM). They examined all the possibilities for positioning the two actuators in the chains, but no formulation was deduced. Recently a new structural synthesis method for generating possible mechanisms with various choices of both R and P pairs and mechanism inversion of planar mechanisms was presented in [9]. It integrates Assur groups (AGs) as elements in the newly developed group-based adjacency matrix. The latter is proposed with the diagonal elements representing three fundamental elements: frame link, driving link, and augmented link AG (AAG). Considering all types of the AGs in the extended adjacency matrix, group permutation and combination are used, and connection forms are generated including variation of the driving link and mechanism inversion. In a previous work [11] a method based on the exploitation of *sub-chain symmetries* was presented. This enabled the number of different possibilities for fixing a frame onto a kinematic chain to be calculated. This *sub-chain symmetries* approach has been applied [12] in order to propose a formulation to enumerate the number of the possibilities to position the actuators for planar driving mechanisms in robotics. However, the formulation has some restriction because it does not take into account the limitation of the number of actuators in a *loop*. Consequently, it is important to be able to fully identify and calculate the number of different possibilities of attaching the actuators in a mechanism resulting from a given kinematic chain. In this context, the aim of this work is to propose a method for calculating all unique inversions (different mechanisms derived from the same linkage by fixing different links), while eliminating the isomorphic ones. Using the symmetries of the chains and the concept of combinatorial analysis, the possibilities for the positioning of the actuators which yield chains with a specified number of links (N), joints (J) and degrees of mobility (M) are assessed. Here only mechanisms with closed loops and single-degree-of-freedom joints of number (J) are concerned. For each type of linkage, the enumeration considers all the possible inversions (selection of fixed link), and for each inversion, considers all the possible placements of the actuator(s) according to the position of the frame. This enumeration only counts actuated mechanisms that are topologically distinct. The results of this study may be useful for robot designers, allowing them to limit the field of research to the possible solutions regarding the position of the actuators of a robot.

2 Preliminary: Basic Notions, Terminology, Definitions and Restrictions

Today, some industrial robots have chains with one or more closed loops, forming a planar or spatial link pin-jointed mechanism. The number of actuators in a *closed loop* must be compatible with its degrees of freedom. This means that the number of actuators is equal to the degrees of freedom of the loop. In other words only one actuator is allowed for loop α, at most two actuators for loop β, and so on [13]. An example of a chain with two degrees of freedom ($M = 2$), nine links ($N = 9$) and eleven joints ($J = 11$) is presented in Fig. 1. The number of independent loops (fundamental circuits) for a planar linkage, is given by the renowned Euler's formula $L = J - N + 1$. After identifying L independent loops, a graph is drawn where each node is a loop and two nodes are connected by e edges if the loops they represent have e joints in common (Fig. 1). Note that this representation of a linkage is not unique since the independent loops are not uniquely defined neither.

In the notation $G_{M\text{-}N}^{i}$: i is the position (number) of the linkage within the list. If there are common joints between two loops the latter are called *neighbour loops* (β and α of Fig. 1 for example), otherwise they are called *distinct loops*. A *mechanism* is a linkage with one link said to be fixed and an *input link*. This may be indicated in the type graph by renumbering, so that the fixed link is link 1. An actuated mechanism has two types of joints: *passive* and *actuated*. To distinguish between them, an adjacency matrix is defined for an actuated mechanism as a matrix A with elements $a_{i,j}$ such that: $a_{i,j} = 1$, if links i and j are connected by a passive joint; $a_{i,j} = 2$, if they are connected by an actuated joint; $a_{i,j} = 0$, otherwise. There are several mechanisms that correspond to a kinematic chain. Their number corresponds to the number of the different possibilities for attaching the frame. As there are chains with symmetries (simple, double, central) the number of possible mechanisms issuing from these chains may decrease as some solutions will be isomorphic depending on the position of the frame and the actuators. Here, only planar pin-jointed mechanisms often applied in robotic conceptual design are investigated.

chain	linkage	loops	graph
$G_{2\text{-}9}^{15}$	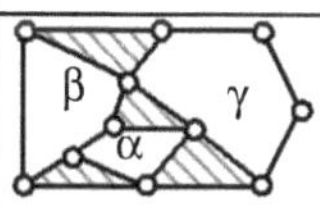	β-α-γ	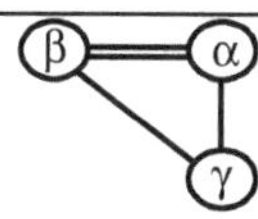

Fig. 1 Linkage represented by its kinematic *chain*, *loops* and *graph*

3 Method for Calculating the Number of Actuator Mechanisms

It is assumed that only mechanisms of unexceptional mobility, where Tchebychev-Grübler's condition holds, are considered and where exactly M joints of a M-DOF linkage are actuated. The position of the end-effector of the robot is not taken into account. Only the position of the frame is retained, and the *isomorphism* of the chain depending on the position of the actuator(s). All the joints have the same status and can accommodate one actuator. The latter can be located indifferently on the joints of any link of the kinematic chains whatever its position is relative to the frame. The proposed method is based on the concepts of *loop* and *graph* using the *symmetries* of the chain and the *combinatorial analysis*. This allows the calculation of the number of possibilities for positioning the actuators of a kinematic chain, by taking into account the position of the frame. Concerning the position of the actuators in a mechanism if multiple solutions appear this is the result of symmetries, which may be *axial* (single, double…) or *central*. Note that type graph $G2$ is *isomorphic* to type graph $G1$ if there exists a renumbering of the links of $G2$ whose adjacency matrix is identical to the adjacency matrix of $G1$. Otherwise, the two graphs are *non-isomorphic* or simply *distinct*. The renumbering of links is represented as an NxN permutation matrix, say P, where element $p_{i,j} = 1$ if link i is renumbered as link j, otherwise $p_{i,j} = 0$. Then, after renumbering, the new adjacency matrix is P^TAP. Thus, isomorphism between linkages with adjacency matrices A and B means that there exists a permutation matrix P such that $P^TAP = B$. This notion of isomorphism extends to actuated mechanisms, where the entries of A and B can be 0, 1, or 2. Note that the proper definition of topological symmetry is the renumbering of a graph such that the adjacency matrix remains unchanged. In other words, a symmetry is a permutation $P \neq I$ such that $P^TAP = A$.

3.1 *One Degree of Freedom Chains*

For chain G_{1-8}^{3} (Fig. 2a) with no axis of symmetry contains $L = 3$ loops and $p = 3$ *couples of neighbour loops*. In this case there are L possibilities to form different combinations between the positions (situations) of the actuator. In fact all joints J of the chain may be occupied by the actuator. Consequently the number of different possibilities for fixing the motor onto a chain (called *motor choice* M_c) is:

$$M_c = N \cdot J \tag{1}$$

Some chains as G_{1-8}^{12} in Fig. 2b, have one axis of symmetry. As the possibilities for attachment of the actuators are compared to the position of the frame, the latter may be cut or not cut by the axis of symmetry. Thus M_c is:

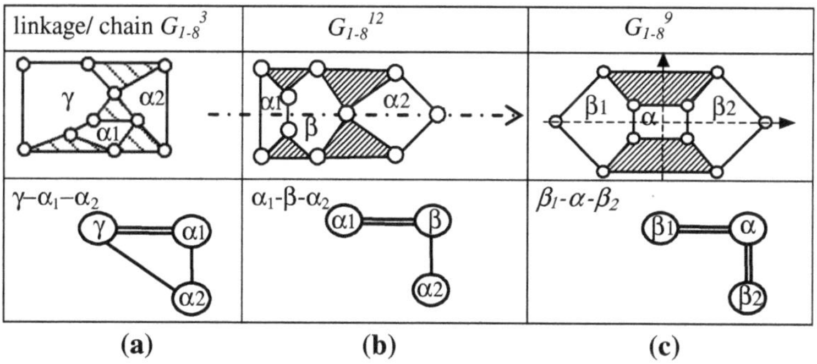

Fig. 2 Chains with: **a** no axis of symmetry; **b** single symmetry; **c** double symmetry (M = 1)

$$M_c = N_{sp}J + N_{ss}(J + J_c)/2 \quad (2)$$

where: $N_{sp} = (N - N_c)/2$ is the number of *symmetrical pairs* in the chain (here $N_{sp} = 3$), N_c is the number of links cut by the axis of symmetry (here $N_c = 1$) and N_{ss} the number of the *symmetrical pairs in the sub-chain* (here $N_{ss} = 1$).

Some chains have double axes of symmetry (G_{1-8}^{9} in Fig. 2c). As in the case with single symmetry, the frame may be cut or not cut by the axes of symmetry. Consequently, the number of the possibilities to attach the motor is:

$$M_c = N_{dsp}J + (J_{cx} + J_{ncx}/2) + (J_{cy} + J_{ncy}/2) \quad (3)$$

where: $N_{dsp} = (N - Nc)/4$ is the number of *double symmetrical pair* (here $N_{dsp} = 1$), J_{cx}, J_{cy} and J_{ncx}, J_{ncy} are respectively the number of joints cut and not cut by the axes of symmetry (***Ox***) and (***Oy***).

3.2 Two Degrees of Freedom Chains

Conventionally, the position of the first actuator is arbitrarily chosen relatively to the frame, the second actuator is positioned onto the remaining joints of the mechanism. For a *chain with no axis of symmetry* (Fig. 3a) there are "*J choose two*" $J(J - 1)/2$ possibilities to form different combinations between the positions of the two actuators for a given position of the frame. From this number, $(6n_\alpha - N_{cl})$ possibilities are withdrawn since the α loops do not allow two actuators to be positioned (n_α is the number of the α loops in the graph and N_{cl} is the number of *common binary links* between two neighbour α loops). Since the chain has no symmetry, there are N possibilities for positioning the frame. Thus M_c is given by:

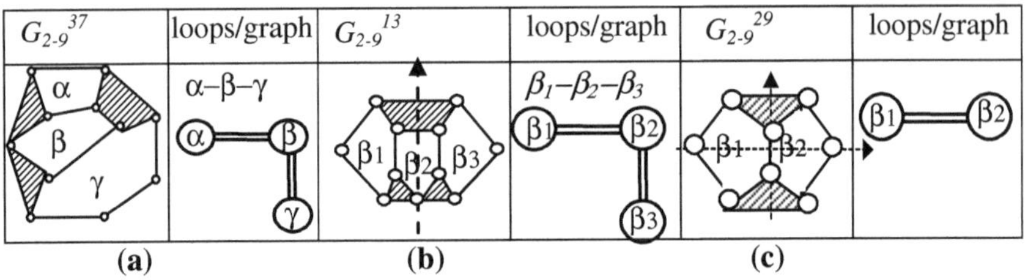

Fig. 3 Chains with: **a** no axis of symmetry; **b** single symmetry; **c** double symmetry (M = 2)

$$M_c = N[J(J-1)/2 - (6n_\alpha - N_{cl})] \tag{4}$$

Consider a chain with single symmetry ($G_{2\text{–}9}^{13}$, Fig. 3b) whose graph contains L loops and p *couples of neighbour loops*. As there are L possibilities of forming different combinations between the positions of the actuator, M_c is:

$$M_c = 1/2\{N[J(J-1)/2 - (6n_\alpha - N_{cl})] + N_c J_\alpha\} \tag{5}$$

where: J_α is the number of symmetrical pairs of joints (situated symmetrically according the axis of symmetry) which do not belong to an α loop and N_{cl} is the number of *common binary links* between two neighbour α loops.

For the *chains with double symmetry* (Fig. 3c) there are $\sum_{1}^{L} i$ possibilities for the formation of different combinations between the positions of the two actuators. Thus, the general expression for M_c is:

$$M_c = 1/4(Nc+1)[J(J-1)/2 - (6n_\alpha - N_{cl})] + (N_{cx} + N_{cy})/2\{1/2[J(J-1)/2 + J_\alpha] - (6n_\alpha - N_{cl})\} + 1/4(J_\alpha + 2J_c) \tag{6}$$

4 Computer Program and Examples of Application

The method described above has been implemented for standard personal computers operating under Silicon Graphics/Linux. The computer program is written in standard C^{++} language. It generates the list of the number of possibilities that exist for each case for mechanisms: with and without symmetries, with one and two degrees of mobility (*DOM*). For each linkage, first a sub-program accepts (reads) the number of *DOF*, links and joints, and creates and writes the base-files for that case. Another sub-program uses these base-files as input files. Then both of them enumerate all unique kinematic inversions (driving mechanisms) for 1 and 2 degrees of freedom (M = 1, 2).

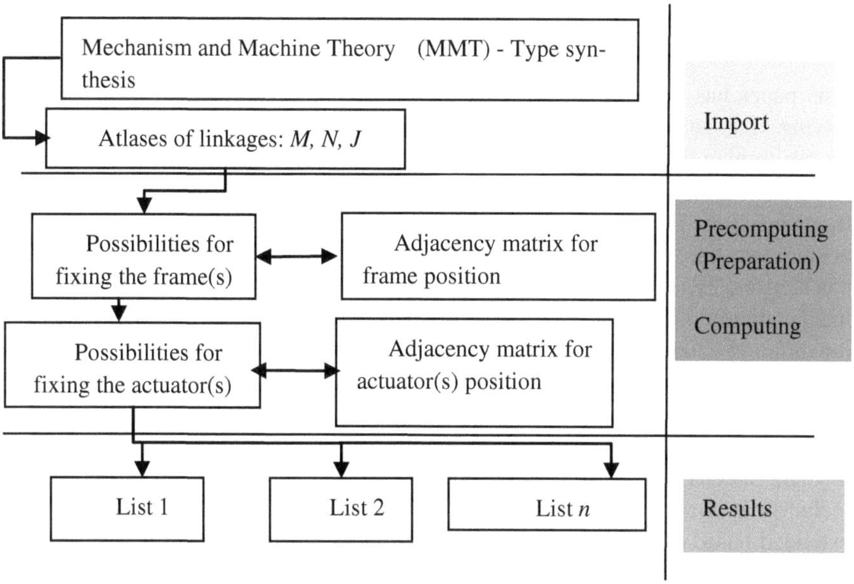

Fig. 4 Pipeline of the model processing scheme with the simplified block diagram for mobility one and two driving mechanisms enumeration

The simplified block diagram of the computer program integrated in the model processing workflow is presented in Fig. 4. In the model processing workflow the atlases of linkages are acquired through files coming from CAD software.

Some results for M_c for different mechanisms with one and two *DOM* and different numbers of links and joints are presented here below:

- for the *two* G_{1-6}^{i} linkages with one degree of mobility, six links and seven joints, the number M_c of distinct actuated (driving) mechanisms are 18 and 11 for Stephenson and Watt chains respectively,
- for the *sixteen* G_{1-8}^{i} linkages with one degree of mobility, eight links and ten joints, the number M_c ranges from 10 to 80,
- for the *two-hundred-thirty* G_{1-10}^{i} linkages with one degree of mobility, ten links and thirteen joints, M_c ranges from 31 to 128,
- for the *three* G_{2-7}^{i} linkages with two degrees of mobility, seven links and eight joints, the values of M_c are 28, 78 and 79,
- for the *thirty-five* G_{2-9}^{i} linkages of mobility two, nine links and eleven joints, the number M_c of distinct actuated mechanisms ranges from 171 to 441.

The *CPU* time to calculate M_c for the 230 chains with 1 degree of mobility, ten links and thirteen joints is 3.53 s (Intel Pentium *P4C:* 2.8 GHz computer running Linux).

5 Conclusions

This paper has presented a method for restricted morphological enumeration for driving kinematic mechanisms with regard to the number of different possibilities for positioning the actuators. The method is based on the exploitation of symmetries and on the concept of loop and combinatorial analysis by taking into account the frame position. The principal point considered relates to the different possibilities of positioning one or two actuators according to the degree of mobility of the mechanism. By successive immobilization of one of the links, the alternative independent motor mechanisms were obtained, from a structural point of view and for different cases of symmetries. Thus, an analytical method enabling the validation of the results obtained by applying the method for chains with one and two degrees of mobility is proposed. Currently the proposed method is restricted to planar pin-jointed mechanisms. The future goal is to extend the method to spatial linkages mechanisms. For this purpose the atlases of spatial topologies will be included at the number enumeration level and the computer program will be extended based on mobility and connectivity laws of the Mechanism and Machine Theory.

References

1. Lee, K., Tortorelli, D.A.: CAD system for designing robotic manipulators, In: Proceedings of IEEE International Conference on Robotics and automation, Saint-Louis, pp. 376–380. doi: 10.1109/ROBOT.1985.1087336 (1985)
2. Gonzales-Palacois, M., Angeles, J.: USyCaMs, a software package for the interactive synthesises of cam mechanisms. In: Proceedings of 1-st Integrated Design and Manufacturing in Mechanical Engineering I.D.M.M.E.'96, vol. 1, pp. 85–94. Nantes France (1996)
3. Gogu, G.: Mobility of mechanisms: a critical review, Elsevier, IFToMM. J. Mech. Mach. Theory **40**, 1068–1097 (2005)
4. Crossley, F.R.E.: A contribution to Grubler's theory in the number synthesis of plane mechanisms. Trans ASME J. Eng. Ind. **86**, 1–8 (1964)
5. Woo, L.S.: Type synthesis of plane linkages. ASME J. Eng. Ind. Ser. B **89**, 159–172 (1967)
6. Freudenstein, F., Dobrjanskyj, L.: On a theory for the type synthesis of mechanisms. In: Proceedings of the 11-th International Congress of Applied Mechanics, pp. 420–428. Springer, Berlin (1964)
7. Mruthyunjaya, T.S.: A computerized methodology for structural synthesis of kinematic chains: part 1-formulation. Mech. Mach. Theory **19**(6), 487–495 (1984)
8. Manolescu, N., Tudosie, I.: Structural and kinematic synthesis of planar kinematic chain (PKC) and mechanisms (PM) with variable chain during the work. In: Proceedings of the 7-th World Congress, The Theory of Machines and Mechanisms, 1, Sevilla, pp. 45–48 Spain (1987)
9. Li, S., Wang, H., Dai, J.S.: Assur-Group inferred structural synthesis for planar mechanisms. Mech. Robot. **7**, 041001 (2015). doi:10.1115/1.4029116
10. Davies, T., Crossley, F.R.E.: Structural analysis of plan linkages by Frankes condensed notation. Pergamon Press, J. Mech. **1**, 171–183 (1966)

11. Mitrouchev, P.: Sub-chain symmetry approach for morphological choice of planar mechanisms in robotics. Eur. J. Mech. (A Solids) **25**(1), 166–188 (2006)
12. Mitrouchev, P.: Formulation for actuators' number enumeration for main planar chains in robotics. Eur. J. Mech. (A Solids) **27**(4), 622–646 (2008)
13. Mitrouchev, P.: Symbolic structural synthesis and a description method for planar kinematic chains in robotics. Eur. J. Mech. (A Solids), ISSN: 0997-7538, Ed. Elsevier, N°20(2001), 777–794

Part II
Humanoid and Legged Robotics

Design and Experiments on a New Humanoid Robot: TIDOM

A. Eon, P. Seguin, M. Arsicault and S. Zeghloul

Abstract This paper is focusing on a new humanoid robot developed by the ROBIOSS team from the Pprime Institute. After a short presentation of its mechanical and control design, a gait generation algorithm based on COG planning is proposed in order to synthesize basic test motions. Two preliminary experiments are carried out on the prototype of the locomotor system, in order to validate the control scheme as well as the global design of the locomotor system of this fully electrically actuated robot.

Keywords Humanoid robots · Mechanical design · Control · Gait generation

1 Introduction

The design and control of anthropomorphic walking robots is a great issue for researchers all over the world [1]. The mechanical design of a humanoid robot, especially defining the actuators requirements (speed, torque, power), is very complex. This is due to several aspects. First of all, these requirements are task-dependant (the robot will not use its actuators in the same way to sit on a chair or to walk dynamically). Furthermore, a biped robot is a complex machine that involves an important number of Degrees Of Freedom (DOF) which leads to highly coupled phenomena. Finally, designers should take care of mass repartition and also of geometric dimensions to build an anthropomorphic humanoid robot. This last "key point" is very important in order to design robots as close as possible to humans.

Several platforms able to walk already exist such as HRP-4 from the AIST and Kawada Industries [3], BHR5 from Beijing Institute of Technology [7].

A. Eon (✉) · P. Seguin · M. Arsicault
Department GMSC, Pprime Institute CNRS - University of Poitiers - ENSMA, UPR 3346 Poitiers, France
e-mail: antoine.eon@univ-poitiers.fr

S. Zeghloul
Institut PPRIME, UPR 3346, University of Poitiers, Poitiers, France

S. Zeghloul et al. (eds.), *Robotics and Mechatronics*,
Mechanisms and Machine Science 37, DOI 10.1007/978-3-319-22368-1_6

The ROBIOSS team from the Pprime Institute, which has already designed the BIP2000 walking robot [5] in the past, has decided to create a new humanoid robot, TIDOM. This paper presents the mechanical and control design of its locomotor system. In order to validate the global design some preliminary gait experiments obtained by COG planning are tested on a physical prototype.

2 Presentation of TIDOM

TIDOM is a new humanoid robot, with a skeleton made of aluminum alloy to reduce its weight, with 26 DOF, 12 for the locomotor system and 14 for the upper part. It will weigh approximately 16 kg for 800 mm tall. This paper is focusing on its locomotor system (Fig. 1), already fully available. This device is weighting 10.8 kg and is 495 mm tall from its feet to its top plane.

2.1 Mechanical Design

The mechanical design of the locomotor system of TIDOM (Fig. 2) is quite different from other humanoid robots [3, 7] which are usually designed with a serial chain from the ankle to the hip, formed by 3 rotational joints whose axis are parallel one to the other. On TIDOM, a rotational joint was built on the knee to allow the rotation of the thigh on its axis. This new design is interesting for mass distribution, because, the motor ensuring the actuation of the rotational motion of the thigh could be placed on a lower position.

Fig. 1 TIDOM

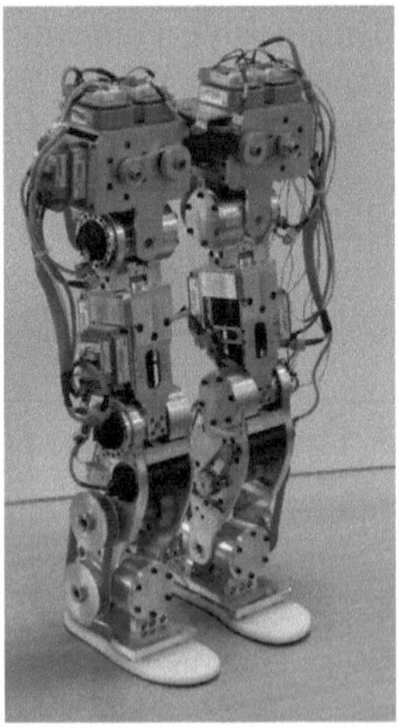

Fig. 2 Kinematic architecture

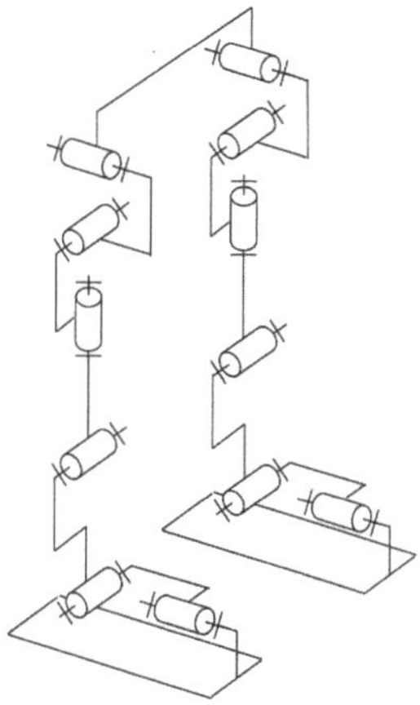

The method proposed to generate preliminary motions is to plan the COG of the robot in order to ensure a balanced motion. A CAD model was used to estimate all the needed parameters, such as the position of the center of gravity, the mass, the inertia, the length of each limb as well as the articular limit on every joints (Tables 1 and 2).

Table 1 Articular limits

Articular coordinate	Joint	Min. angle (°)	Max. angle (°)
q_1/q_{12}	Ankle: Abd./Add.	−45	25
q_2/q_{11}	Ankle: Flex./Ext.	−26	25
q_3/q_{10}	Knee: Flex./Ext.	−70	0
q_4/q_9	Hip: Rotation	−40	43
q_5/q_8	Hip: Flex./Ext.	−10	33
q_6/q_7	Hip: Abd./Add.	−20	20

Table 2 Segment properties

Rigid body	Mass (kg)		Length characteristics
S_i	m_i		
S_1/S_{13}	Foot	0.43	$L_{1\_longitudinal}$ = 185.3 mm ; $L_{1\_transversal}$ = 79.1 mm
S_2/S_{12}	Ankle	0.56	H_2 = 63.5 mm
S_3/S_{11}	Shank	1.22	L_3 = 160 mm
S_4/S_{10}	Thigh	1.08	L_4 = 168.2 mm
S_5/S_9	Inf. hip	0.35	
S_6/S_8	Sup. hip	0.55	L_{hip} = 136 mm (distance between hip centers)
S_7	Pelvis	2.41	

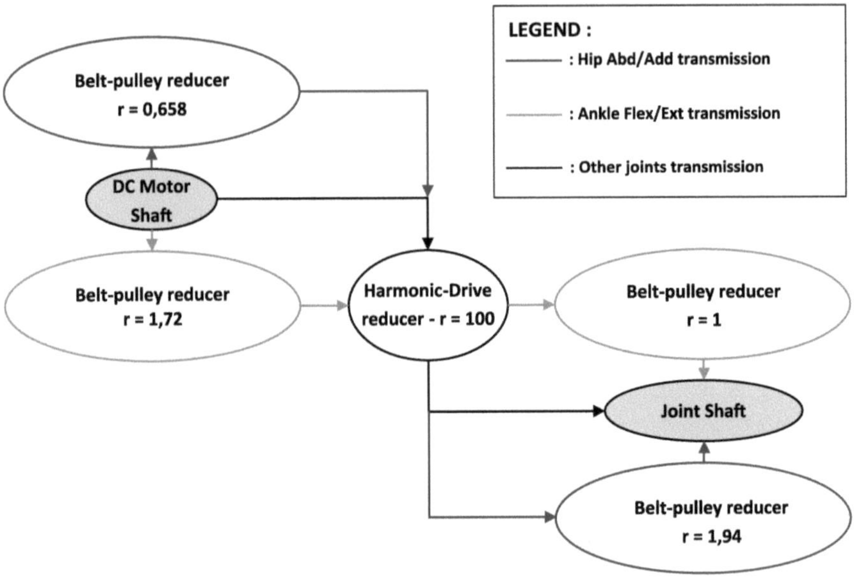

Fig. 3 TIDOM's different transmissions

The motion transmission between motors and limbs is depending on the considered joint as shown in Fig. 3. Most of the joints are actuated with a DC motor coupled to a harmonic drive reducer on its shaft (black arrows in Fig. 3). For the hip and the ankle joints, a universal joint is needed. It was then not possible to keep this design to actuate the two required joints (flexion/extension and abduction/adduction). This would have led to large transversal dimensions. As a result, the proposed solution was to move the second motor in the limb, as well as its reducer, let us call it "moved-transmission". Two belt/pulleys transmissions are used to adapt the power of the motor to the joint axis (blue and red arrows in Fig. 3) leading to different articular performances on the considered joint axis (Table 3). There are two different types of Faulhaber 24 V DC motor embedded in TIDOM: 3257 and 3863 references developing respectively 83 and 214 W. The 3863 DC

Table 3 Articular performances

	DOF: 1,3,4,5	DOF: 2	DOF: 6
Global reduction ratio	100	172	128
Nominal speed (RPM)	52.1	32	43.1
Max. speed (RPM)	70	40.7	43.6
Nominal torque (Nm)	3	3	5.8
Max. repetitive torque (Nm)	5.1	5.1	9.9
Impulse torque (Nm)	8.1	8.1	15.7

motors are used for the "moved-transmission" joints in order to support the increased friction for these particular axes due to multiple revolute joints and reducers, and also to increase the articular performances for the abduction/adduction joint of the hip.

2.2 *Control Architecture*

The control architecture (Fig. 4) is based on distributed axis controllers (one per axis), embedded on the robot and placed close to the motors. These controllers are linked to an industrial Programmable Logic Controller (PLC) with 2 CAN buses (one per leg, to increase the bandwidth). The supervision is made by a distant PC connected to the PLC over an Ethernet link.

The PLC is referenced X20CP3486 from B&R Automation with an Intel Celeron 650 processor inside and an industrial real-time OS called Automation Runtime.

The axis controllers are Elmo Whistle referenced SOL-WHI A10/60E01. They are extremely compact intelligent digital servo drives and are able to drive motor with a peak power up to 480 W and to control the axis loop (Current, Speed, Position). It receives periodically (10 ms) its orders from the PLC and sends back its state, position and eventual errors over CAN Bus.

TIDOM is equipped with 12 encoders (one per motor, 512 pulses/rev), directly mounted on the motor shaft. This information is used by the axis controller to drive

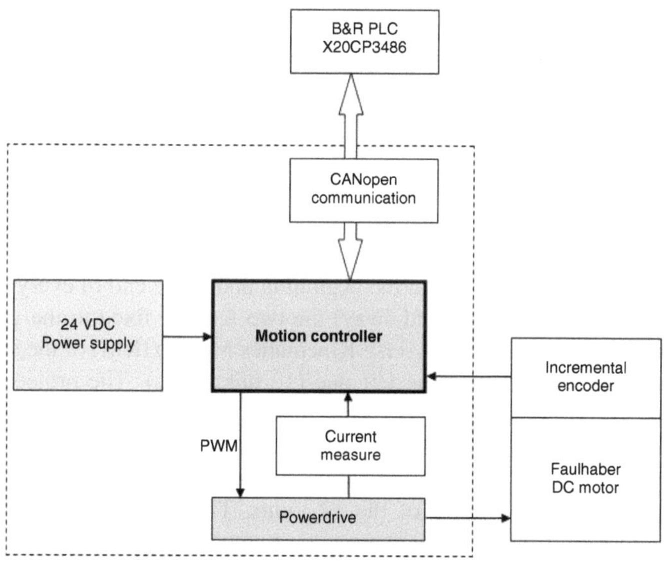

Fig. 4 Control architecture of TIDOM

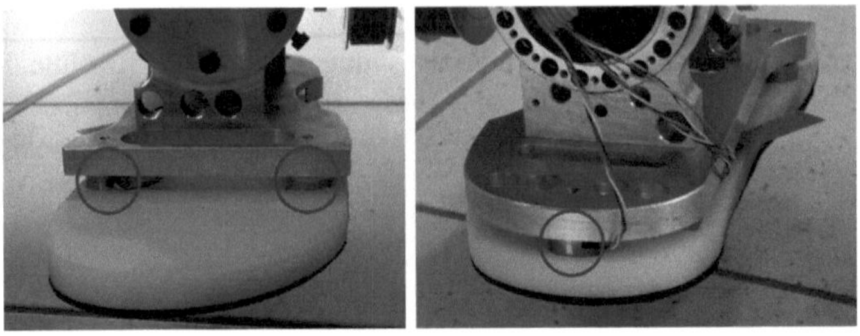

Fig. 5 Force sensors

the motor in close loop (position and speed). The articular displacement is computed from this measure and sent to the PLC over CAN bus.

TIDOM's feet are equipped with 3 ENTRAN ELFM unidirectional force sensors to measure the normal contact force in three different points (Fig. 5). With this, it is possible to compute the measured ZMP of the robot during the motion. The acquisition is made directly on the PLC with adequate additional modules.

3 Preliminary Experiments

The method used to generate motions is based on a particular case of ZMP criterion [6]: the dynamic balance of the motion is ensured if the ZMP is within the support polygon. Instead of using the ZMP, the gait generation is based on the projection on the ground of the COG of the complete multibody model.

3.1 Quasistatic Gait Generation

The first step needed to synthesize a motion consists of defining the reference path of the COG, its projection being located within the support polygon. To avoid torque peaks, these references are built as piecewise polynomial functions of time with null speed and acceleration at the beginning and at the end of every phase. For the motions tested here (squat and sway) the two feet are fixed to the ground.

The next step is to solve the Inverse Kinematics Model (IKM) of the serial chain defined from the right foot to the left one (12 unknowns). The projection of the COG needs to follow the reference paths (max 3 nonlinear equations). There are also 8 nonlinear equations for the postural constraints (6 for the orientation and the position of the left foot, 2 to avoid pitching and rolling of the pelvis S7). The last constraints are the range limits of the 12 joints. The method used to solve this redundant problem is to minimize at every time step k a criterion J given in Eq. (1)

under constraints with fmincon function from Matlab, using the Sequential Quadratic Programming (SQP) algorithm described in [4].

$$J = \frac{1}{2} \cdot \Delta q^t \cdot I_{d_{12\times 12}} \cdot \Delta q \tag{1}$$

where $\Delta q = q_k - q_{k-1}$ and q_k is the articular coordinate vector at time step k.

The solution that minimizes this criterion leads to the one that limits the variation of the joint angles during the motion. Self-collisions and joint limits are checked after the IKM.

3.2 First Experiment: Squat Motion

The first basic task tested on the robot is a squat motion. The COG moves only on the vertical axis (Fig. 6). The robot starts with stretch knees and lowers its COG 10 mm down. The total time of the motion is set to 2 s.

Figure 7 shows the position of the feet on the ground plane as well as the COG reference (in red) and the computed ZMP (in green) for the synthesized motion. The ZMP remains very close to the projection of the COG and stay inside the sustentation polygon. Figure 8 is a snapshot of the motion tested on the robot (from top left to bottom right). The behavior of the robot is coherent with the computed one.

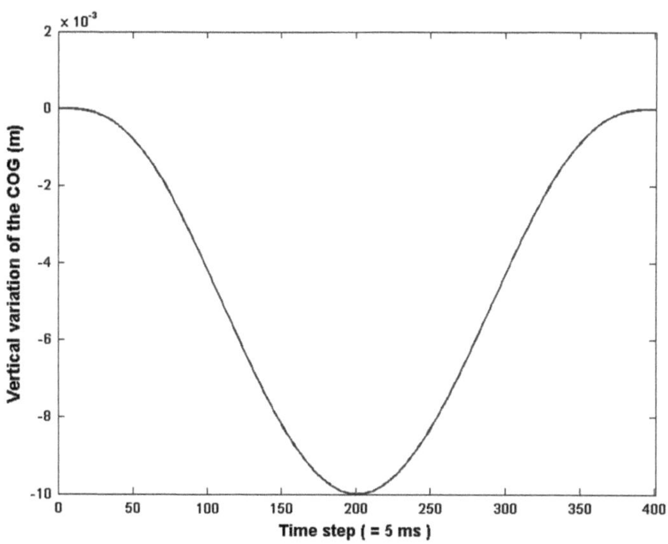

Fig. 6 Squat motion: COG reference trajectory

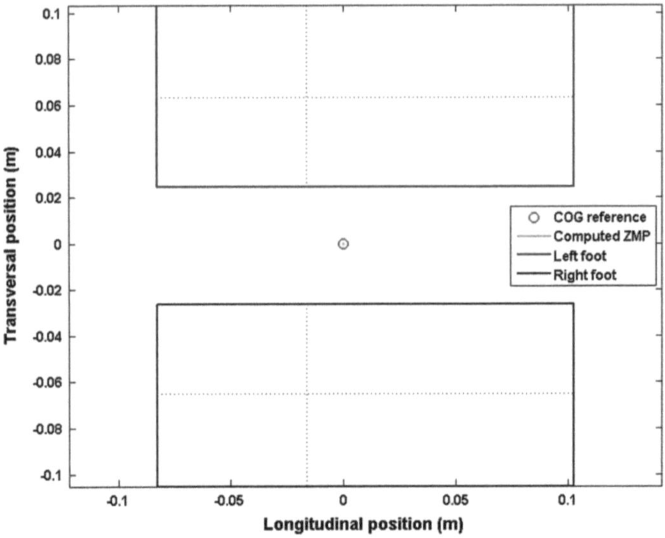

Fig. 7 Squat motion: COG reference trajectory and computed ZMP on the ground

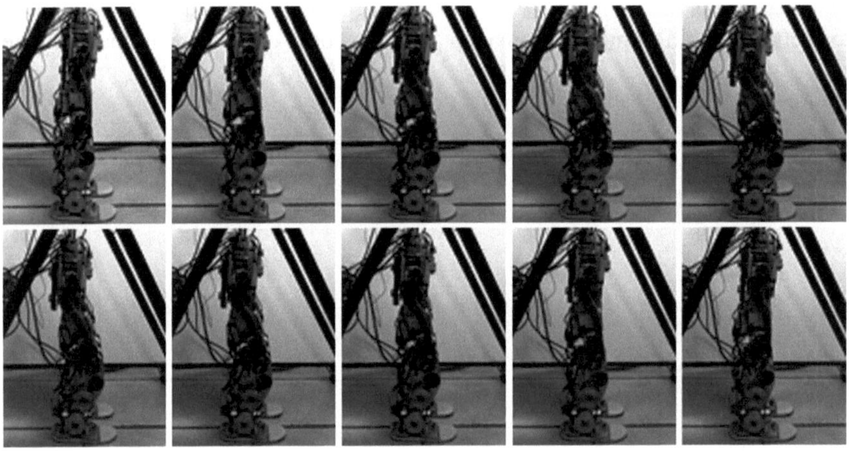

Fig. 8 Squat motion snapshot

3.3 Second Experiment: Sway Motion

In order to test the behavior of the robot in a transversal plane, a sway motion is tested. The COG reference paths are given in Fig. 9.

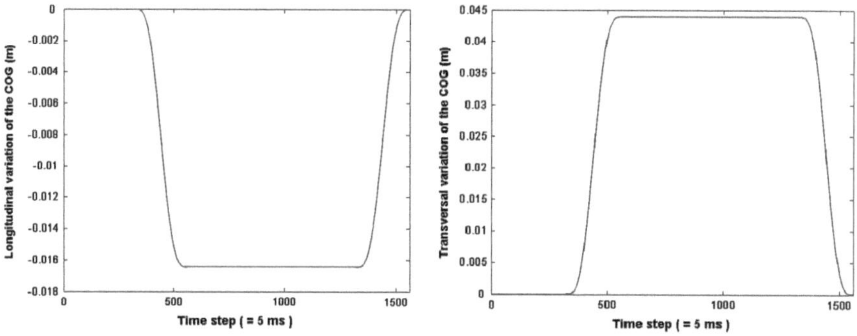

Fig. 9 Sway motion: COG reference trajectory

The robot starts with 30° initial knee flexion and transfers its COG on the left foot. The transfer phases last 1.2 s. The robot keeps its COG fix on its left foot during 3.8 s. The maximum transversal speed of the COG is 0.039 m/s.

Figure 10 shows the position of the feet on the ground plane as well as the COG reference (in red) and the computed ZMP (in green) for the synthesized motion. The ZMP overshoots the COG reference trajectory (8.7 mm) but still remains inside the sustentation polygon. This is a typical behavior of humanoid robots during double support transfer phases, due to a large displacement in a short time. To correct this behavior, a ZMP planning method should be used [2]. Figure 11 is a snapshot of the motion tested on the robot (from top left to bottom right). The behavior of the robot is coherent with the computed one.

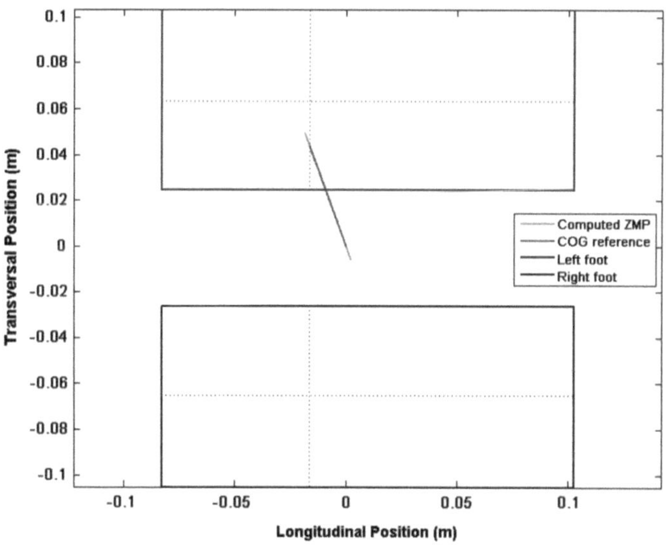

Fig. 10 Sway motion: COG reference trajectory and computed ZMP on the ground

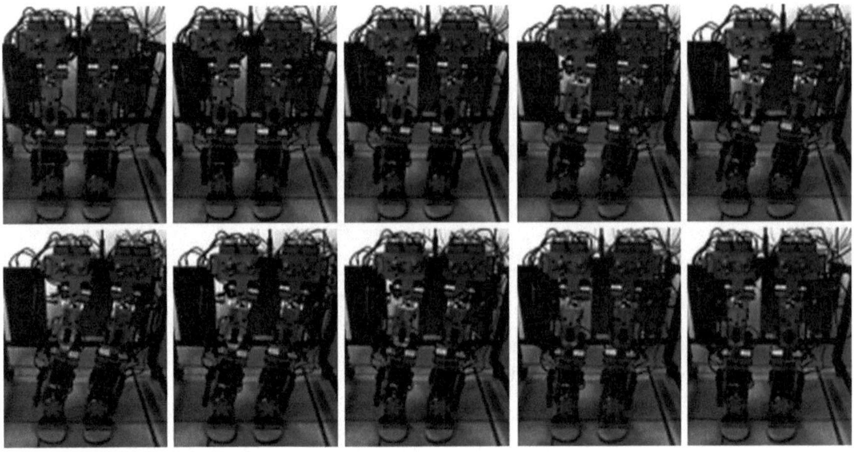

Fig. 11 Sway motion snapshot

4 Conclusions

The first experiments carried out on TIDOM have shown the good behavior of the robot in terms of mechanics, control and balance. These motions are parameterized and additional tests could be carried out with different durations and displacements without any difficulties. The COG planning is sufficient for basic motions, but a more complex motion algorithm will be used to synthesize and to test walking or stepping motions. This will be the next step of this study.

Acknowledgments This work has been sponsored by the French government research program « Investissements d'Avenir » through the Robotex Equipment of Excellence (ANR-10-EQPX-44) and is supported by the Poitou-Charentes region 2007-2013 (program project 10 Images and interactivities), in partnership with the European Union (FEDER/ERDF, European Regional Development Fund).

References

1. Hayder, F.N., et al.: Modeling, stability and walking pattern generators of biped robots: a review. Robotica **32**, 907–934 (2014)
2. Kajita, S., et al.: Biped walking pattern generation by using preview control of zero-moment point. In: Proceedings of IEEE International Conference on Robotics and Automation ICRA2003, Taipei, 2, pp. 1620–1626 (2003)
3. Kaneko, K., et al.: Humanoid robot HRP-4-humanoid robotics platform with lightweight and slim body. In: Proceedings of IEEE/RSJ International Conference on Intelligent Robots and Systems IROS2011, San Francisco, pp. 4400–4407 (2011)
4. Nocedal, J., Wright, S.J.: Numerical Optimization, 2nd edn. Springer Series in Operations Research, Springer Verlag (2006)

5. Sardain, P., et al.: An anthropomorphic biped robot : dynamical concepts and technological design. IEEE Trans. Syst. Man Cybern. **28**, 823–838 (1998)
6. Vukobratovic, M., Borovac, B.: Zero-moment point—thirty five years of its life. Int. J. Humanoid Rob. **1**, 157–173 (2004)
7. Yu, Z., et al.: Design and development of the humanoid robot BHR-5. Adv. Mech. Eng. Article ID852937, 11 p (2014)

Experimental Inspiration and Rapid Prototyping of a Novel Humanoid Torso

D. Cafolla and M. Ceccarelli

Abstract In this paper, problems for design and operation of solutions for a robotic torso are discussed by referring to a background with experiences and designs that have been developed at LARM in Cassino, Italy. A new solution is presented with conceptual views as waist-trunk structure that makes a proper partition of the performance for walking and arm operations as sustained by a torso.

Keywords Humanoid robots · Design · Torso · Rapid prototyping

1 Introduction

Human torso pays an important role in human body in both capability and performance during actions. Its anatomy is quite complex as composed of several systems whose integration ensures a very efficient functioning. Understanding of the anatomy of human torso is very helpful for achieving successful humanoid designs to not only replicate human anatomy but also even for better functioning of human-like operations.

Human-like torso features are essential in humanoid robots. Humanoid robots have been designed and built in the last decades as much as the technology has permitted complex integration of multi-body systems with suitable capabilities in actuators, sensors, control equipment, artificial intelligence, and interfaces [1, 2]. They have been used within lab environments, but even in some implemented applications with the main purposes of exploring design solutions and performance capabilities in mimicking humans and their actions.

D. Cafolla (✉) · M. Ceccarelli
LARM, University of Cassino and South Latium, Cassino, Italy
e-mail: cafolla@unicas.it

M. Ceccarelli
e-mail: ceccarelli@unicas.it

S. Zeghloul et al. (eds.), *Robotics and Mechatronics*,
Mechanisms and Machine Science 37, DOI 10.1007/978-3-319-22368-1_7

Recently, solutions have reached market availability with user-oriented performance, like for example Nao [3] and Pino [4], although they still require high expertise for a proper operation and use. Humanoid robots are still studied and developed by many research groups. The Japanese have been leaders in this research field since the beginning of modern developments of humanoid robots. A first prototype named WABOT-1 started walking since 1972 [5]. This and many other prototypes have been developed at Waseda University, Tokyo [6]. In Japan other relevant prototypes have been also built such as BLR-G2 at Gifu University [7], and ASIMO by Honda Motor Corporation [8], and many other research groups are still working in this field. In fact, the National Institute of Advanced Industrial Science and Technology in Tokyo (AIST) is sponsoring a humanoid project, which has begun in 1998, by supporting 15 research institutes throughout Japan [9], with last HRP-2P solutions. Also in USA and Europe there are many researchers working on humanoid robots. Some interesting prototypes of humanoid robot are: JOHNNIE and ERMES at the University of Munich (Germany) [10, 11]; ARMAR at University of Karlsruhe (Germany) [12]; ELVIS at Chalmers University in Goteburg (Sweden) [13]; ROBONAUT, DART and EVA by NASA (USA) [14]; COCO, COG and KISMET at MIT (USA) [15]; the hydraulic and pneumatic humanoid robots for entertainment by SARCOS (USA) [16]; Humanoid Robot CENTAUR at KIST (South Korea) [17]; The Human Robot at KAIST (South Korea) [18]; JOHNNY WALKER at the University of Western Australia (Australia) [19].

In general a special attention on torso structures in humanoid robots can be understood as focused along two main directions, namely design developments of sophisticated solutions towards fully android behaviors, and design arrangements of limited-skill solutions towards low-cost easy-operation systems. Most interesting advances are worked out in project developments for sophisticated solutions, like for example for the above mentioned Honda Asimo, Wabian, HRP humanoids. But those advances are also used in the second category of approaches where challenges are attached to achieve torso designs and operations with less complicated solutions for a more user-friendly operation and practical implementation.

This paper is aimed to present a torso solution for humanoid robots with a design that is based on a driving parallel manipulator kinematic architectures (PKM).

2 The Attached Problem

In general, humanoid robots are developed to fully mimic the structure and operation capabilities of human beings within research and development projects both with application purposes and study aims.

Human torso is a central platform of human body to which limbs are attached and within which most of the organs are located. Its anatomy is very compact and complex, mainly consisting of a skeleton system and muscle assembly as concerning with biomechanical functions, Fig. 1, [20]. Main requirements and design structures

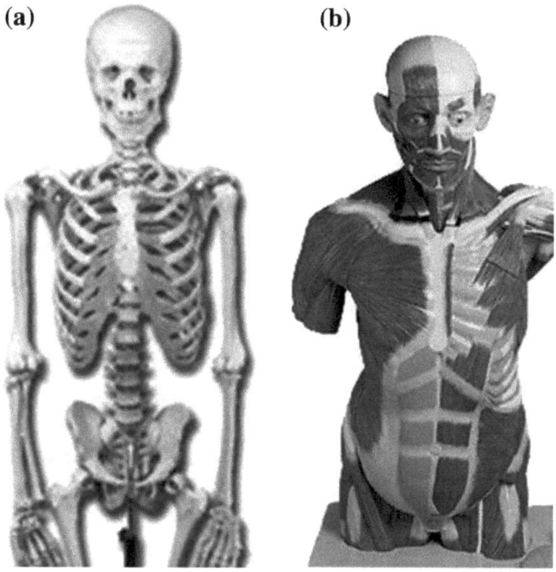

Fig. 1 Anatomy of human torso. **a** Skeleton structure. **b** Muscle system

of torso for a humanoid robot can be summarized as in Fig. 2, in which a segmentation of the system emphasizes primary functions of subsystems as specifically focused on mechanical properties related to trunk and waist. It is remarkable that although the gross motion of the torso can be described with 3 d.o.f.s (degree of freedom), the human torso is a multi-d.o.f. system whose d.o.f. identification can be related both to the gross d.o.f.s and multiple spine d.o.f.s.

There are hundred pairs of muscles and ligaments, complex blood and nervous systems with different functions to make a human torso an important part of the human body with high performance capabilities. These characteristics are those that in general are the main design goals in designing such a torso platform in humanoid robots. This aspect is stressed in the scheme of Fig. 2 where additionally the torso is partitioned in trunk and waist to which different main functions are demanded. The

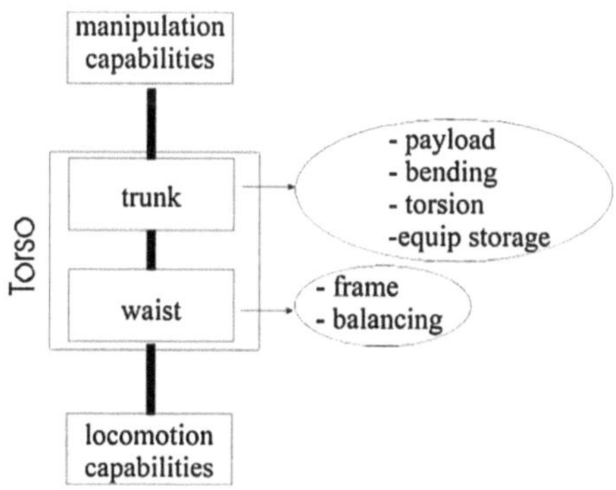

Fig. 2 A scheme for torso main features towards robot design

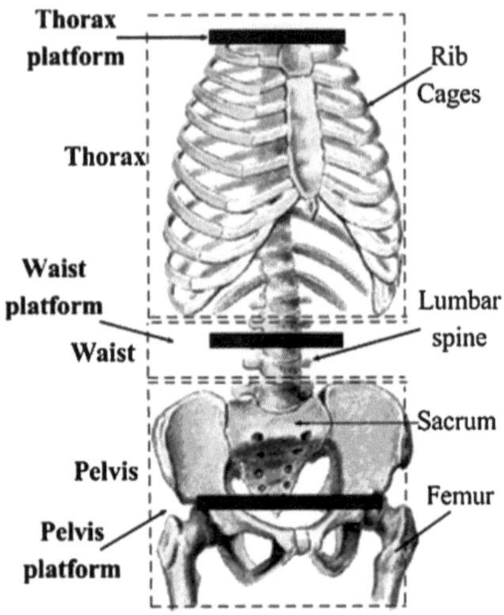

Fig. 3 A scheme for platform identification in human torso [21]

trunk task can be recognized in motion capability and organ/equipment storage, and the waist can be characterized by frame structure and balancing purposes.

By considering gross motions and main features of human torso it is also possible to model the torso system with three main body segments as reference platforms likewise it is proposed in Fig. 3, [21] according to the idea of segmentation of parts in a robotic humanoid structure. The trunk segmentation can be considered as referring to the thorax platform connected to the waist platform whereas the waist segment can be considered as related to the waist platform and pelvis platform.

Alternatively, as it is a widely used solution in current humanoid robots, the torso structure is designed with a serial chain structure with well identified actuators for the corresponding gross motions, like in the case in Fig. 4, [22], where the platform body of the torso is reduced just to the connection link. The advantages of such a design can be considered mainly in a fairly simple mechanical design and direct operation control of those gross motions whereas main drawbacks can recognized in a limited payload capability with poor stiffness.

3 Human Torso Behavior Characterization

In order to work out human torso behavior characterization IMU (Inertial Measurement Unit) sensors have been used. The characteristics of these sensors, Fig. 5, are shown in Table 1, [23]. The measurement tests were carried out to acquire the force acting on the human body during 7 different kinds of test movements that were carried out by 12 subjects. Several parameters of the 12 subjects were taken to consider the forces acting on them such as: their height, their weight, the length of their arms, the length of their legs, the torso diameter and the

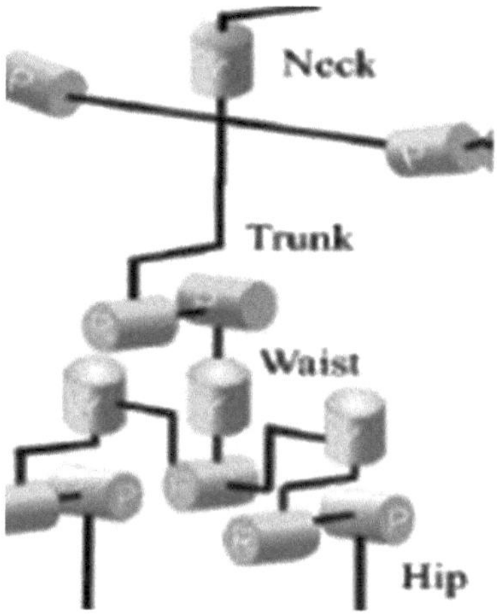

Fig. 4 A design scheme for humanoid torso in Wabian 2 in 2005 [22]

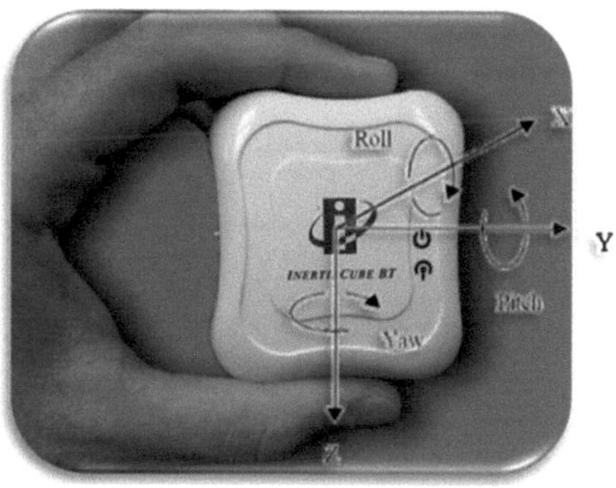

Fig. 5 A used IMU Sensor [23]

waist diameter. Each experiment was divided into 7 parts with different basic motion, namely: turn left and right slowly maintaining the torso in a straight position; turn fast from left to right maintaining the torso in a straight position; lean forward and then backward slowly; lean forward and then backward faster; lean right and left slowly; lean right and left faster; walk forward slowly. The procedure can be shortly summarized in acquiring measurement data from IMUs on a subject performing prescribed motion tests and successively in elaborating the acquired measurements also for interpretation aims. Tests have been carried out with several people and in Fig. 6 characteristic results are reported as referring to a the 12 male and female different subjects of an average of 1.68 m in high and 720 N in weight.

A statistical study have been elaborated to obtain mean acceleration and velocities for a mean people characterization developing a Matlab script. In order to calculate the forces and velocities acting on the shoulder and along the spine segment lengths and segments heights, have been considered. Segment lengths are expressed in percentages of total body height, segment weights expressed in

Table 1 IMU Sensor Characteristics [23]

Performance specifications	Numerical data
Degrees of freedom	3 (Yaw, Pitch, and Roll)
Angular rate	0°–1200° per second
Accuracy (RMS)	1° in yaw, 0.50° in pitch & roll at 25 °C
Angular resolution	0.01° RMS
Update rate	180 Hz
Size	60 × 54 × 32 mm
Weight	67 g

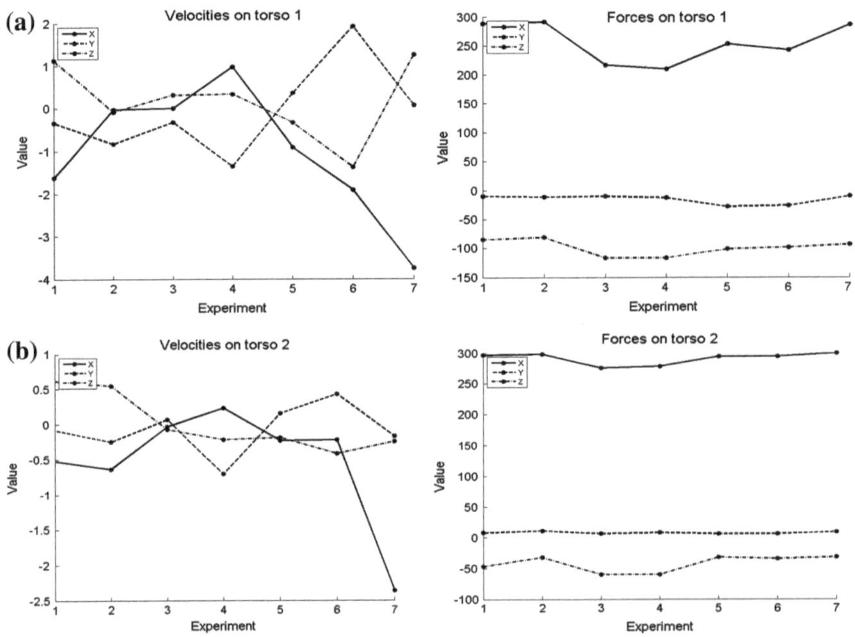

Fig. 6 Measured angular velocities and forces during natural movements of a torso as function of the 7 tests. **a** At the trunk part. **b** At the waist part

percentages of total body weight. The values in the computation represent mean values for the individuals of the tests with a comparison with respect the ones in the scientific literature.

An additional computation has been developed to extrapolate the mean angular velocities, mass and linear acceleration and convert the velocities in deg/s from rad/s using expression (1) and (2). The mean weight of the arm between male and females is about 5 % (represented by the terms into parenthesis) of the total weight of the body, the mass of the Trunk is about 4.5 % (represented by the terms into parenthesis) of the total weight of the body, Eqs. (3) and (4). The final step is to compute the forces acting on the shoulders and on the trunk along the spine, having the mass from (3) and (4) and having the acceleration measured through the experiment, known parameters have been put into Eq. (5) one time for each sensor.

$$C = \frac{180}{\pi} \tag{1}$$

$$V = V_r\, C \tag{2}$$

$$M_a = M_m \left(\frac{0.0512 + 0.0447}{2} \right) \tag{3}$$

$$M_t = M_m \left(\frac{0.468412 + 0.4500}{2} \right) \tag{4}$$

$$F = m\, a \tag{5}$$

Mean values of the seven parts of the experiment are plotted into different charts. On X axes the number of the experiment is represented while on Y axes numerical values are shown. Figure 6a shows the results that have been extrapolated through the use of Sensor 2 positioned on the trunk of the subject. It shows the velocities acting that have a range between (−3.7475; −1.3568; −1.3770) deg/s and (0.9780; 1.9248; 1.2551) deg/s; It also shows the forces acting that have a range between (209.6200; −27.8090; −116.6900) N and (290.9500; −9.4248; −80.7070) N.

Figure 6b shows the results that have been extrapolated through the use of Sensor 4 positioned on the pelvis of the subject. It shows the velocities acting that have a range between (−2.3621; −0.7038; −0.4180) deg/s and (0.2322; 0.4272; 0.6118) deg/s; It also shows the forces acting that have a range between (275.8000; 5.7680; −59.9120) N and (299.7600; 11.0850; −31.7120) N.

4 A New Design with PKM Structure

A design of the torso has been elaborated with the rapid prototyping. The trunk and pelvis have been shelled and divided into two parts in order to make the prototype light and compact as well as to optimize the consumption of the needed material for the 3D printing. A humanoid human-like torso at human scale can be characterized by the a 0.80–1.00 m height, 20–30 cm width, and 300–500 N weight. The shelled trunk is shown in Fig. 7a and the shelled pelvis is shown in Fig. 7b. The two shells have been connected with pins in order improve also the stiffness of the prototype. A shape design has been developed to make the structure similar to a human chest. Thus, a CAD model of the chest have been developed to extrapolate the most important curves and to use them to reshape the platform structure. The entire assembled structure is shown in Fig. 7c. In particular, in the 3D printing process the

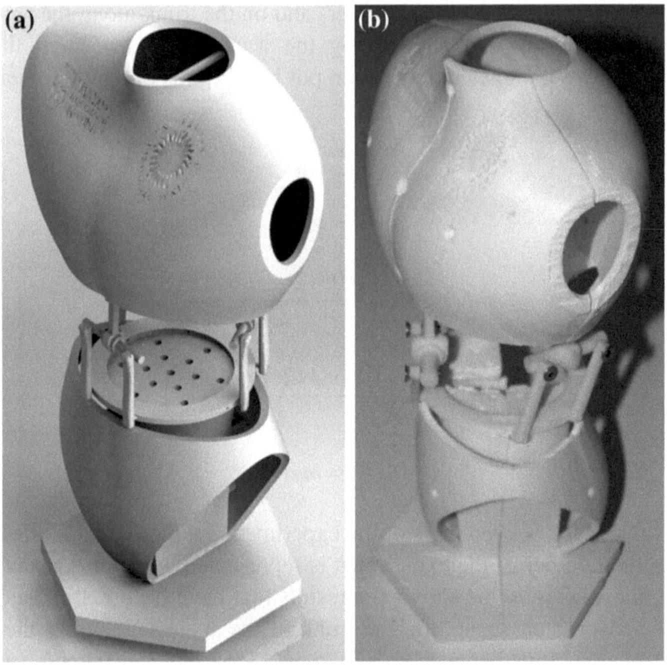

Fig. 7 3D printing prototype of a PKM torso. a Prototype Rendering. b 3D Printed prototype

Replicator G software of 3D printer has been used to properly locate the parts on printer base and the Slic3r routine has been used for positioning several parts within the printing volume to produce several parts in few sessions and to save printing material and support material. Before printing different gcodes were generated to simulate the 3D printing. GCodeSimulator was used to simulate the 3D printing for an estimation of the used materials, final volume, final mass, and final dimensions. Once optimized the model, the entire prototype has been printed and Figs. 7 and 8 show the results with the assembly and driving parallel mechanism for the torso design with human-like appearance and performance. A kinematic model is provided in Fig. 8b and several simulations have been carried out to choose suitable actuators as described in [24] and [25]. The proposed system has 3 d.o.f.s. The location of the actuator has been added in Fig. 8b. The chosen actuators are MG995 Tower Pro servomotors which characteristics are: No-load Operation speed: 0.17 s/60° (4.8 V) 0.13 s/60° (6.0 V); Input Voltage (limits): 6 ~ 20 V; Working Torque: 13 kg/cm at 6.0 V; Rotate Speed: 53–62 R/M; Rotation angle range: 0–180°; Working Voltage: 3.0 ~ 7.2 V; Working current: 100 mA; Structure material: metal gear, dual ball bearings; Dimensions: 5.5 × 3.7 × 1.8 cm; Weight: 60 g.

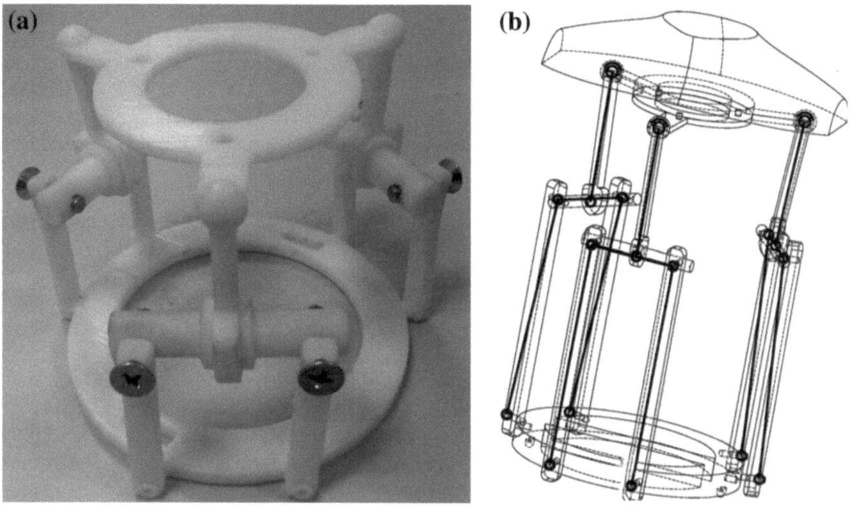

Fig. 8 The PKM mechanism for torso design in Fig. 7. **a** The 3D printed prototype. **b** A kinematic scheme

5 Conclusions

Torso design pays and important role in human-like operation of humanoid robots. The characteristics of human torso can be partitioned in waits and trunk as structures that perform mainly supporting actions for leg walking and arm manipulations, respectively. A solution is presented as based on PKM architectures as result of an experimental characterization of basic motion of human torso by using inertial magnetic sensing tools. A prototype was designed and a Rapid Prototyping simulation has been carried out by using different software to optimize a final 3D printed prototype. The built 3D prototype fully respects the design targets with light structure of 10.38 kg when compared with human structure, with low-cost and easy to manufacture features, and with easy operation to be controlled because of its pure revolute joints.

References

1. Android World homepage: Bipedal projects. www.androidworld.com/index.htm (2014)
2. Rosheim, M.E.: Robot Evolution. Wiley, New York (1994)
3. Kilner, C., et al.: Mechatronic design of NAO humanoid. In: IEEE International Conference on Robotics and Automation, Kobe (2009)
4. St. Pino: http://orionrobots.co.uk/Pino. 2014
5. Koganezawa, K., Takanishi, A., Sugano, S. (eds.): Development of Waseda Robot, 3rd edn. Publisher Ichiro Kato Laboratory, Tokyo (1991)
6. HRI Takanishi lab: http://www.takanishi.mech.waseda.ac.jp/top/research/wabian/2014

7. Furusho, J., Sano, A.: Sensor-based control of a nine-link biped. Int. J. Rob. Res. (Special Issue on Legged Locomotion) **9**(2), 83–98 (1990)
8. Sakagami, Y., et al.: The intelligent ASIMO: system overview and integration, intelligent robots and systems. In: IEEE/RSJ International Conference, vol. 3, pp. 2478–2483 (2002)
9. AIST homepage: http://www.aist.go.jp/sangi/29.html (2014)
10. Pfeiffer, F., et al.: Sensor and control aspects of biped robot "Johnnie". Int. J. Humanoid Rob. **1**(3), 481–496 (2004)
11. University of Muenchen homepage: www.unibw-muenchen.de/hermes (2014)
12. Karlsruhe University homepage: http://wwwipr.ira.uka.de/~asfour/armar/armar.html (2014)
13. Chalmers University homepage: http://humanoid.fy.chalmers.se/elvis.html (2014)
14. http://robonaut.jsc.nasa.gov/2014
15. MIT homepage: www.ai.mit.edu/projects/humanoid-robotics-group/index.html (2014)
16. Sarcos homepage: http://www.sarcos.com (2014)
17. KIST homepage: http://intmob-robot.kist.re.kr/field.html (2014)
18. KAIST homepage: http://mind.kaist.ac.kr/3_re/HumanRobot/HumanRobot.htm (2014)
19. UWA homepage: http://www.ee.uwa.edu.au/~braunl/eyebot/robots/android.html (2014)
20. Virginia, C.: Bones and Muscles: An Illustrated Anatomy. Wolf Fly Press, New York (1999)
21. Liang, C., Ceccarelli, M.: Design and simulation of a waist–trunk system for a humanoid robot. Mech. Mach. Theory **53**, 50–65 (2012)
22. Aiman Musa, M.O., et al.: Development of a humanoid robot having 2-DOF waist and 2-DOF trunk. In: IEEE-RAS International Conference on Humanoid Robots Humanoids 2005, Tsukuba, pp. 333–338 (2005)
23. InterSense: http://www.intersense.com/ (2014)
24. Cafolla, D., Ceccarelli, M.: Design and simulation of humanoid spine. In: Proceedings of the 5th European Conference on Mechanism Science (EUCOMES 2014) Guimarães, Portugal, pp. 585–593 (2014)
25. Cafolla, D., Ceccarelli, M.: Design and FEM analysis of a novel humanoid torso. In: Proceedings of the 5th International Symposium on Multibody Systems and Mechatronics (MUSME 2014) Huatulco, Mexico, pp. 477–488 (2014)

Design of Robots Used as Education Companion and Tutor

Albert Causo, Giang Truong Vo, I-Ming Chen and Song Huat Yeo

Abstract Robot has a special appeal in education as it can enhance interaction between content-user like students, content-producer like teachers, and the actual material content. Deciding on the form of the robot to use in exploring, testing and implementing educational ideas can be challenging for researchers and educators since form and function could affect user-interaction and determine implementation success. This paper presents a review of published works on robots used in education. Robot design is analyzed with special focus on form, actuation and mobility, interaction capability, and sensing. Robots are grouped into three categories based on the forms they take: humanoid, semi-humanoid, and pet-like.

Keywords Robot design · Education · Humanoid · Semi-humanoid · Pet-like robot

1 Introduction

The use of multi-media tool including robotics has become increasingly popular in recent years, helped with children being more accepting and playing more with technologically advanced devices [1]. Robots have been used to study cognition,

A. Causo (✉) · G.T. Vo · I.-M. Chen · S.H. Yeo
School of Mechanical and Aerospace Engineering, Robotics Research Centre,
Nanyang Technological University, Singapore, Singapore
e-mail: acauso@ntu.edu.sg

G.T. Vo
e-mail: tgvo@ntu.edu.sg

I.-M. Chen
e-mail: michen@ntu.edu.sg

S.H. Yeo
e-mail: myeosh@ntu.edu.sg

S. Zeghloul et al. (eds.), *Robotics and Mechatronics*,
Mechanisms and Machine Science 37, DOI 10.1007/978-3-319-22368-1_8

language learning, social interaction, and domain specific subjects such as computer, mathematics, and sciences [2].

Robots in education could be categorized according to the target user, venue of learning, subject domain or robot role [2, 3, 4]. The target users could be students, ranging from pre-school to secondary students. A robot appeals to younger users because it allows total physical response (TPR), which is difficult to accomplish with multi-media tools [5]. Learning venue is another way of classifying robots which is not limited to curricular activities (learning) and extra-curricular activities (at home or under guidance of parents or instructors).

Early applications of robots in education was in the field of computer science and programming. While non-technical domains include music [6] and language learning [7, 8]. Some robots have also been employed as socially assistive agents to encourage cognitive and social development of children [9].

Shin and Kim [10] have classified the role of robots in education into learning tool, learning companion, and learning instructor. Mubin et al. [2] also referred to the three roles as tool, peer, and tutor. As a tool, the robot is a technology aid which students can use to learn the course material. As a peer, robots provide assistance like prompting or providing feedback to students in going though the course material. The most challenging role of a robot is as a tutor, as it requires better communication and autonomy technology than what is currently available.

As robotics in education continue to advance, it is important to provide an overview of the robots that have been used. Reviews such those by Mubin et al. [2] and Goodrich and Schultz [4] cover the applications, successes, and challenges of the field.

However, deciding on the design of the robot is not the easiest task. Guidelines on robot design are not readily available and borrowing design concepts from other professional fields, such as toy design and psychology, is not enough.

This paper aims to present a review of robots with physical embodiment and capable of providing user-interaction that are used as education companion and tutor, a subset of socially assistive robots. The papers reviewed were found on IEEE Xplore Digital Library, ACM Digital Library, Science Direct, SpringerLink, ERIC, and Academic Search Premier. Section 2 presents the design aspects reviewed which include form, actuation and mobility, interaction capability and sensing. Section 3 discusses the most popular robots available. The paper is summarized in Sect. 4.

2 Design Criteria

Robots with similar functionality tend to have similar design and functionality. This section discusses the similarities in robot design, as applied in education robots.

2.1 Form

We categorized the robots into three: humanoid, semi-humanoid, and pet-like. A humanoid robot's resemblance to humans makes it attractive as a platform for education. For example, Nao [11, 12] and Asimo [13] look like a small child, and encourage children to interact. While humanoid robots have bi-pedal mobility, the lower body of semi-humanoid robot use wheels to move around. Examples include Tiro [6], Robovie [14], and Papero [15]. Saya [16, 17] is also classified as a semi-humanoid due to its lack of lower body mobility.

Pet-like robots may take the form of pets, animals, or fantasy characters. Pleo is a dinosaur [18, 19] and Dragonbot is a small-sized dragon character [20]. Pet-like robots are usually designed for specific tasks in mind: iCat is meant to convey emotional feedback to the user [21].

The form of the robot could require some design features to be added, for example, fur like materials for pet-like robots and silicon for others. Without additional "skin", robots would have it's default cover of metal and plastic.

2.2 Actuation and Mobility

These education robots cover a big range of degrees-of-freedom (DOF). Humanoids, semi-humanoids, and legged pet robots tend to have more than 20 DOFs. Leonardo has 69 DOFs [22] with most of the DOFs for the control of facial expressions.

For humanoid robots, mobility is provided by two legs while semi-humanoids rely on wheels, which provide maneuverability and stability. An advantage of humanoid and semi-humanoid robots is their arms, which could be used for gesturing and providing feedback to the user.

2.3 Interaction Capability

To be learning peer or tutor, robots must be able to interact with the users, of which, speech recognition and speech synthesis are the most important capabilities. Instructions and feedback are given or obtained by the user through robust speech processing software.

Another key factor in interaction is facial expression or a way of conveying the robot's "emotion". Robots like Leonardo [22] has a face whose skin can be controlled by actuators, enabling emotion to be displayed. Other robots rely on electronic face, like Dragonbot [20], which uses a digital display for eyes.

While gesture recognition mostly depends on software, generating appropriate gesture to accompany the robot's audio is mainly a hardware issue. In this regard, humanoids and semi-humanoids have an advantage over pet robots. Gesturing for communications and providing directions is possible if the robots have arms.

2.4 Sensors

Full robot autonomy is difficult to achieve without the robots being able to sense its environment and users. The sensors can be grouped into two broad categories: sensors for navigation and sensors for interaction. Navigation sensors include laser range finders or ultrasonic sensors to avoid collision and aid in navigation path planning. Pressure sensors and touch sensors on the feet and other body parts are also installed in some robots to help with collision avoidance.

For communications, microphone and speakers are required components of the robot, either as inherent part or as external devices connected to the robot. For the vision system of the robot, standard video cameras are usually built in and become the physical eyes of the robots. To augment the capability to detect people and distance, stereo cameras are sometimes included in the robot, or an external one is placed behind the robot.

3 Robots in Focus

3.1 Humanoid Robots

3.1.1 Nao Robot

Nao Robot (see Fig. 1b) was developed and launched by Aldebaran Robotics in 2004 [12]. It is a humanoid robot that features up to 25 degrees of freedom (DOF) which helps NAO re-create smooth human-like motions like walking and

Fig. 1 Humanoid Robots (*left* to *right*): Asimo and Nao

dancing. It had been used as a care-receiver in a vocabulary learning context, where the children learned English along the way as they taught Nao robot [11]. Even though NAO could be operated autonomously, it could also be tele-operated using the Wizard-of-Oz technique. The Nao robot is equipped with an Intel Atom CPU running at 1.6 GHz and uses Linux operating system. It has built-in cameras, microphone and speakers, and can perform voice recognition, speech synthesis and moving object detection.

3.1.2 Asimo Robot

ASIMO (see Fig. 1) is a humanoid robot developed by Honda [13] and designed to be a multi-functional mobile assistant. In the educational field, it was employed to study interactive learning between children and robot as a companion [13]. In the latest model (2014), Asimo is 130 cm tall, weights 50 kg and features 57 DoF in total. It has extensive functions which can recognize voice, face, gesture and moving objects.

3.2 Semi-humanoid Robots

3.2.1 Tiro Robot

The teaching assistant robot, Tiro (Fig. 2) helped teachers as an educational media in class, and as children's classmate for music [6] or English learning [5]. Teachers could also use a remote-control to interact with the robot if they were across the classroom. Tiro displayed and explained the learning material while the teacher roamed around the classroom and attended to any individuals who needed extra help. Tiro had been used for class management such as checking students' attendance, getting attention, and acting as a timer for activities and as tutor such as providing lesson objectives, conversation scripts, English chants and dancing, storytelling and role playing, praising and cheering up. Tiro stands 130 cm high and weights 70 kg, with 8 DoF (3 DoF on each arm and 2 wheels). It has an LED panel

Fig. 2 Semi-humanoid robots (from *left* to *right*): Tiro, Robovie R3, PaPeRo, Maggie, Saya

to display emotion with 60 templates. It also has cameras for localization and user recognition, as well as 16 pairs of ultra-sonic sensors.

3.2.2 Robovie R3

The Robovie R3 (Fig. 2) stands 108 cm tall, weighs 35 kg, 17 DOF (eyes × 4, neck × 3, 2 arms × 4, 2 wheels), and 11 tactile sensors throughout its body. It is equipped with 2 USB cameras for eyes, 2 mono microphones for ears, a speaker for a mouth, and can be equipped with a laser range finder in its base for obstacle detection. It also communicates with a local network which includes other robots, cameras, and mobile phone services. Robovie offers voice-based casual English conversation, and nonverbal communication. It serves hundreds of interactive behaviors such as shaking hands, hugging, playing rock-paper-scissors, exercising, greeting, kissing, singing, briefly conversing, and pointing to objects in the surroundings. Those services made children interact with Robovie over the long-term [14].

3.2.3 Papero

The PaPeRo (Fig. 2) by NEC is 38.5 cm tall, weighs 5 kg [15], has 4 DoFs (2 for the neck and 2 wheels), and has 2 cameras for facial recognition and microphones for voice recognition. It also has ultra-sonic sensors for object detection and avoidance in order to roam around looking for people and start conversation using speech synthesis software. PaPeRo can also connect to the Internet to look for information when asked.

3.2.4 Maggie

Maggie (Fig. 2) is another wheeled robot with 6 DOF (2 for head, 1 for each arm and 2 wheels). In total, it has 12 sets of bumper, IR, sonar and tactile sensors, as well as 2 camera, a laser scanner and a tablet.

3.2.5 Saya Robot

First developed as a receptionist robot in 2004 [16], Saya was tested as teacher (tutor) in a real Tokyo classroom in 2009 with primary school students. With expressive human-like face featuring 27 DoFs, the robot can display several emotions and give basic instruction through its speech recognition and synthesis.

3.3 *Pet Robots*

3.3.1 Pleo the Dinosaur

Pleo (Fig. 3) was initially unveiled in 2006, with the latest iteration 'Pleo rb' released in 2011. It is a sophisticated robotic toy with features such as light, color and object detection, beat detection to listen and dance to music. In order to facilitate such functions, it has a cameras, two microphones and sensors including: 12 touch sensors (head, chin, shoulders, back, feet), 4 ground foot sensors for surface detection, 14 force-feedback sensors (1 per joint), orientation tilt sensor for body position, and infrared mouth sensor for object detection inside the mouth. Pleo has been used as a care-receiver to study the interaction and social skill development of primary children [19].

3.3.2 EMuu

eMuu (Fig. 3) is an expressive, one-eyed robotic head which can display its emotions based on social interaction with human.

3.3.3 Probo

Probo Robot (Fig. 3) was created by Vrije Universiteit Brussel to enable human-robot interaction by face-to-face communication, with special focus on children. Probo's communicative skills lie initially on facial expressions: the robot has 20 high-precision motors to actuate the ears, eyebrows, eyelids, eyes, trunk, mouth and neck. For Probo, a shared control between the operator and the reactive systems of the robot is targeted.

3.3.4 The Huggable

The Huggable (Fig. 3) is a new type of robotic companion being developed at the MIT Media Lab for healthcare, education, and social communication applications. It has a full body sensitive skin with over 1500 sensors, quiet back-drivable actuators, video cameras in the eyes, microphones in the ears, an inertial

Fig. 3 Pet robots (from *left* to *right*): Pleo, eMuu, Probo, Aibo, The Huggable, Dragonbot, Leonardo, iCat

measurement unit, a speaker, and an embedded PC with 802.11 g wireless networking. In educational application, the Huggable serves as a semi-autonomous robotic communication avatar controlled by a teacher via the internet to interact with a student in an educational activity.

3.3.5 DragonBot

DragonBot (Fig. 3) is a platform built by MIT to support long-term interactions between children and robots. The robot's design is based on squash and stretch principles of animation, which creates more natural and organic motion and allows for a range of expressive body movements, while keeping the actuator count low. The robot runs entirely on an Android phone, which displays an animated virtual face. Additionally, the phone provides sensory input (camera and microphone) and fully controls the actuation of the robot (motors and speakers). Most importantly, the phone always has an Internet connection, so a robot can harness cloud-computing paradigms to learn from the collective interactions of multiple robots. A custom tele-operation interface that runs either on a tablet or laptop computer allows researchers or caregivers a range of controls, from speaking and acting as the robot to merely triggering pre-recorded speech and behavior [20]. To support long-term interaction, DragonBot is a "blended-reality" character, if the phone is removed from the robot, a virtual avatar would appear on the screen so that the user could still interact with the virtual character. DragonBot was specifically designed to be a low-cost platform that can support longitudinal human-robot interactions "in the wild."

3.3.6 Leonardo

Leonardo (Fig. 3) is another robot created by MIT Personal Robots Group to facilitate the study of human-robot interaction and collaboration. Featuring 69 DoFs, Leonardo is not only able to display expressions but also perform gestures and manipulation. Aside from facial and voice recognition, the robot can detect temperature, proximity and pressure [22].

4 Summary

This paper shows a brief review of different robot designs used in education research and pedagogy implementations. The robot criteria were compared based on four major criteria: form, actuation and mobility, interaction capability, and sensors.

Robots in the education field could be humanoid robot like Nao or as pet-like characters like MIT's Dragonbot. Though the robots vary not only in shape and size but also in functions, there are a few points that display the common features of such robots:

- Like most social robots, they have good appearance to encourage the users to interact, especially young users like students of kindergarten and primary schools. Many of them have fur or textile skins to cover the mechanical frame, like the Huggable and Probo, to facilitate physical interaction.
- While biped humanoid robots such as Nao has begun to have more applications, immobile animal-like and pet robots still dominate in the educational context, perhaps due to safety and reliability concern.
- All of the robots have embedded processor which can connect to the Internet wirelessly to look for information if needed. They also possess at least two perception modes to interact with human. Most of them can perform facial and voice recognition and express emotion either by LED or mechanical means.

Acknowledgments We would like to acknowledge the Singapore Millennium Foundation for supporting this work.

References

1. Beran, T.N., Ramirez-Soriano, A., Kuzyk, R., Fior, M., Nugent, S.: Understanding how children understand robots: perceived animism in child-robot interaction. Int. J. Hum. Comput. Stud. **69**(7–8), 539–550 (2011)
2. Mubin, O., Stevens, C.J., Shahid, S., Mahmud, A.A., Dong, J.: A review of the applicability of robots in education. Technol. Educ. Learn. **1**, 1–7 (2013)
3. Hendler, J.: Robots for the rest of us : designing systems "out of the box". In: Robots for Kids: Exploring New Technologies For Learning, pp. 2–7. The Morgan Kaufmann Publishers, San Mateo (2000)
4. Goodrich, M.A., Schultz, A.: Human robot interaction: a survey. Found. Trends in Hum-Comput. Interact. **1**(3), 203–275 (2007)
5. Han, J., Kim, D.: R-Learning services for elementary school students with a teaching assistant robot. In: Proceedings 4th ACM/IEEE International Conference Human Robot Interaction (HRI), New York, pp. 255–256 (2009)
6. Han, J.H., Kim, D.H., Kim, J.W.: Physical learning activities with a teaching assistant robot in elementary school music class. J. Convergence Inf. Technol. **5**(5), 1406–1410 (2009)
7. Kanda, T., Shimada, M., Koizumi, S.: Children learning with a social robot. In: Proceedings 7th ACM/IEEE International Conference on Human-Robot Interaction, Boston, Massachusetts, USA, pp. 351–358 (2012)
8. Kanda, T., Hirano, T., Eaton, D., Ishiguro, H.: Interactive robots as social partners and peer tutors for children: a field trial. Hum-Comput. Interact. **19**, 61–84 (2004)
9. Tapus, A., Mataric, M.J., Scassellati, B.: Socially assistive robotics. IEEE Robot. Autom. Mag. **14**(1), 35–42 (2007)
10. Shin, N., Kim, S.: Learning about, form and with robots: students' perspectives. In: Proceedings 16th IEEE International Symposium Robot and Human interactive Communication (RO-MAN), Jeju, Aug 2007, pp. 1040–1045 (2007)
11. Fumihide, T., Matsuzoe, S.: Children teach a care-receiving robot to promote their learning: field experiments in a classroom for vocabulary learning. J. Hum-Robot Interact. **1**(1), 78–95 (2012)

12. Shamsuddin, S., Yussof, H., Ismail, L., Hanapiah, F.A.: Initial response in HRI—a case study on evaluation of child with autism spectrum disorders interacting with a humanoid robot NAO. In: International Symposium Robotics and Intelligent Sensors (IRIS), vol. 41, pp. 1448–1455 (2012)
13. Okita, S.Y., Ng-Thow-Hing, V., Sarvadevabhatla, R.: Learning together: asimo developing an interactive learning partnership with children. In: Proceedings 18th IEEE International Symposium Robot and Human interactive Communication (RO-MAN), pp. 1125–1130 (2009)
14. You, Z., Shen, C., Chang, C., Liu, B., Chen, G.: A robot as a teaching assistant in an english class. In: Proceedings 6th International Conference Advanced Learning Technologies, July 2006, Kerkrade, pp. 87–91 (2006)
15. Osada, J.: Scenario design for children care robot. In: Proceedings Robot Design Forum, Daejeon, Korea, Nov 2005, pp. 29–36 (2005)
16. Hashimoto, T., Kato, N., Kobayashi, H.: Development of educational system with the android robot SAYA and evaluation. Int. J. Adv. Robot. Syst. **8**(3), 51–61 (2011)
17. Hashimoto, T., Kobayashi, H., Polishuk, A., Verner, I.: Elementary science lesson delivered by robot. In: Proceedings 8th ACM/IEEE International Conference on Human-Robot Interaction (HRI), Tokyo, March 2013, pp. 133–134 (2013)
18. Janssen, J., Wal, C., Neerincx, M., Looije, R.: Motivating Children to Learn Arithmetic with an Adaptive Robot Game. In: Mutlu, B., Bartneck, C., Ham, J., Evers V., Kanda, T. (eds.) Social Robotics, Lecture Notes in Computer Science. Springer, Berlin, pp. 153–162 (2011)
19. Heerink, M., D'az, M., Albo-Canals, J., Angulo, C., Barco, A., Casacuberta, J., Garriga, C.: A field study with primary school children on perception of social presence and interactive behavior with a pet robot. In: Proceedings 21st IEEE International Symposium Robot and Human interactive Communication (RO-MAN), Paris, Sept 2012, pp. 1045–1050 (2012)
20. Kory, J., Breazeal, C.: Storytelling with Robots: Learning companions for preschool children's language development. In: Proceedings 23rd IEEE International Symposium on Robot and Human Interactive Communication (RO-MAN), Washington, DC, 2014, pp. 643–648 (2014)
21. Saerbeck, M., Schut, T., Bartneck, C., Janse, M.D.: Expressive robots in education varying the degree of social supportive behavior of a robotic tutor. In: Proceedings ACM Conference on Human Factors in Computing Systems, Georgia, Atlanta, 2010, pp. 1613–1622 (2010)
22. Breazeal, C., Berlin, M.: Spatial scaffolding for sociable robot learning. In: Proceedings 23rd AAAI Conference on Artificial Intelligence, 2008, pp. 1268–1273 (2008)

Walking of a Biped Robot Balanced with a Reciprocating Torso

Víctor De-León-Gómez, J. Alfonso Pámanes and Víctor Santibáñez

Abstract The results of walking tests on a five degree of freedom (DOF) biped robot are presented in this paper. The robot is developed at the Instituto Tecnológico de La Laguna in Mexico and uses a reciprocating mass as a balancing device. The gait pattern proposed for the biped is based on cycloidal motions that avoid impacts of the feet with the soil during the walking. The parameters of the synchronized motions of torso and limbs keep balanced the robot and avoid saturations of motor torques. Simulation of walking allow to verify the suitable behavior of the Zero Moment Point (ZMP) before to accomplish the tests. The obtained results of the experiments finally validate the efficacy of the proposed walking pattern.

Keywords Biped robot · Gait pattern · ZMP · Cycloidal motion · Biped motion planning

1 Introduction

The application of humanoid robots for tasks in environments normally used by human beings has been largely considered by experts and interaction of these machines with people is already a reality. Thus, biped locomotion appears as a natural way of displacement for humanoids. Not any special installations are necessary to make feasible the collaboration of this kind of robots anywhere (i.e. in hospitals, cars, trains, airports, etc.) if biped walking is applied. Accordingly, a number of studies on biped robots has been addressed to accomplish stable walking

V. De-León-Gómez · J. Alfonso Pámanes (✉) · V. Santibáñez
Graduate and Research Division, Instituto Tecnológico de La Laguna,
C.P. 27000 Torreón, Coahuila, Mexico
e-mail: alfonso.pamanes@gmail.com

J. Alfonso Pámanes
Department of Mechanical Engineering, Instituto Tecnológico de La Laguna,
C.P. 27000 Torreón, Coahuila, Mexico

S. Zeghloul et al. (eds.), *Robotics and Mechatronics*,
Mechanisms and Machine Science 37, DOI 10.1007/978-3-319-22368-1_9

on several surfaces, slopes, steps, and stairs. Some prototypes have been applied for experimental validation of the proposed approaches. A planar five bar biped robot, named Rabbit, has been used for evaluation of controllers related to walking and running, impact analysis, limit cycles and hybrid systems [7, 30, 31]. The robot WEBIAN with 52 DOF is capable of walking at 0.21 m/s [10, 26, 27], and has been used for analysis of stability of control algorithms [17, 32]. This kind of analysis has been applied in the ASIMO humanoid, which has 57 DOF and abilities for walking, running, jumping, turning, climbing and descending stairs [12]. The HRP-2, with 30 DOF and 58 kg [13, 16] has been used in several laboratories for a number of studies on control, motion planning, and interaction man-robot, among other. The PetMan humanoid, from Boston Dynamics, walks in a natural way and is capable of flexing and jumping, getting an excellent gait even if perturbations occur [24]. Basically, walking control of biped robots can be managed in both ways, by using time dependent and non-dependent algorithms. The present paper deals with a biped robot controlled by time dependent algorithms that apply a pre-specified trajectory. Several authors have developed contributions on this subject; in [25] the BLR-G2 robot was controlled by a decoupled control for the frontal and sagittal planes. In [15] a PID control for trajectory tracking was applied to the Meltran II robot. On the other hand, for controlling a 5 DOF biped robot on [19], a calculated torque control was considered, and studies of simulation were reported for trajectory tracking with force control augmented of a 20 axis biped robot [9]. Other approaches for development of controllers for trajectory tracking are based on the notion of the ZMP [29], which is often used as criterion of walking stability [11, 13, 14, 17, 18, 23, 32].

Our biped robot, named Bip-ITLag, uses trajectories based on cycloidal laws for motions of the free foot and hip on the sagittal plane. These kind of motions have been earlier used for 3D walking of the Bioloid humanoid of 18 DOF [3, 22]. In other works some advances on the development of the Bip-ITLag have been presented [1, 2, 4–6, 20, 21, 28]. Some simulation studies or preliminary experimental analysis with the robot fixed from the hip were discussed in those papers. This article presents the results obtained during the first real-time walking tests of the Bip-ITLag robot. A description of the design of this biped is presented in Sect. 2. The equations that describe the specified motions of the hip and foot are presented in Sect. 3, and the equations for the forward and inverse kinematics of the robot are considered in Sect. 4. In Sect. 5 the computation of the ZMP is included for the proposed trajectories. Results of the developed walking tests with the Bip-ITLag biped robot are showed in Sect. 6. The conclusions and comments on future work are presented in the final section.

2 Biped Robot Description

The Bip-ITLag is 131 cm high and weights 92 kg. The displacement of the robot is constrained to remain in the sagittal plane by walking on the horizontal plane. Each femur and tibia of the robot uses a parallelogram linkage in order to avoid an

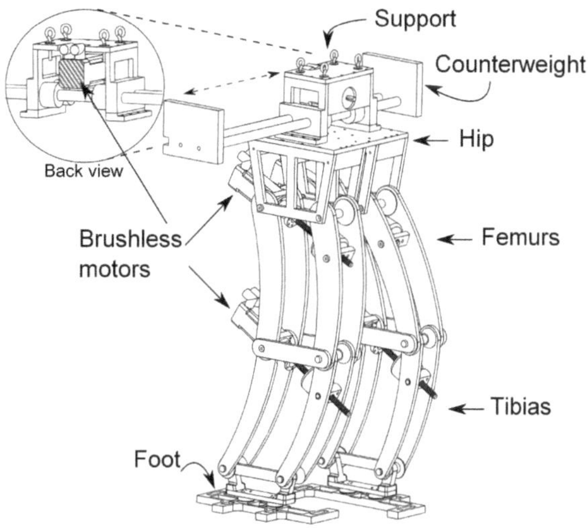

Fig. 1 ITLag biped robot

additional DOF to control the orientation of the hip; thus, this one is constrained to be parallel to the supporting foot, as showed in Fig. 1. The actuators of the legs are brushless DC motors that are coupled to screw-nuts mechanisms, positioned in such a way that the motors torque is amplified to produce the rotations of both, tibia and femur. As appreciated in the aforementioned figure, the torso is a reciprocating mass which is displaced on the frontal plane in order to keep balanced the robot during the walking. The motion is generated by a brushless DC motor and transmitted through a timing belt to the torso. The design of the balancing mechanism can be also appreciated in Fig. 1; the counterweight bars are 91.44 cm long, and the weight of the whole balancing mechanism is 27.70 kg (8.68 kg of the frame and 19.02 kg of the counterweight) (Fig. 2).

3 Trajectory Planning

The trajectory planning for motions of the foot and hip should be achieved in such a way that stable walking be obtained. The cycloidal trajectories applied in this study give null accelerations at the beginning and at the end of displacements, and consequently dynamic forces of the links are not significant at this instant. Then, when the motion is in progress, smooth changes of accelerations occurs, and major shaking forces are avoided in the whole mechanism. This suitable behavior of the dynamic forces contributes to the stability of walking. The cycloidal trajectories are based on proposals made on [3, 22], where the trajectory parameters were suitably chosen by taking into account the size of the robot. In the case of the Bip-ITLag the parameters must be adjusted. The robot should to begin in a posture such that the jambs are almost vertical; then, when the first step is beginning the hip moves down

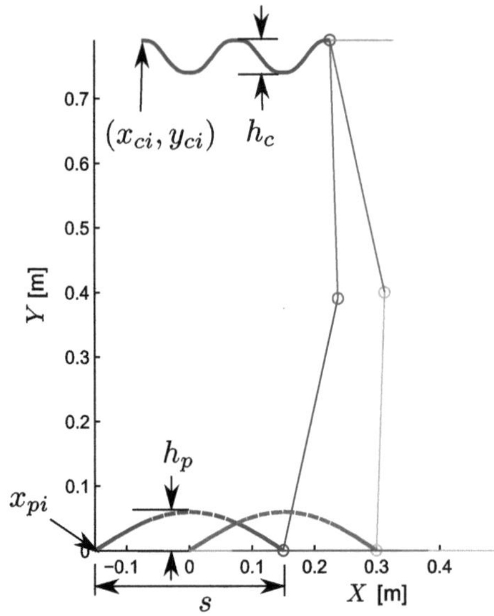

Fig. 2 Path described by the hip and feet of the robot

helping to reduce torque consumption. For the biped robot Bip-ITlag the cycloidal trajectories are defined by the following equations.

For the free foot

$$x_p = \begin{cases} x_{pi} + s\left[\frac{t}{T_1} - \frac{1}{2\pi}\sin\left(\frac{2\pi t}{T_1}\right)\right] & \text{if} \quad 0 < t \leq T_1 \\ x_{pi} + s & \text{if} \quad T_1 < t \leq T \end{cases}$$

$$y_p = h_p \sin\left(\left(\frac{x_p}{s} + \frac{1}{2}\right)\pi\right) \qquad \text{if} \quad 0 < t \leq T$$

$$\dot{x}_p = \begin{cases} \frac{s}{T_1}\left[1 - \cos\left(\frac{2\pi t}{T_1}\right)\right] & \text{if} \quad 0 < t \leq T_1 \\ 0 & \text{if} \quad T_1 < t \leq T \end{cases}$$

$$\dot{y}_p = \frac{h_p \pi \dot{x}_p}{s}\cos\left[\pi\left(\frac{x_p}{s} + \frac{1}{2}\right)\right] \quad \text{if} \quad 0 < t \leq T$$

$$\ddot{x}_p = \begin{cases} \frac{2\pi s}{T_1^2}\left[\sin\left(\frac{2\pi t}{T_1}\right)\right] & \text{if} \quad 0 < t \leq T_1 \\ 0 & \text{if} \quad T_1 < t \leq T \end{cases}$$

$$\ddot{y}_p = \frac{h_p \pi \ddot{x}_p}{s}\cos\left[\pi\left(\frac{x_p}{s} + \frac{1}{2}\right)\right] - \dot{x}_p\left[\frac{h_p \pi^2 \dot{x}_p}{s^2}\sin\left[\pi\left(\frac{x_p}{s} + \frac{1}{2}\right)\right]\right] \quad \text{if} \quad 0 < t \leq T$$

for the hip

$$x_c = \begin{cases} x_{ci} + v_c t & \text{if} \quad 0 < t \le T_1 \\ x_{ci} + \frac{s}{2} & \text{if} \quad T_1 < t \le T \end{cases}$$

$$y_c = \begin{cases} y_{ci} + h_c\left[\frac{2t}{T_1} - \frac{1}{2\pi}\sin\left(\frac{4t\pi}{T_1}\right)\right] & \text{if} \quad 0 < t \le \frac{T_1}{2} \\ y_{ci} + \\ h_c\left[1 - \left(\frac{2t-T_1}{T_1} - \frac{1}{2\pi}\sin\left(\frac{4t\pi}{T_1}\right)\right)\right] & \text{if} \ \frac{T_1}{2} < t \le T_1 \\ y_{ci} & \text{if} \quad T_1 < t \le T \end{cases}$$

$$\dot{x}_c = \begin{cases} v_c & \text{if} \quad 0 < t \le T_1 \\ 0 & \text{if} \quad T_1 < t \le T \end{cases}$$

$$\dot{y}_c = \begin{cases} \frac{2h_c}{T_1}\left[1 - \cos\left(\frac{4\pi t}{T_1}\right)\right] & \text{if} \quad 0 < t \le \frac{T_1}{2} \\ -\frac{2h_c}{T_1}\left[1 - \cos\left(\frac{4\pi t}{T_1}\right)\right] & \text{if} \quad \frac{T_1}{2} < t \le T \\ 0 & \text{if} \quad T_1 < t \le T \end{cases}$$

$$\ddot{x}_c = 0 \qquad \text{if} \quad 0 < t \le T$$

$$\ddot{y}_c = \begin{cases} \frac{8\pi h_c}{T_1^2}\left[\sin\left(\frac{4\pi t}{T_1}\right)\right] & \text{if} \quad 0 < t \le \frac{T_1}{2} \\ -\frac{8\pi h_c}{T_1^2}\left[\sin\left(\frac{4\pi t}{T_1}\right)\right] & \text{if} \quad \frac{T_1}{2} < t \le T_1 \\ 0 & \text{if} \quad T_1 < t \le T \end{cases}$$

and for the balancing mechanism

$$r_d = \begin{cases} r_{ini} + \varepsilon\Delta_{r1}\left[\frac{t}{T_1} - \frac{1}{2\pi}\sin\left(\frac{2t\pi}{T_1}\right)\right] & \text{if} \quad 0 < t \le T_1 \\ r_{ini} + \varepsilon\Delta_{r1} + \\ \varepsilon\Delta_{r2}\left[\frac{t-T_1}{T_2} - \frac{1}{2\pi}\sin\left(\frac{2\pi(t-T_1)}{T_2}\right)\right] & \text{if} \quad T_1 < t \le T \end{cases}$$

$$\dot{r}_d = \begin{cases} \frac{\varepsilon\Delta_{r1}}{T_1}\left[1 - \cos\left(\frac{2\pi t}{T_1}\right)\right] & \text{if} \quad 0 < t \le T_1 \\ \frac{\varepsilon\Delta_{r2}}{T_2}\left[1 - \cos\left(\frac{2\pi(t-T_1)}{T_2}\right)\right] & \text{if} \quad T_1 < t \le T \end{cases}$$

$$\ddot{r}_d = \begin{cases} \frac{2\pi\varepsilon\Delta_{r1}}{T_1^2}\left[\sin\left(\frac{2\pi t}{T_1}\right)\right] & \text{if} \quad 0 < t \le T_1 \\ \frac{2\pi\varepsilon\Delta_{r2}}{T_2^2}\left[\sin\left(\frac{2\pi(t-T_1)}{T_2}\right)\right] & \text{if} \quad T_1 < t \le T \end{cases}$$

where $T_2 = T - T_1$ represents time left on the trajectory, and $\varepsilon = \pm 1$ is used for changing counterweight motion direction every step.

Table 1 Trajectory tracking parameters proposed for the Bip-ITLag robot

Parameter	Description	Value	Unit
T	Step period	6.0	[s]
T_1	Simple support period	3.0	[s]
s	Step size	0.3	[m]
x_{pi}	Foot's initial position	−0.15	[m]
h_p	Foot's maximum height	0.03	[m]
x_{ci}	Hip's initial X coordinate	0.075	[m]
y_{ci}	Hip's initial height	0.79	[m]
h_c	Hip's maximum height	−0.01	[m]
$v_c = s/T$	Hip velocity	0.05	[m/s]
r_{ini}	Counterweight initial position	−0.28	[m]
Δ_{r1}	1st counterweight displacement	0.028	[m]
Δ_{r2}	2nd counterweight displacement	0.252	[m]

All the parameters used by the precedent equations are specified in Table 1. The set of final parameters was established by completing a number of simulations until obtaining a suitable behavior of the robot, with a stable walking and with no saturation of the torques of motors.

4 Kinematic Modelling

The equations used for transforming the hip and free foot proposed motions to joint motions of the robot are presented in this section. The vector of specified operational coordinates is $\boldsymbol{x}_d = [x_c\ y_c\ x_p\ y_p]^T$, where $(x_p,\ y_p)$ and $(x_c,\ y_c)$ represent the position of the free foot and the hip in Cartesian coordinates respectively, and coordinates $(x_s,\ y_s)$ give the current position of the foot fixed on the floor. Both positions are referred to the frame shown in Fig. 3. On the other hand, the joint variables are the components of the vector $\boldsymbol{q}_d = [q_{d1}\ q_{d2}\ q_{d3}\ q_{d4}]^T$ where q_{di}, with $i = 1\ldots4$, represent the angle of the robot's i link, as shown in Fig. 3. The joint variables determine the instantaneous posture of the robot during the walking. The values of the parameters showed in Fig. 3, that define the geometry of the robot, are given in Table 2.

The forward kinematic model of a serial robot of n DOF is a function that relates $\boldsymbol{q}_d$ and $\boldsymbol{x}_d$. This function allow us to find the relationship between operational coordinates and the joint variables of the robot.

$$\boldsymbol{x}_d = \boldsymbol{f}(\boldsymbol{q}_d), \tag{1}$$

Fig. 3 ITLag biped robot's leg schematic representation

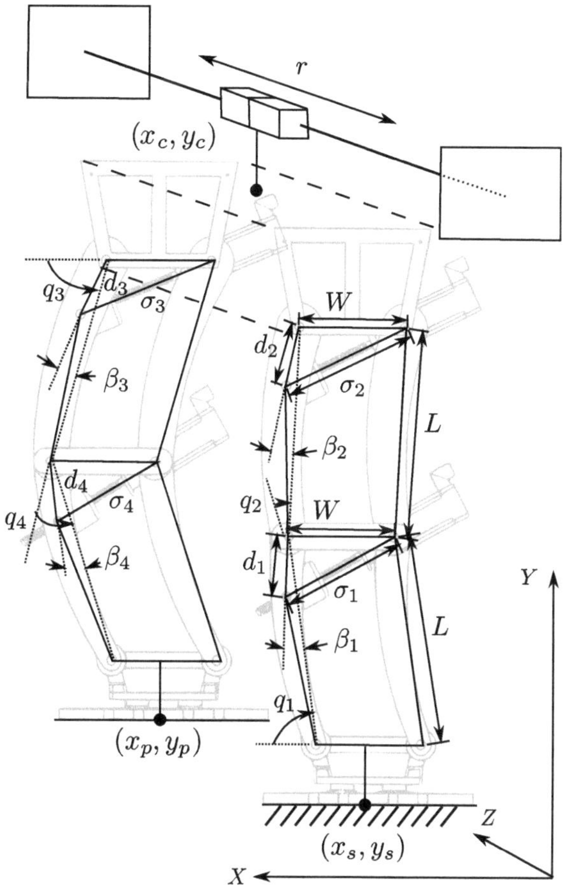

For the Bip-ITLag, the forward kinematics equations are given by

$$\begin{bmatrix} x_c \\ y_c \\ x_p \\ y_p \end{bmatrix} = \begin{bmatrix} L(\cos(q_{d1}) + \cos(q_{d1} + q_{d2})) + x_s \\ L(\sin(q_{d1}) + \sin(q_{d1} + q_{d2})) + y_s \\ L(\cos(q_{d1}) + \cos(q_{d1} + q_{d2}) + (\cos(q_{d3}) + \cos(q_{d3} + q_{d4}))) + x_s \\ L(\sin(q_{d1}) + \sin(q_{d1} + q_{d2}) + (\sin(q_{d3}) + \sin(q_{d3} + q_{d4}))) + y_s \end{bmatrix}$$

On the other hand, the inverse kinematics can be expressed by the generic vector equation $\boldsymbol{q}_d = \boldsymbol{f}^{-1}(\boldsymbol{x}_d)$ And the equations of the inverse kinematics for the Bip-ITLag are:

Table 2 Geometric parameters of the Bip-ITLag biped robot

Parameter	Description	Value	Unit
L	Link length	0.40	[m]
W	Link width	0.21	[m]
β_1, β_4	Curvature angle of the tibia	0.1977	[rad]
β_2, β_3	Curvature angle of femur	0.1791	[rad]
d_1, d_4	Distance from the screw joint of the tibia to knee	0.09	[m]
d_2, d_3	Distance from the screw joint of the femur to hip	0.115	[m]

$$q_{d1} = \operatorname{atan2}((y_c - y_s), (x_c - x_s)) - \frac{1}{2}\varepsilon_1 \cos^{-1}\left(\frac{(x_c - x_s)^2 + (y_c - y_s)^2 - 2L^2}{2L^2}\right)$$

$$q_{d2} = \operatorname{atan2}((y_c - y_s), (x_c - x_s)) + \frac{1}{2}\varepsilon_2 \cos^{-1}\left(\frac{(x_c - x_s)^2 + (y_c - y_s)^2 - 2L^2}{2L^2}\right) - q_{d1}$$

$$q_{d3} = \pi - \operatorname{atan2}\left((y_c - y_p), (x_c - x_p)\right) + \frac{1}{2}\varepsilon_3 \cos^{-1}\left(\frac{(x_c - x_p)^2 + (y_c - y_p)^2 - 2L^2}{2L^2}\right)$$

$$q_{d4} = \operatorname{atan2}\left((y_c - y_p), (x_c - x_p)\right) - \frac{1}{2}\varepsilon_4 \cos^{-1}\left(\frac{(x_c - x_p)^2 + (y_c - y_p)^2 - 2L^2}{2L^2}\right) - q_{d3}$$

where

$$\varepsilon_i = \begin{cases} +1 \text{ Human posture (knee forward)} \\ -1 \text{ Bird posture (knee backwards)} \end{cases}$$

Now, in order to determine the relationship between operational and joint velocities the time derivative of Eq. (1) is obtained:

$$\dot{\boldsymbol{x}}_d = J(\boldsymbol{q}_d)\dot{\boldsymbol{q}}_d \tag{2}$$

where $J(\boldsymbol{q}_d)$ is the jacobian matrix defined as

$$J(\boldsymbol{q}_d) = \frac{\partial \boldsymbol{f}(\boldsymbol{q}_d)}{\partial \boldsymbol{q}_d}.$$

Then, by isolating $\dot{\boldsymbol{q}}_d$ from (2) we determine the desired vector of joint velocity:

$$\dot{\boldsymbol{q}}_d = J^{-1}(\boldsymbol{q}_d)\dot{\boldsymbol{x}}_d. \tag{3}$$

On the other hand, getting the time derivative of Eq. (2) we obtain

$$\ddot{x}_d = \dot{J}(\boldsymbol{q}_d)\dot{\boldsymbol{q}}_d + J(\boldsymbol{q}_d)\,\ddot{\boldsymbol{q}}_d\,. \tag{4}$$

Finally, isolating $\ddot{\boldsymbol{q}}_d$ from (4) we have

$$\ddot{\boldsymbol{q}}_d = J^{-1}(\boldsymbol{q}_d)[\ddot{x}_d - \dot{J}(\boldsymbol{q}_d)\dot{\boldsymbol{q}}_d] \tag{5}$$

that represent the desired joint accelerations.

As observed in the previous section, for the counterweight, the position, speed and acceleration desired (r_d, $\dot{r}_d$ y $\ddot{r}_d$) are directly assigned based in cycloid motions.

5 ZMP as Walking Stability Criterion

The walking of a robot can be divided into two phases, simple support phase (SSP) and double support phase (DSP). The SSP occurs while the robot is doing a step and it is supported by only one foot. The DSP take place since the instant when the feet are touching the floor until one of them launches. During the walking, the forces and torques in joints generated by the motion of the biped robot are reflected in the support foot (or feet). In this foot (or feet), if the equilibrium of the robot is kept, the reaction of the floor must to produce the required moment and forces. Thus, the horizontal torques that can be exercised by the floor over the support foot (or feet) are in X and Z directions and are generated by the vertical component of the restrictive force of the floor. The theoretical point where this force acts has been called by Vukobatrović Zero Momento Point (ZMP) [29]. Consequently, to keep the robot balanced during the walking it is necessary that the ZMP stay inside the support polygon of the robot's feet. This polygon is the footprint in SSP or the convex area between footprints in DSP.

For the walking of the Bip-ITLag, with the launching of the oscillating foot, at the beginning of the SSP, the control starts for trajectory tracking of variables corresponding to the oscillating foot and the hip. At the same time, the counterweight is kept on the same side of the support foot in such a way that the ZMP is into the support polygon corresponding to this foot. When the step finishes, the DSP starts and the support polygon growth; then, the counterweight is displaced toward the opposite side by applying a cycloidal motion in such a way that the equilibrium of the robot is retained. This process is repeated at each step.

The ZMP coordinates of the Bip-ITLag can be obtained as

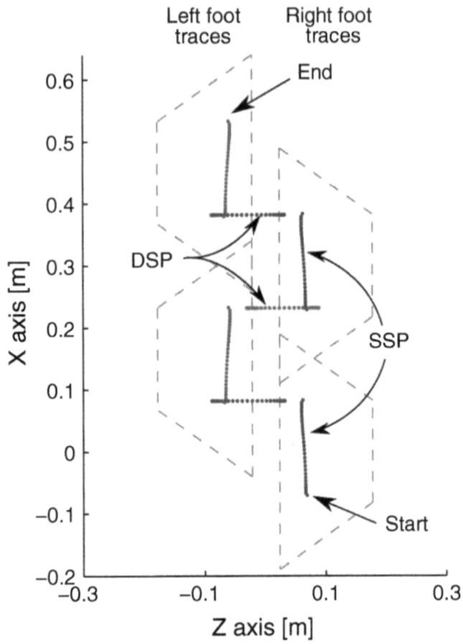

Fig. 4 Paths of the ZMP obtained with the proposed motions for the robot

$$x_{zmp} = \frac{-\tau_{Tz} - h_t f_{Tx}}{f_{Ty}} \tag{6}$$

$$z_{zmp} = \frac{\tau_{Tx} - h_t f_{Tz}}{f_{Ty}} \tag{7}$$

where h_t represent the height of the ankle corresponding to the support foot, i.e. the vertical distance from the soil to the joints of the support foot with the tibia. f_{Tx}, f_{Ty} and f_{Tz} are, respectively, the x, y and z components of the force applied by the robot on the support foot through these joints; and τ_{Tx} and τ_{Tz} are, respectively, the x and z components of the moment applied by the robot on the same joints. Such forces and moments appear as consequence of the inertial and gravitational effects of the motion.

Figure 4, shows the path of the ZMP obtained in a simulation process by using the best values of the parameters of motion given in Table 1. It can be observed that the generated path is always inside the support area, which means that the robot will keep a stable walking.

Table 3 Controller gains for tracking task

Controller	α	k_p	k_d	k_i	γ
μ_1 Tibia 1	1	110	2	35	100
μ_2 Femur 1	1	100	2	20	90
μ_3 Femur 2	1	100	2	20	90
μ_4 Tibia 2	1	100	2	30	90
f_{cp} Counterweight	300	40	25	1	30

6 Experimentation

The experimental tests on the biped robot Bip-ITLag were completed in real time by using the toolbox Real-Time Windows Target by Simulink, Matlab ©. Two Sensoray 626 data acquisition boards were used for communication and application of the control signals. Five Kollmorgen drivers model S603 were used with the AKM32D brushless motors of the same brand. The sampled time was 3 ms.

The law of control used for tracking the proposed joint trajectories is

$$\boldsymbol{u} = K_p \mathbf{tanh}(\alpha\tilde{\boldsymbol{q}}) + K_d\dot{\tilde{\boldsymbol{q}}} + \mathbf{Sat}\left(K_i \int \tilde{\boldsymbol{q}} dt, \gamma\right) \tag{8}$$

where $\boldsymbol{u} = [\mu_1\ \mu_2\ \mu_3\ \mu_4\ f_{cp}]^T$ is the torque vector in the rotation and counterweight applied force axis, $K_p = \text{diag}[k_{pi}]$, $K_d = \text{diag}[k_{di}]$, $K_i = \text{diag}[k_{ii}]$, $\mathbf{tanh}(\boldsymbol{\vartheta}) = [\tanh(\vartheta_1)\ \tanh(\vartheta_2)\ \cdots\ \tanh(\vartheta_n)]$ and furthermore $\mathbf{Sat}(\boldsymbol{v}) = [\text{sat}(v_1)\ \text{sat}(v_2)\ \cdots\ \text{sat}(v_n)]$, where $\text{sat}_i(v_i) = \text{sign}(v_i)\min(|v_i|, \gamma_i)$.

The parameters for the controller and gains used are defined in Table 3.

The parameters of trajectories used for achieving the experimental walking are those obtained for simulation and computation of the ZMP in the precedent Section. The period of walking applied, which gives the best stability of the robot, is 6 s (3 s. for SSP and 3 s. for DSP). The control signals of torque supplied to the motors during the tests are showed in Figs. 5, 6 and 7.

A sequence of postures while the robot walking during the tests is appreciated in Fig. 8. Under the initial posture the rear foot is the left one, the counterweight is positioned on the right side and, therefore, the first step is achieved by using the left foot. When the first step finishes, a DSP start and then the counterweight moves to the left side in order to balance the robot during the next step. A video of the robot during the walking could be found in [8].

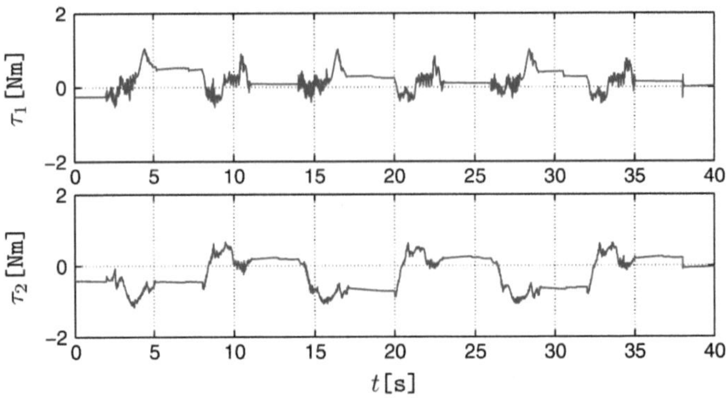

Fig. 5 Applied torque control signal to tibia and femur motors of *right* leg during tracking task

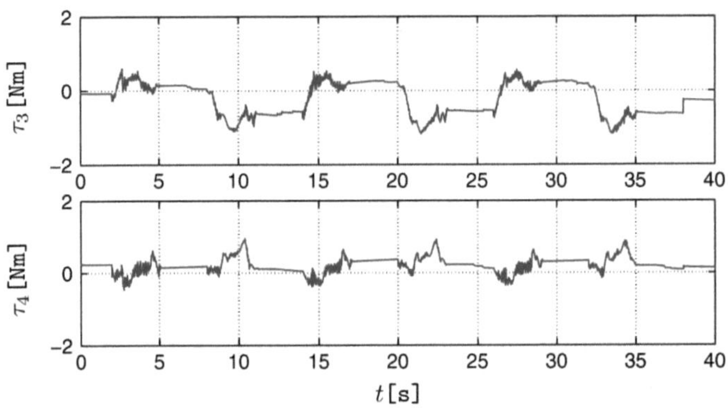

Fig. 6 Applied torque control signal to tibia and femur motors of *left* leg during tracking task

Fig. 7 Applied torque control signal to balancing motor for doing the tracking task

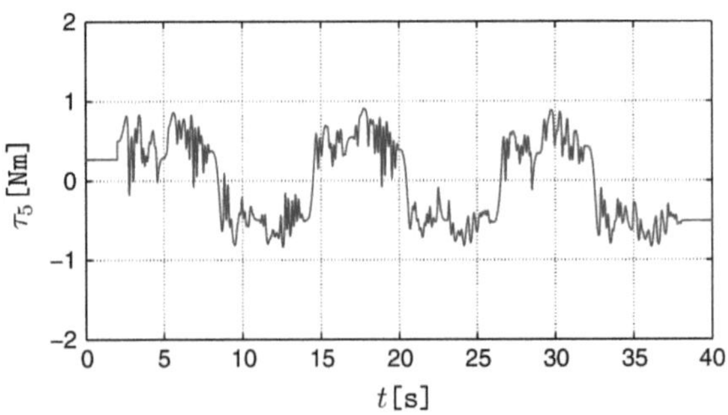

Fig. 8 Walking phases of the ITLag robot biped; first step is given with the *left* leg

7 Conclusions

The results of walking tests in real time of the biped robot Bip-ITLag were presented in this paper. This robot was designed and built at the Mechatronics and Control Laboratory of the Instituto Teconológico de La Laguna in Mexico. The walking pattern of the robot is based on cycloidal trajectories of the free foot and the hip in order to avoid impacts, and significant shaking forces, during the walking. The parameters used for the walking of the robot were previously found by achieving studies of simulation, and applying the ZMP as criterion of stability. The control applied for completing the walking is a PID saturated one. By taken into account the results obtained from the applied torques, it is feasible to propose a quicker walking. However, we consider as future work, the previous incorporation of force sensors in the feet for real ZMP linear feedback, in such a way that a more confident scheme of control be obtained.

References

1. Alvarez, E.: Modelado dinámico de un robot bípedo: Aplicación al diseno de sus mecanismos y al análisis de la marcha. Master's thesis, Instituto Tecnológico de la Laguna, Torreón, México (2006)
2. Alvarez, E., Pámanes, J.A., Arias, L.E.: Balancing of a 4 dof biped robot for a walking pattern based in cycloidal motions. In: Memorias del IX Congreso Mexicano de Robótica. Monterrey, México (2007)
3. Arias, L.E., Pámanes, J.A.: Especificación de movimientos generales de la pelvis y el pie libre de robots bípedos para marcha sin impacto. In: IX Congreso Iberoamericano de Ingeniería Mecánica. Las Palmas de Gran Canaria, Espana (2009)
4. Campos, C., Campa, R., Llama, M.: Modeling and real-time motion control of a 4-dof planar parallelogram-link biped mechanism. In: Proceedings of the European Control Conference, pp. 3341–3346. Budapest, Hungary (2009)
5. Campos, C., Campa, R., Llama, M., Pámanes, A.: Modelling, simulation and analysis of a 5-dof planar parallelogram-link biped mechanism. Int. Rev. Model. Simul. (I.RE.MO.S.) **4**, 3337–3352 (2011)
6. Carrera, I.: Control de servomotores para un robot bípedo de 4 g.d.l a través de una tarjeta controladora de multiejes. Master's thesis, Instituto Tecnológico de la Laguna, Torreón, México (2006)
7. Chevallereau, C., Abba, G., Aoustin, Y., F. Plestan, E.R.W., Canudas, C., Grizzle, J.: Rabbit: a testbed for advanced control theory. IEEE Control Syst. Mag. **23(5)**, 57–79 (2003)
8. De-León-Gómez, V., Pámanes, J.A., Santibanez, V.: [video] bipitlag primeros pasos (first walk). Instituto Tecnologico de la Laguna, Lab. de Mecatrónica y Control (2013). https://www.youtube.com/watch?v=3nba04aOPbQ
9. Fujimoto, Y., Kawamura, A.: Simulation of an autonomous biped walking robot including environmental force interaction. IEEE Robot. Autom. Magazine, **5**(2), 33–42 (1998)
10. Hashimoto, S., Narita, S., Kasahara, H., Shirai, K., Kobayashi, T., Takanishi, A., Sugano, S., et al.: Humanoid robots in waseda university-hadaly-2 and wabian. Auton. Robots **12**(1), 25–38 (2002)
11. Hirai, K., Hirose, M., Haikawa, Y., Takenake, T.: The development of honda humanoid robot. In: The 1998 IEEE International Conference on Robotics and Automation, pp. 1321–1326. Leuven, Belgium (1998)
12. Honda: Asimo the world's most advanced humanoid robot (2014). http://asimo.honda.com/asimo-specs/
13. Kajita, S., Kanehiro, F., Kaneko, K., Fujiwara, K., Yokoi, K., Hirukawa, H.: A realtime pattern generator for biped walking. In: The 2002 IEEE International Conference on Robotics and Automation, pp. 31–37. Washington, D.C., USA (2002)
14. Kajita, S., Kanehiro, F., Kaneko, K., Yokoi, K., Hirukawa, H.: The 3d linear inverted pendulum mode: a simple modeling for a biped walking pattern generation. In: the 2001 IEEE/RSJ International Conference on Intelligent Robots and Systems, pp. 239–246. Maui, HI, USA (2001)
15. Kajita, S., Tani, K.: Experimental study of biped dynamic walking. IEEE Control Syst. Mag. **16**(1), 13–19 (1996)
16. Kaneko, K., Kanehiro, F., Kajita, S., Yokoyama, K., Akachi, K., Kawasaki, T., Ota, S., Isozumi, T.: Design of prototype humanoid robotics platform for hrp. In: The 2002 IEEE/RSJ

International Conference on Intelligent Robots and Systems, pp. 2431–2436. Lausanne, Switzerland (2002)
17. Lim, H., Yamamoto, Y., Takanishi, A.: Control to realize human-like walking of a biped humanoid robot. In: The IEEE International Conference on Systems, Man and Cybernetics, Computational Cybernetics and Simulations, pp. 3271–3276. Nashville, TN, USA (2000)
18. Loffler, K., Gienger, M., Pfeiffer, F., Ulbrich, H.: Sensors and control concept of a biped robot. IEEE Trans. Industr. Electron. **51**(5), 972–980 (2004)
19. Mitobe, K., Mori, N., Aida, K., Nasu, Y.: Nonlinear feedback control of a biped walking robot. In: The 1995 IEEE International Conference on Robotics and Automation, pp. 2865–2870. Nagoya, Japan (1995)
20. Moquet, S.: Conception d'un robot bipede, projet international de fin d'études effectué á l'institut technologique de la laguna. Tech. rep, Francia (2004)
21. Ortiz, J.: Contribución al diseno mecánico de un robot caminante bípedo. Tech. rep., Instituto Tecnologico de la Laguna, Torreón, México (2005)
22. Pámanes, J.A., Nunez, J.V.: Análisis del caminado de un robot bípedo para un patrón de marcha basado en movimientos cicloidales. In: Memorias del III Congreso Mexicano de Robótica (COMRob2001) de la AMRob, pp. 62–67. Querétaro, México (2001)
23. Park, J.H., Kim, K.D.: Biped robot walking using gravity-compensated inverted pendulum mode and computed torque control. In: the 1998 IEEE International Conference on Robotics and Automation, pp. 3528–3533. Leuven, Belgium (1998)
24. Raibert, M.: Dynamic legged robots for rough terrain. In: 10th IEEE-RAS International Conference on Humanoid Robots (Humanoids), p. 1. Nashville, TN, USA (2010)
25. Sano, A., Furusho, J.: Realization of natural dynamic walking using the angular momentum information. In: The 1990 IEEE International Conference on Robotics and Automation, pp. 1476–1481. Cincinnati, OH, USA (1990)
26. Takanobu, H., Tabayashi, H., Narita, S., Takanishi, A., Guglielmelli, E.: Dario., P.: Remote interaction between human and humanoid robot. Auton. Robots **25**(4), 371–385 (1999)
27. Thuilot, B., Goswami, A., Espiau, B.: Bifurcation and chaos in a simple passive bipedal gait. In: The 1997 IEEE International Conference on Robotics and Automation, pp. 792–798. Albuquerque, NM, USA (1997)
28. Velázquez, A.: Implantación de controladores en tiempo real para un robot bípedo de 4 g.d.l. Master's thesis, Instituto Tecnológico de la Laguna, Torreón, México (2007)
29. Vukobratović, M., Borovac, B.: Zero-moment point - thirty five years of its life. Int. J. Humanoid Rob. **1**(1), 157–173 (2004)
30. Westervelt, E.R., Buche, G., Grizzle, J.W.: Experimental validation of a framework for the design of controllers that induce stable walking in planar bipeds. Int. J. Robot. Res. **23**(6), 559–582 (2004)
31. Westervelt, E.R., Buche, G., Grizzle, J.W.: Inducing dynamically stable walking in an underactuated prototype planar biped. In: The 2004 IEEE International Conference on Robotics and Automation, pp. 4234–4239. New Orleans, USA (2004)
32. Yamaguchi, J., Soga, E., Inoue, S., Takanishi, A.: Development of a bipedial humanoid robot: control method of whole body cooperative dynamic biped walking. In: The 1999 IEEE International Conference on Robotics and Automation, pp. 368–374. Detroit, MI, USA (1999)

Part III
Parallel Manipulators

Determining the Reachable Workspace for 6-DOF Delta Manipulators

C.K. Huang and K.Y. Tsai

Abstract This paper presents analytical methods that use the characteristics of workspace boundaries to predict the equations for developing workspaces of 6-DOF Delta manipulators. All the equations can be determined by some extreme points where all the limbs are at their joint limits. The study shows that the manipulators have two different types of workspaces. How to develop Delta manipulators with a better workspace is also investigated.

Keywords 6-DOF · Delta · Parallel manipulators · Reachable workspace

1 Introduction

The reachable workspace of a parallel manipulator is a set of points that can be reached by the tool center point (TCP) on the moving platform. A point is in the workspace if the point can be approached by at least one orientation of the platform. Discretization methods are commonly employed to develop the workspace for 6-DOF manipulators. The methods are easy to implement, and they are applicable to different types of parallel manipulators. However, searching for the exact boundary of the workspace using a discretization method is very time-consuming, and the result obtained can be just a theoretical workspace, as some of its subspaces might not be reachable through a continuous motion starting from the initial assembly configuration. The exact workspace boundary can be developed using analytical methods, but the process is extremely complicated because the boundary is not a smooth surface. It has been shown that the workspace boundary of a

C.K. Huang · K.Y. Tsai (✉)
School of Mechanical of Engineering, National Taiwan University of Science and Technology, Taiwan, Republic of China
e-mail: kytsai@mail.ntust.edu.tw

C.K. Huang
e-mail: D10103001@mail.ntust.edu.tw

S. Zeghloul et al. (eds.), *Robotics and Mechatronics*,
Mechanisms and Machine Science 37, DOI 10.1007/978-3-319-22368-1_10

Stewart-Gough manipulator consists of many patches generated by different sets of four constraint equations, so the boundary surface can be developed by solving four equations [1]. The difficult part is searching for the equations to generate each patch of the boundary surface. In a previous work, the characteristics of workspace boundaries were employed to develop the compatible reachable workspace [2]. This paper shows that some characteristics can be employed to develop workspaces for different types of 6-DOF parallel manipulators.

The workspace of parallel manipulators has been intensively investigated over the last three decades [3–5]. The studies, in general, are focused on 6-DOF Stewart-Gough manipulators. This paper presents methods for developing workspace for 6-DOF Delta parallel manipulators. Some characteristics of workspace boundaries of Stewart-Gough manipulators are employed to predict the equations for generating all the boundary patches. The characteristics of workspace boundaries are studied and how to develop Delta manipulators with a better workspace is investigated. The proposed method is applicable to different types of 6-DOF parallel manipulators.

2 Characteristics of Workspace Boundaries

The characteristics of the workspace boundaries of 6-DOF symmetrical Stewart-Gough parallel manipulators that can be employed to develop the workspace for other types of manipulators are summarized in this section. Figure 1 shows a typical workspace boundary of a symmetrical Stewart-Gough parallel manipulator. The boundary surface consists of several patches generated by different sets of four constraint equations (numbers k and k′ denote that the *k*th limb is at the upper limit and lower limit respectively). The boundaries of those patches are bifurcation curves with five limbs at limited positions, and the bifurcation curves intersect at extreme points with six limbs at limited positions. Apart from the {6, 0} upper extreme point A and the {0, 6} lower extreme point F, there are five types of extreme points: $M_i = \{5, 1\}$, $N_i = \{4, 2\}$, $P_i = \{3, 3\}$, $N_i' = \{2, 4\}$ and $M_i' = \{1, 5\}$, for i = 1, 2, …, 6, where the first number and the second number in each bracket denote the number of limbs at their upper limits and lower limits, respectively.

The extreme points, M_i (including M_i and M_i'), N_i (including N_i and N_i') and P_i, in the figure are termed type-1, type-2 and type-3 extreme points, respectively. There are only six type-3 extreme points because P_i and P_i' are coincident for Stewart-Gough manipulators. This work uses sets of limbs at their limited positions to denote the constraint equations for developing boundary patches, bifurcation curves and extreme points. The six numbers for some extreme points on the upper boundary in Fig. 1a are A = (1, 2, 3, 4, 5, 6), M_1 = (1, 2, 3, 4, 5, 6′), M_2 = (1, 2, 3′, 4, 5, 6), M_3 = (1, 2′, 3, 4, 5, 6), N_1 = (1, 2, 3′, 4, 5, 6′), N_2 = (1, 2′, 3′, 4, 5, 6), P_1 = (1, 2′, 3′, 4, 5, 6′). As can be seen, the limbs at the lower limits from two neighboring type-1 extreme points M_1 and M_2 (and the remaining four limbs at the

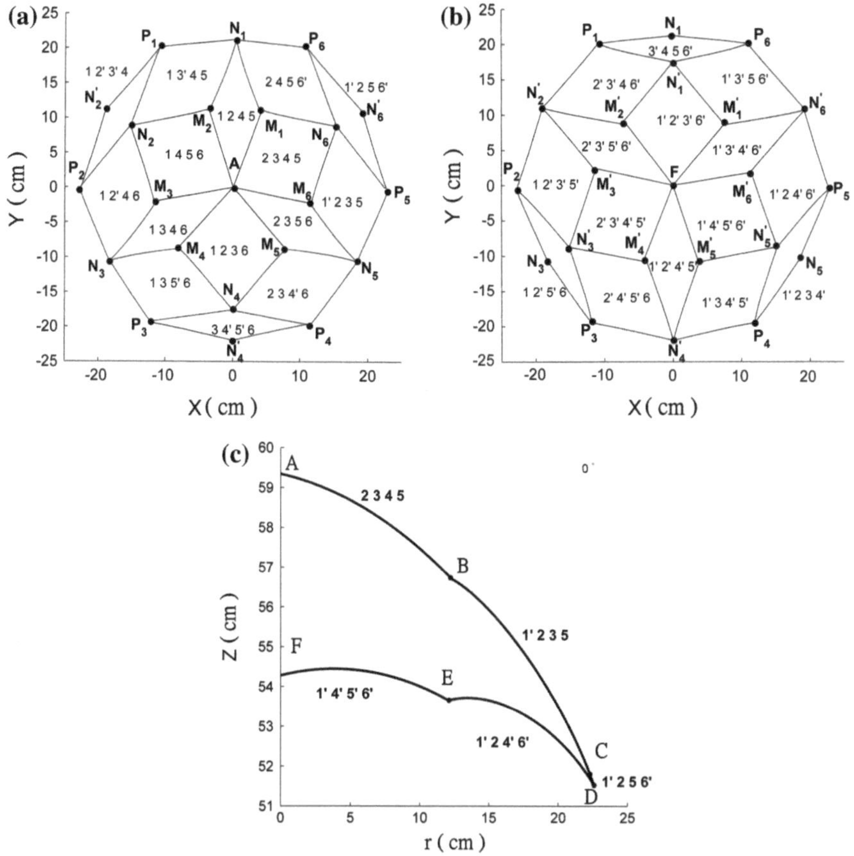

Fig. 1 Boundary surface of a Stewart-Gough manipulator. **a** Upper boundary. **b** Lower boindary. **c** Cross section at $\phi = 0°$

upper limits) on the upper boundary generate the type-2 extreme point N_1 between them and there are two bifurcation curves that connect the type-2 point to the two type-1 points. Furthermore, the three limbs at the lower limits from two neighboring type-2 extreme points N_1 and N_2 on the upper boundary generate the type-3 extreme point P_1 between the two type-2 extreme points and there are two bifurcation curves from the type-3 point to the two type-2 points. The extreme points on the lower boundary have the same properties.

The five numbers obtained by the intersection of two sets of six numbers from extreme points A and M_i, from F and M_i' or from a type-i extreme point and its neighboring type-(i + 1) point for i = 1, 2 give the five limbs for generating the bifurcation curve that connects the two corresponding extreme points, and the four numbers from the intersection of two sets of six numbers from two opposite extreme points (two extreme points that are not connected by a bifurcation curve) for a patch determine the four limbs for generating the patch. Thus, the intersection numbers (1,2,3,4,6′) from M_1 (1,2,3,4,5,6′) and N_1 (1,2,3,4,5′,6′) give the five limbs that generate the bifurcation curve between the two points, and numbers (1,3,4,5′6′) from the intersection of N_1 and P_1 determine the five limbs that generate the bifurcation curve from N_1 to P_1. The intersection numbers (1, 2, 4, 5) from M_1 and M_2 (or A and N_1) determine that limbs 1, 2, 4 and 5 at their upper limits generate patch $AM_1N_1M_2$. This work employs the above properties to predict the equations for developing the boundary curve on a cross-section of the workspace. Once the equations for a boundary surface are determined, the boundary curve on a cross-section shown in Fig. 1c can be developed by solving four equations in five variables [2].

3 6-DOF Delta Manipulators

Figure 2 shows a 6-DOF Delta manipulator with six RSS limbs. Coordinate frames $F_0 \equiv x_0 - y_0 - z_0$, $F' \equiv x' - y' - z'$ and $F_i \equiv x_i - y_i - z_i$ denote the fixed reference frame on the base, the moving frame on the platform and the fixed coordinate frame for limb i respectively. The coordinates of points C_i and B_i related to $\mathbf{F_0}$ can be determined by

$$\begin{bmatrix} \mathbf{c_i} \\ 1 \end{bmatrix} = \mathbf{T_i} \begin{bmatrix} {}^{i}\mathbf{c_i} \\ 1 \end{bmatrix} \quad \text{and} \quad \begin{bmatrix} \mathbf{b_i} \\ 1 \end{bmatrix} = \mathbf{T}' \begin{bmatrix} \mathbf{b}'_{\mathbf{i}} \\ 1 \end{bmatrix} \tag{1}$$

where ${}^{i}\mathbf{c_i} = [l_1 \cos\theta_1 \quad l_1 \sin\theta_1 \quad 0]^T$ represents the coordinates of point C_i with respect to F_i, $\mathbf{b}'_{\mathbf{i}}$ denotes the coordinates of point B_i with respect to F', $\mathbf{T}'$ is the homogeneous matrix that transforms coordinates from F' to F_o, $\mathbf{T_i}$ is the constant homogeneous matrix that transforms coordinates from F_i to F_o, θ_i denotes the ith

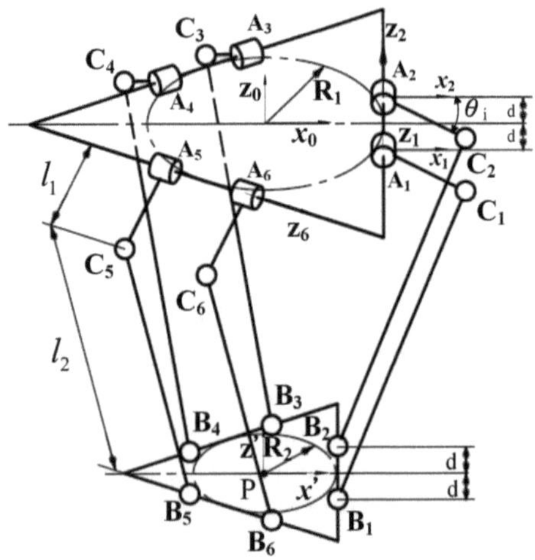

Fig. 2 A 6-DOF Delta manipulator

actuated joint displacement. The distance between points B_i and C_i equal to link length l_2 gives

$$(\mathbf{c_i} - \mathbf{b_i})^{\mathbf{t}}(\mathbf{c_i} - \mathbf{b_i}) = l_2^2 \quad i = 1, 2, 3, .., 6 \tag{2}$$

Twelve constraint equations for the workspace boundary can be obtained by substituting θ_{imax} or θ_{imin} into the equations.

The extreme points for many manipulators with different shapes are developed and the results show that type-3 extreme points P_i and P_i' are not coincident for some manipulators. For the manipulator with $R_1 = 20$, $R_2 = 7$, $d = 10$, $l_1 = 20$, $l_2 = 80$ and $-30° \le \theta_i \le -10°$ for i = 1, 2, …, 6, the extreme points and bifurcation curves are shown in Fig. 3. With $P_i = P_i'$ for i = 1, 2, …, 6, the number and the relative positions of the extreme points are exactly the same as those shown in Fig. 1a, b for a Stewart-Gough manipulator. Figure 4 shows the extreme points and bifurcation curves of another manipulator with the same design parameters, except that $R_1 = 25$ and $R_2 = 15$.

A closed boundary surface cannot be obtained after connecting type-2 and type-3 extreme points on the upper and lower boundary surfaces. Checking the intersection numbers from extreme points P_i, N_i' and N_{i+1}' shows that there is an additional bifurcation curve from P_i to one of the two neighboring points N_i' and N_{i+1}' for manipulators with $P_i \ne P_i'$. Likewise, there is an extra bifurcation curve from P_i' to one of the two neighboring points N_i and N_{i+1}. All the bifurcation curves can be developed by the following steps.

1. Solve different sets of six constraint equations to obtain M_i and M_i' and develop the upper bifurcation curves from the (6, 0) extreme point A to M_i and the lower bifurcation curves from the (0, 6) extreme point G to M_i' for i = 1, 2, …, 6.
2. Let k = 1. Repeat (i) let $M_7 = M_1$ and $M_7' = M_1'$ if k = 6; (ii) use M_k and M_{k+1} to develop N_k and the bifurcation curves from M_k and M_{k+1} to N_k; (iii) use M_k' and

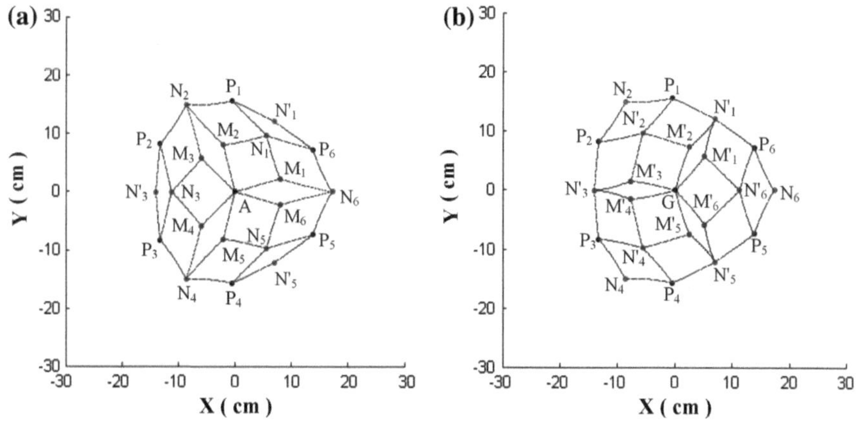

Fig. 3 Boundary surface of a 6-DOF Delta manipulator. **a** Upper boundary. **b** Lower boundary

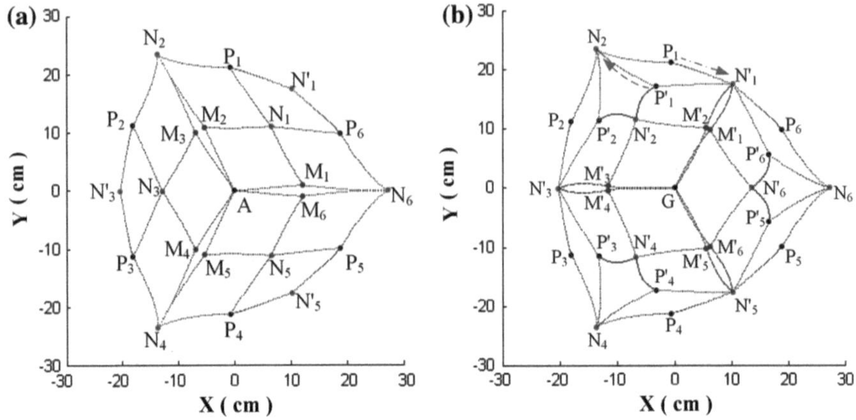

Fig. 4 Boundary surface with $P_i \neq P'_i$. **a** Upper boundary. **b** Lower boundary

M'_{k+1} to develop N'_k and the bifurcation curves from M'_k and M'_{k+1} to N'_k; and (iv) let k = k + 1, until k = 7.

3. Let j = 1. Repeat (i) let $N_7 = N_1$ and $N'_7 = N'_1$ if j = 6; (ii) use N_k and N_{k+1} to develop P_k and the bifurcation curves from N_k and N_{k+1} to P_k; (iii) use N'_k and N'_{k+1} to develop P'_k and the bifurcation curves from N'_k and N'_{k+1} to P'_k; and (iv) let j = j + 1, until j = 7.
4. Exit if $P_i = P'_i$ for i = 1, 2, …, 6.
5. Let i = 1. Repeat (i) let $N_7 = N_1$ and $N'_7 = N'_1$ if i = 6; (ii) if the intersection of P_i and N'_i gives five numbers, then use the five limbs related to the five numbers to develop the bifurcation curve from P_i to N'_i; otherwise, use the five limbs related to the five numbers from the intersection of P_i and N'_{i+1} to develop the bifurcation curve from P_i to N'_{i+1}; (iii) if the intersection of P'_i and N_i gives five numbers, then use the five limbs related to the five numbers to develop the bifurcation curve from P'_i to N_i; otherwise, use the five limbs related to the five numbers from the intersection of P'_i and N_{i+1} to develop the bifurcation curve from P'_i to N_{i+1}; and (iv) let i = i + 1, until i = 7. Exit.

It is much easier to develop the workspace of a manipulator with $P_i = P'_i$ using the method for Stewart-Gough manipulators, so this work focuses on the development of the workspace for a manipulator with $P_i \neq P'_i$. The workspace boundary has one property that is the same as that of a Stewart-Gough manipulator: the extreme points, N_i and N'_i, are on the same cross-section and the boundary surface is symmetrical about the vertical planes through N_1N_4, N_2N_5 and N_3N_6. Thus, the boundary can be represented by the cross-sections between two neighboring type-2 extreme points. The sets of equations for boundary segments change if the cross-section passes an extreme point. If the cross-sections are developed from N_1 to N_2 in Fig. 4, then the equations change at M'_2, P_1, P'_1, and M_2.

The cross-sections in the two very narrow regions from N_1 to M_2' and from M_2 to N_2 can be neglected. A section between M_2' and P_1 intersects patches $AM_1N_1M_2$, $M_2N_1P_1N_2$, $N_1P_6\,N_1'\,P_1$, $P_1'\,N_1'P_1N_2$, $M_2'N_1'\,P_1'\,N_2'$ and $GM_2'N_2'M_3'$, so the equations for generating the six boundary segments in the clockwise direction can be determined by the four intersection numbers from six pairs of extreme points (M_1,M_2), (N_1, N_2), (P_6, P_1), (P_1, P_1'), (N_1', N_2') and (M_2', M_3'), respectively.

A cross-section between P_1 and P_1' does not intersect patch $N_1P_6\,N_1'\,P_1$, and the segments on a boundary curve between P_1', and M_2 are generated by (M_1,M_2), (N_1, N_2), (P_1, P_1'), (P_1', P_2'), (N_1', N_2') and (M_2', M_3'). The boundary surfaces of the two manipulators are shown in Figs. 5 and 6.

Compared with that shown in Fig. 5, the workspace with $P_i \neq P_i'$ has some extra patches on the top of the boundary and away from the central line, as shown in Fig. 6b. Those very narrow regions can cause serious problems in control or trajectory planning. Thus, a practical design should have a workspace with $P_i = P_i'$.

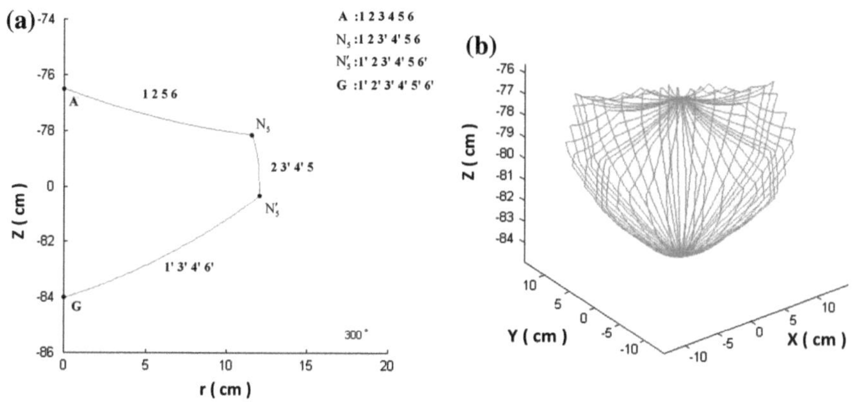

Fig. 5 Workspace with $P_i = P'_i$. **a** Cross section at $\phi = 300°$. **b** Isometric view

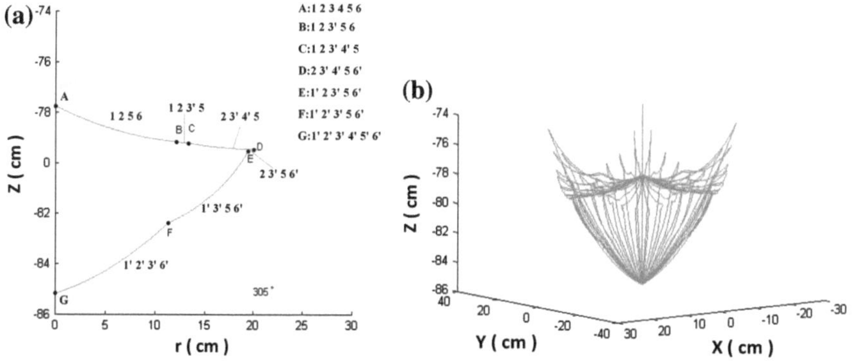

Fig. 6 Workspace with $P_i \neq P'_i$. **a** Cross section at $\phi_1 = 305°$. **b** Isometric view

A lot of workspaces with different shapes are developed and the results show that manipulators with a smaller ratio of R_2/R_1 have the workspace with $P_i = P_i'$, where R_2 and R_1 (defined in Fig. 2) determine the size of the platform and the base, respectively. In general, a workspace with $P_i = P_i'$ can be obtained if $R_2/R_1 < 0.5$

4 Conclusions

Methods involving the characteristics of workspace boundaries are presented to develop the reachable workspace for 6-DOF Delta manipulators. The study on 6-DOF Delta manipulators shows that the manipulators have two types of boundary surfaces, one of which has a workspace of better shape and is easier to develop. Manipulators for generating a specific type of workspace can be identified by the size of the platform and the base. The methods are applicable to other types of symmetrical manipulators.

Acknowledgments Financial support under grant NSC101—2221—E011—014 from the National Science Council of Taiwan is gratefully acknowledged.

References

1. Tsai, K.Y., Lee, T.K., Huang, K.D.: Determining the workspace boundary of 6-DOF parallel manipulators. Robotica **24**(5), 605–611 (2006)
2. Tsai, K.Y., Lo, I-T., Lin, P.J.: Compatible reachable workspaces of symmetrical stewart-gough parallel manipulators. Mech. Mach. Theory (2014, in press)
3. Merlet, J.P.: Determination of 6D workspaces of Gough-type parallel manipulator and comparison between different geometries. Int. J. Robot. Res. **18**(9), 902–916 (1999)
4. Gosselin, C.: Determination of the workspace of 6-DOF parallel manipulators. J. Mech. Trans. Automat. Des. **112**(3), 331–336 (1990)
5. Masory, O., Wang, J.: Workspace evaluation of Stewart platform. Adv. Robot. **9**(4), 443–461 (1995)

A Reconfiguration Strategy of a Parallel Delta-Type Robot to Improve the Kinematic Performance

A.L. Balmaceda-Santamaría and E. Castillo-Castaneda

Abstract In this work a reconfiguration strategy of a Delta-type robot to improve its kinematic performance is presented. The strategy considers an extra actuator that modifies locations and dimensions of the fixed platform. The direct kinematic analysis of the reconfigurable manipulator is determined to obtain the robot Jacobian, since this is used to analyse kinematic performance indices, such as Minimum singular value, Manipulability index and Global conditioned index.

Keywords Parallel robot · Reconfiguration · Kinematics · Performance · Screw theory

1 Introduction

A reconfiguration concept was presented in [1] as the change of the characteristics of the robot in operation, distinguished in two types: static and dynamic. The main approaches that have been proposed for reconfiguration of parallel mechanisms such as the Gough-Stewart platform are modular and have a variable geometry design [2–5]. Many reconfigurable parallel robots with variable geometry are presented, among them is a double planar parallel reconfigurable robot [6], a modular fully isotropic parallel manipulator [7], a reconfigurable parallel mechanism for machine tools applications [8] and a family of reconfigurable parallel mechanisms with different reconfigurable limbs [9].

Modular reconfigurable parallel mechanisms have been proposed too, e.g. [10] and [11], in fact the most of the researches in the reconfigurable robotics is based on the modularity, however, modular robots have a high level mechanical complexity

A.L. Balmaceda-Santamaría · E. Castillo-Castaneda (✉)
Unidad Queretaro, CICATA, Instituto Politecnico Nacional, Mexico City, Mexico
e-mail: ecastilloca@ipn.mx

A.L. Balmaceda-Santamaría
e-mail: albert.balmaceda@gmail.com

S. Zeghloul et al. (eds.), *Robotics and Mechatronics*,
Mechanisms and Machine Science 37, DOI 10.1007/978-3-319-22368-1_11

in their structures. On the other hand, the reconfiguration of robots focused on variable geometry, could be a low cost alternative when it comes to constructing them. In this work the reconfiguration of the 3-DOF Delta-type robot Parallix LKF-2040 based on the concept of variable geometry is reported. Such robot was designed as a didactic version [12]. A new reconfiguration for this robot, in order to improve kinematic performance indices of the robot Parallix is proposed.

2 Description of the Delta-Type Robot

The Parallix LKF-2040 robot shown in the Fig. 1, is a translational manipulator that was developed at IPN—CICATA.

The robot Parallix comprises 3 stationary motors disposed on the robot base through brackets. The motor axes are coupled to a kinematic chain at the bracket level. This mechanism has a structure of the famous Delta robot, however the mechanism of Fig. 1 is a robot 3-RUU.

2.1 Proposed Reconfiguration Strategy

Figure 2 shows the mechanical concept of reconfigurable robot Parallix proposed in this work.

The reconfigurable robot is composed by a reconfigurable fixed platform with a fixed element (6) and three mobile elements in form of framing square (7a, 7b, 7c). These framing squares simultaneously modify the radius and height between the axes of the motors (8a, 8b, 8c) and the center of the fixed element (6). In the center of the fixed platform there is a screw (1) driven by a fourth actuator (8d). A nut

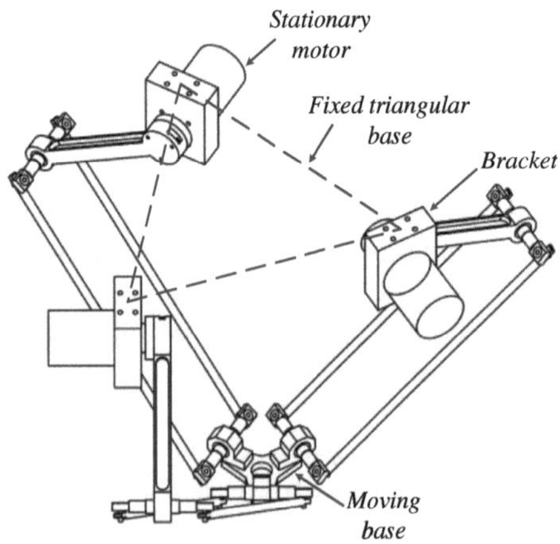

Fig. 1 Configuration of the parallel manipulator Parallix LKF-2040

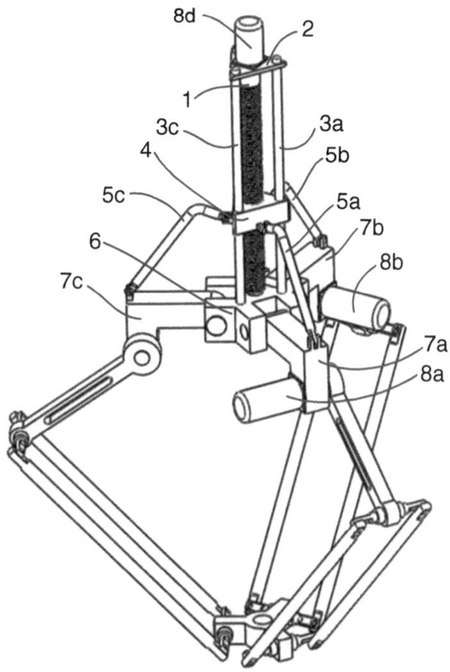

Fig. 2 Reconfiguration strategy proposed (from patenting document at the IMPI-Mexico: MX/a/2013/006781)

(4) coupled to the screw, slides along three guide bars (3a, 3b, 3c). The nut moves the framing squares through coupling bars (5a, 5b, 5c). The set of components screw (1), nut (4), motor (8d) and coupling bars (5a, 5b, 5c) establish a system used here as example of a linear displacement mechanism. It should be noted that this configuration is highly versatile, since the mechanism can be reconfigured also with decoupled movements of framing squares from the rest of the mechanism and be able to return to the original configuration and mechanical characteristics of the robot Parallix LKF-2040.

3 Direct Kinematic Analysis of the Reconfigurable Mechanism

In Fig. 3, the geometrical parameters of the reconfigurable manipulator Parallix are shown.

The position analysis for the reconfiguration mechanism is performed based on the method proposed in [13]. For this analysis, it is first necessary to calculate the generalized coordinate α_4 (see Fig. 3) as follows

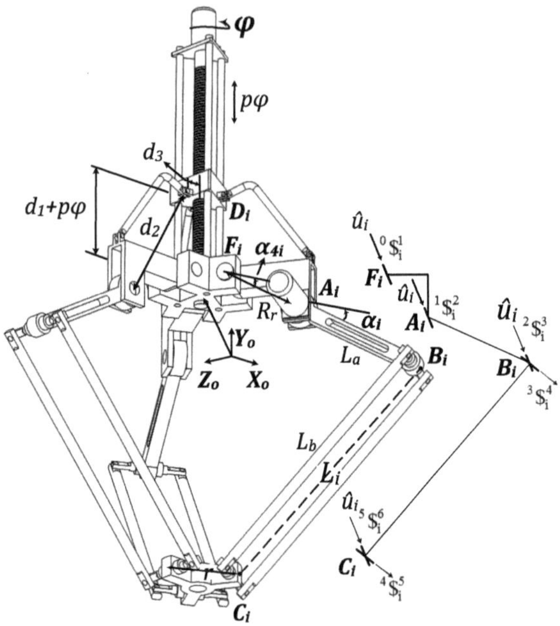

Fig. 3 Geometrical parameters of the mechanism

$$\left(R_f + Rr\cos(\alpha_4) - d_3\right)^2 + (d_1 + p\varphi - Rr\sin(\alpha_4))^2 = d_2^2 \tag{1}$$

therein d_1 is an offset between the point D_i and the base plane of nut, d_2 is the distance between point D_i and the point A_i, d_3 is the perpendicular distance between the screw axis and the point D_i and p is the thread pitch of the screw. Then solving and simplify the Eq. (1) we have

$$k_1\sin(\alpha_4) + k_2\cos(\alpha_4) = k_3 \tag{2}$$

where $k_1 = -2Rr(d_1 + p\varphi), k_2 = 2Rr\left(R_f - d_3\right), k_3 = d_2^2 - R_f^2 - Rr^2 - d_3^2 + 2R_fd_3 - d_1^2 - p^2\varphi^2 - 2d_1p\varphi$. Now, solving the Eq. (2) for α_4 yields

$$\sin(\alpha_4) = \frac{2k_1k_3 \pm \sqrt{4k_1^2k_3^2 - 4(k_1^2 + k_2^2)(k_3^2 - k_2^2)}}{2(k_1^2 + k_2^2)} \tag{3}$$

Thus, with the Eq. (3) can be calculated the generalized coordinate α_4 generated by the linear displacement system used as a technological option in this work.

Let $X_oY_oZ_o$ be the fixed reference system in the center of the fixed platform of the manipulator (see Fig. 3). The coordinates of the point $F_i = (Fx_i, Fy_i, Fz_i)$, as well as $A_i = (Ax_i, Ay_i, Az_i)$ that depends on the rotation φ of the motor, namely $\alpha_{4i}(\varphi) = \alpha_4(\varphi)$ since α_{4i} is the same angle for all framing squares and $B_i = (Bx_i, By_i, Bz_i)$ are calculated by projections as follows

$$F_i = \begin{bmatrix} R_f\cos(\theta_i) & 0 & R_f\sin(\theta_i) \end{bmatrix}$$
$$A_i = \begin{bmatrix} R_r\cos(\alpha_{4i})\cos(\theta_i) + Fx_i & R_r\sin(\alpha_{4i}) & R_r\cos(\alpha_{4i})\sin(\theta_i) + Fz_i \end{bmatrix}$$
$$B_i = \begin{bmatrix} La\cos(\alpha_i)\cos(\theta_i) + Ax_i & By_i & La\cos(\alpha_i)\sin(\theta_i) + Az_i \end{bmatrix}$$

where θ_i is the orientation angle of each kinematic chain; $R = R_f + R_r$, where R_f is the distance from the fixed reference system $X_oY_oZ_o$ to each point F_i. And evidently, $By_i = La\sin(\alpha_i) + Ay_i$ is equal for all kinematic chains of the mechanism. Unless otherwise specified, $i = 1, 2, 3$ in the rest of this document. Whereas the moving platform is an equilateral triangle $C_1C_2C_3$ of side r and center P. It is established that $C_1 = (X, Y, Z)$ and hence the coordinates of C_2 and C_3 can be computed according to

$$C_{2,3} = \left(X + r\cos\left(\theta_{2,3}\right), Y, Z + r\sin\left(\theta_{2,3}\right)\right) \tag{4}$$

where $\theta_{2,3}$ is the orientation angle for the kinematic chain 2 and 3, respectively. Also, in order to calculate the coordinates X, Y and Z we consider three closure equations $(\boldsymbol{C}_i - \boldsymbol{B}_i) \bullet (\boldsymbol{C}_i - \boldsymbol{B}_i) = Lb^2$ that generates

$$X^2 + Y^2 + Z^2 + K_{i1}X + K_{i2}Y + K_{i3}Z + K_{i4} = 0 \tag{5}$$

therein $K_{ij}(j = 1, 2, 3, 4)$ are obtained according to the parameters and generalized coordinates of the mechanism. Furthermore, the Eq. (5) can be expressed for X and Y in terms of Z as follows

$$X = W_1Z + W_2, Y = W_3Z + W_4 \tag{6}$$

where the coefficients $W_1, \ldots, W_4$ are calculated according to the coefficients K_{ij}. And finally, by substituting Eq. (6) in any of the Eq. (5) produces a second-order equation with Z as unknown variable as follows

$$aZ^2 + bZ + c = 0 \tag{7}$$

therein $a = 1 + W_1^2 + W_3^2$, $b = K_{11}W_1 + 2W_1W_2 + K_{13} + 2W_3W_4 + K_{12}W_3$, $c = K_{11}W_2 + W_4^2 + W_2^2 + K_{14} + K_{12}W_4$. From Eq. (7), two solutions Z_1 and Z_2 that depend on the generalized coordinates can be obtained. Thus, X and Y coordinates can be recursively computed with the Eqs. (7) and (6). Finally, the point P is calculated with the Eq. (8) as below

$$P = \left(X + \frac{r}{3}(\cos(\theta_2) + \cos(\theta_3)),\ Y,\ Z + \frac{r}{3}(\sin(\theta_2) + \sin(\theta_3))\right) \tag{8}$$

Velocity analysis determines the velocity state of the moving platform with respect to the fixed platform [13], when the joint rates $\dot{\alpha}_4 = \dot{q}_4$ and $\dot{\alpha}_i = \dot{q}_i$ are known. In this work $\dot{q}_4$ is the derivative from Eq. (3) and is used as an independent variable in the velocity analysis.

Let $V = [0,\ \boldsymbol{v}]^T$ be the velocity state of the moving platform observed from the fixed platform, where $\boldsymbol{v}$ is the velocity vector of any point on the moving platform. The velocity state $\boldsymbol{V}$ may be expressed in screw form through the limbs of the robot as follows

$$\dot{q}_4\ {}^0\$_i^1 + \dot{q}_i\ {}^1\$_i^2 +_2 \omega_3^i\ {}^2\$_i^3 +_3 \omega_4^i\ {}^3\$_i^4 +_4 \omega_5^i\ {}^4\$_i^5 +_5 \omega_6^i\ {}^5\$_i^6 = \boldsymbol{V} \tag{9}$$

where i denote the ith kinematic chain. In order to obtain an input-output velocity equation of the reconfigurable manipulator, consider that a line in Plücker coordinates, $L_i = [lx_i \quad ly_i \quad lz_i \quad Lx_i \quad Ly_i \quad Lz_i]^T$, passes through the points B_i and C_i (Fig. 3), reciprocal to all the screws in the same limb, excepting the screw associated to the actuated revolute joint. Therefore, the systematic application of the Klein form, which is denoted as $\{*;*\}$, of the line L_i to both sides of Eq. (9) and reducing terms, yields the input-output velocity equation as

$$v = \mathbf{A}^{-1}\left(\mathbf{B}'\mathbf{q}_4 + \mathbf{B}\mathbf{q}\right) \tag{10}$$

therein $\mathbf{A} = [lx_i \quad ly_i \quad lz_i]^T$ is the active Jacobian matrix of the manipulator, while $\dot{\mathbf{q}}_4 = [\dot{q}_4 \quad \dot{q}_4 \quad \dot{q}_4]$ and $\dot{\mathbf{q}} = [\dot{q}_1 \quad \dot{q}_2 \quad \dot{q}_3]$ are the first-order generalized coordinates of the mechanism. $\mathbf{B}' = diag\left[\{{}^0\$_1^1; L_1\} \quad \{{}^0\$_2^1; L_2\} \quad \{{}^0\$_3^1; L_3\}\right]$ and $\mathbf{B} = diag\left[\{{}^1\$_1^2; L_1\} \quad \{{}^1\$_2^2; L_2\} \quad \{{}^1\$_3^2; L_3\}\right]$ are the passive Jacobians.

4 Performance Indices

In this section the Minimum Singular Value (MSV), Manipulability Index (MI) and Global Condition Index (GCI) are considered as performance indices, in order to show the advantages of the reconfigurable manipulator. A Cartesian trajectory of the center P of the moving platform is generated by solving the forward displacement analysis by considering $\alpha_{1,3}(t) = 0.25\sin(t)$ and $\alpha_2(t) = 0.25\cos(t)$, where t is $0 \le t \le 2\pi$, arbitrarily selected (see Fig. 4).

The MSV proposed in [14] is calculate as follow

$$\text{MSV} = min(J^T J) \tag{11}$$

therein the Jacobian matrix $\mathbf{J}$ of the manipulator is calculated from the Eq. (10) as

$$v = \left(\mathbf{A}^{-1}\mathbf{B}\right)\dot{\mathbf{q}} = \mathbf{J}\mathbf{q} \tag{12}$$

since $\dot{q}_4$ is zero, because the reconfiguration is considered in this paper as a particular case with alternating movements between the framing squares and the rest of the mechanism. The Fig. 5a shows the performance of the manipulators in the

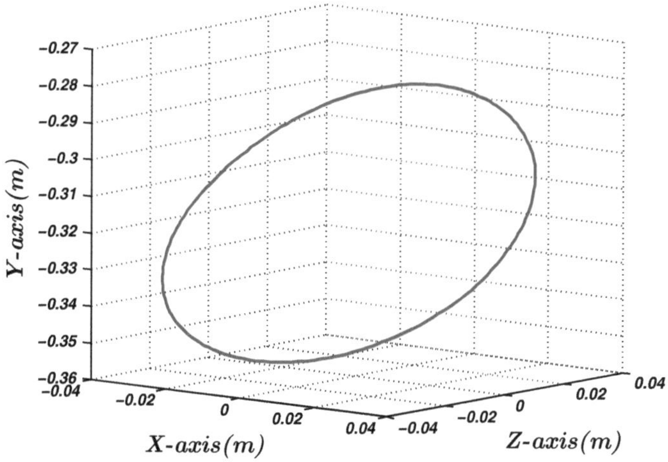

Fig. 4 Trajectory of the moving platform

trajectory of Fig. 4 calculated with Eq. (11). It is evident that the proposed reconfigurable mechanism obtained the best performance in that trajectory with values nearest to zero.

The MI proposed in [15] measures the capability of a robot in any configuration of generate velocities in the end-effector and the closeness to a singular configuration. Similarly, using the Eq. (12) to compute the matrix **J**, the MI $= \sqrt{\det(JJ^T)}$ and its inverse is defined as follows

$$\mathrm{IMI} = 1/\sqrt{\det(\mathbf{J}\mathbf{J}^T)} \qquad (13)$$

To have a better appreciation of the MI, its inverse (IMI) is used, since it has a range between $0 \leq \mathrm{IMI} \leq \infty$. So that from the Eq. (13), if IMI $= 0$ indicates a better capability to generate movements in the end-effector, but when values of IMI tending to infinity indicate that the manipulator is in a singular configuration. Figure 5b shows the behavior of the IMI computed with Eq. (13) and it is evident that the reconfigurable manipulator significantly improves the manipulability with respect to the robot Parallix.

The GCI is proposed in [16], means the distribution of the Condition number throughout the manipulator workspace. This index can be expressed as shown below

$$\mathrm{GCI} = \sum_{i=1}^{n_w} k_i^{-1}/n_w \qquad (14)$$

therein n_w is the number of points of the discretized workspace and the numerator is the sum of the reciprocal of the Condition number obtained for each the points that

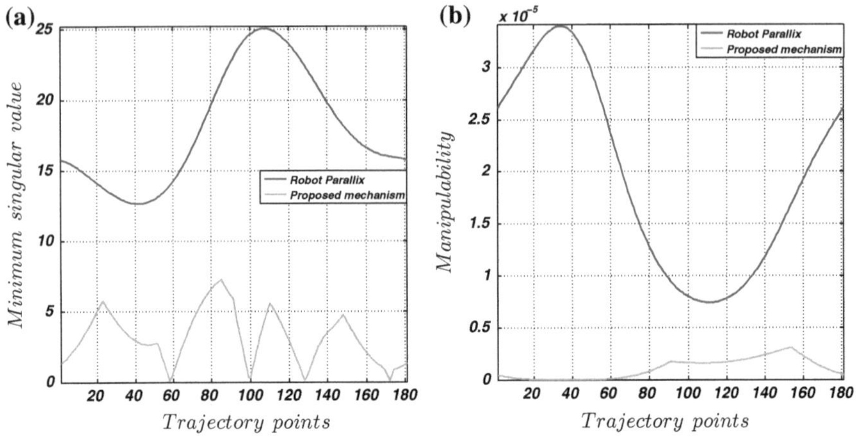

Fig. 5 Performance indices of the parallel manipulators. **a** Minimum singular value. **b** Manipulability index

comprise the entire manipulator workspace. From the Eq. (14), if the GCI = 0 it is say that the manipulator workspace is bad-conditioned, but if GCI = 1 then the manipulator have a well-conditioned workspace. The values obtained of the GCI for robot Parallix is 0.1146 and for the proposed reconfigurable mechanism is 0.1734. It is worth to note that the reconfiguration is advantageous for the robot Parallix. Therefore, by selecting a convenient value of α_{4i} it is possible to enhancement the kinematic performance of the original configuration, namely when $\alpha_{4i} = 0°$. In addition, as performance indices mentioned above are improved by the reconfiguration, all other kinematic performance indices dependent on matrix **J** of the manipulator are indirectly improved.

5 Conclusions

In this paper a new reconfiguration of a Delta-type robot is proposed to improve the kinematic indices. The reconfiguration proposal is in patenting process at IMPI-Mexico. The inverse displacement and velocity analyses are presented. The calculation of performance indices for a trajectory defined is presented, to show that performance indices are improved with respect to the Parallix original configuration. The results evidence that by selecting a suitable configuration of the fixed platform, it is possible to obtain a better kinematic performance for the Delta-type manipulator. It is worth notice that any performance indices related with the Jacobian matrix can be improved thanks to this new reconfiguration.

References

1. Krefft, M. et. al.: A reconfigurable parallel robot with high flexibility. In: Proceeding of ISR Robotik 2006 Joint Conference Robotics München (2006)
2. Du Plessis, L.J., Snymar, J.A.: An optimally re-configurable planar Gough-Stewart machining platform. Mech. Mach. Theory **41**, 334–357 (2006)
3. Kumar, S.G., Nagarajan, T., Srinivasa, Y.G.: Characterization of reconfigurable Stewarte platform for contour generation. Robot. Comput.-Integr. Manuf. **25**, 721–731 (2009)
4. Chun-Ta, Chen.: Reconfiguration of a parallel kinematic manipulator for the maximum dynamic load-carrying capacity. Mech. Mach. Theory **54**, 62–75 (2012)
5. Borràs, J. et. al.: A reconfigurable 5-DoF 5-SPU parallel platform. ASME/IFToMM International Conference on Reconfigurable Mechanisms and Robots 617—662 (2009)
6. Simaan, N., Shoham, M.: Stiffiness synthesis of a variable geometry six-degrees-of-freedom double planar parallel robot. Int. J. Robot. Res. **22**(9), 757–775 (2003)
7. Gogu, G.: Isogliden-TaRb: a family of up to five axes reconfigurable and maximally regular parallel kinematic machines. J. Manuf. Syst. **36**(5), 419–426 (2007)
8. Zhang, D., Shi, Q.: Novel design and analysis of a reconfigurable parallel manipulator using variable geometry approach. Practical Applications of Intelligent Systems. Adv. Intell. Soft Comput. **124**, 447–457 (2012)
9. Ye, W., et al.: A new family of reconfigurable parallel mechanism with diamond kinematotropic chain. Mech. Mach. Theory **74**, 1–9 (2014)
10. Haage, M. et al.: Reconfigurable parallel kinematic manipulator for flexible manufacturing. In: 13th IFAC Symposium on Information Control Problems in Manufacturing, vol. **13**, part 1, pp. 145–150 (2009)
11. Zhiming, J., Phillip, S.: Design of a reconfigurable platform manipulator. J. Robot. Syst. **15**(6), 341–346 (1998)
12. Gonzalez-Hernandez, A., Castillo-Castaneda, E.: Stiffness estimation of a parallel manipulator using image analysis and camera calibration techniques. Robotica **31**, 657–667 (2013)
13. Gallardo-Alvarado, J., Balmaceda-Santamaría, A.L., Castillo-Castaneda, E.: An application of screw theory to the kinematic analysis of a Delta-type robot. J. Mech. Sci. Technol. **28**(9), 3785–3792 (2014)
14. Klein, C.A., Blaho, B.E.: Dexterity measures for the design and control of kinematically redundant manipulators. Int. J. Robot. Res. **6**(2), 72–83 (1987)
15. Yoshikawa, T.: Manipulability of robotic mechanism. Int. J. Robot. Res. **4**(2), 3–9 (1985)
16. Gosselin, C., Angeles, J.: A global performance index for the kinematic optimization of robotic manipulators. J. Mech. Des. **113**(3), 220–226 (19991)

Workspace and Singularity Analysis of a Delta like Family Robot

R. Jha, D. Chablat, F. Rouillier and G. Moroz

Abstract Workspace and joint space analysis are essential steps in describing the task and designing the control loop of the robot, respectively. This paper presents the descriptive analysis of a family of delta-like parallel robots by using algebraic tools to induce an estimation about the complexity in representing the singularities in the workspace and the joint space. A Gröbner based elimination is used to compute the singularities of the manipulator and a Cylindrical Algebraic Decomposition algorithm is used to study the workspace and the joint space. From these algebraic objects, we propose some certified three dimensional plotting describing the the shape of workspace and of the joint space which will help the engineers or researchers to decide the most suited configuration of the manipulator they should use for a given task. Also, the different parameters associated with the complexity of the serial and parallel singularities are tabulated, which further enhance the selection of the different configuration of the manipulator by comparing the complexity of the singularity equations.

Keywords Delta-like robot · Cylindrical algebraic decomposition · Workspace · Gröbner basis · Parallel robot

R. Jha (✉) · D. Chablat
Institut de Recherche en Communications et Cybernétique de Nantes,
UMR CNRS 6597, Nantes, France
e-mail: Ranjan.Jha@irccyn.ec-nantes.fr

D. Chablat
e-mail: Damien.Chablat@irccyn.ec-nantes.fr

F. Rouillier
INRIA Paris-Rocquencourt, Institut de Mathématiques de Jussieu,
UMR CNRS 7586, Paris, France
e-mail: Fabrice.Rouillier@inria.fr

G. Moroz
INRIA Nancy-Grand Est, Universit´e de Lorraine, LORIA, UMR CNRS 7503,
Villers-lès-Nancy, France
e-mail: Guillaume.Moroz@inria.fr

S. Zeghloul et al. (eds.), *Robotics and Mechatronics*,
Mechanisms and Machine Science 37, DOI 10.1007/978-3-319-22368-1_12

1 Introduction

The workspace can be defined as the volume of space or the complete set of poses which the end-effector of the manipulator can reach. Many researchers published several works on the problem of computing these complete sets for robot kinematics. Based on the early studies [12, 21], several methods for workspace determination have been proposed, but many of them are applicable only for a particular class of robots. The workspace of parallel robots mainly depends on the actuated joint variables, the range of motion of the joints and the mechanical interferences between the bodies of mechanism. There are different techniques based on geometric [11, 15], discretization [3, 4, 9], and algebraic methods [5, 7, 17, 22] which can be used to compute the workspace of parallel robot. The main advantage of the geometric approach is that, it establish the nature of the boundary of the workspace [19]. Also it allows to compute the surface and volume of the workspace while being very efficient in terms of storage space, but when the rotational motion is included, it becomes less efficient.

Interval analysis based methods can be used to compute the workspace but the computation time depends on the complexity of the robot and the requested accuracy [9]. Discretization methods are usually less complicated and can easily take into account all kinematic constraints, but they require more space and computation time for higher resolutions. The majority of numerical methods which is used to determine the workspace of parallel manipulators includes the discretization of the pose parameters for computing workspace boundaries [3]. There are other approaches, which are based on optimization algorithms [20] for fully serial or parallel manipulators, analytic methods for symmetrical spherical mechanisms [2]. In [1] a method for computing the workspace boundary for manipulators with a general structure is proposed, which uses a branch-and-prune technique to isolate a set of output singularities, and then classifies the points on such set according to whether they correspond to motion impediments in the workspace. A cylindrical algebraic decomposition (CAD) based method is illustrated in [5, 6], which is used to model the workspace and joint space for the 3 RPS parallel robot.

This paper presents the results which are obtained by applying algebraic methods for the workspace and joint space analysis of a family of delta-like robot including complexity information for representing the singularities in the workspace and the joint space. The CAD algorithm is used to study the workspace and joint space, and a Gröbner based elimination process is used to compute the parallel and serial singularities of the manipulator. The structure of the paper is as follows. Section 2 describes the architecture of the manipulator, including kinematic equation and joint constraints associated with the manipulators. Section 3 discusses the computation of parallel as well as serial singularities and their projections in workspace and joint-space. Section 4 presents a comparative study on the shape of the workspace of different delta-like robots. Section 5 finally concludes the paper.

2 Manipulators Architecture

The manipulator architecture under the study is a three degree of freedom parallel mechanism which consists of three identical legs, the different arrangements of these legs give rise to family of delta like robot. Several types of delta-like robot were studied, few of them are Orthoglide [9, 18], Hybridglide, Triaglide [13] and UraneSX [9]. The position vectors of end points of *i*th leg are $\mathbf{P}^j$ and $\mathbf{B}_i^j$, also $\mathbf{A}_i^j$ and $\mathbf{B}_i^j$ are the position vectors of end points of *i*th actuator where *j* represents the manipulator type from the family of delta like robot (*j* = 1—Orthoglide, 2—Hybridglide, 3—Triaglide, 4—UraneSX). $\boldsymbol{\rho}_i^j$ represents the prismatic joint variables whereas $\mathbf{P}^j$ represents the position vector of the tool center point which is shown in Eq. (1).

$$||\mathbf{A}_i^j\mathbf{B}_i^j|| = \rho_i^j \quad \mathbf{P} = [xyz]^T \quad \text{with} \quad i = 1,2,3 \quad \text{and} \quad j = 1,2,3,4 \tag{1}$$

The kinematic equations of the family of delta like robot can be generalized as $||\mathbf{P} - \mathbf{B}_i^j|| = \mathbf{L}_i^j$. All the computations and analysis are done for $\mathbf{L}_i^j = \mathbf{L}$ and by imposing the following constraints on joint variables. Without joint limits, all the robots admit two assembly modes and eight working modes.

$$0 < \rho_1 < 2L \quad 0 < \rho_2 < 2L \quad 0 < \rho_3 < 2L \tag{2}$$

The *Orthoglide* mechanism is driven by three actuated orthogonal prismatic joints. A simpler virtual model can be defined for the Orthoglide, which consists of three bar links connected by the revolute joints to the tool center point on one side and to the corresponding prismatic joint at another side. Several assembly modes of these robots depends upon the solutions of direct kinematic problem is shown in Fig. 1a. The point $\mathbf{P}_i$ represents the pose of corresponding robot. However more than one value of *i* for the point $\mathbf{P}_i$ shows the multiple solutions for the DKP. The constraint equations for the Orthoglide are

$$(x-\rho_1)^2 + y^2 + z^2 = L^2; x^2 + (y-\rho_2)^2 + z^2 = L^2; x^2 + y^2 + (z-\rho_3)^2 = L^2$$

The *Hybridglide* mechanism consists of three actuated prismatic joints, in which two actuators are placed parallel and third one perpendicular to others two. Also the three bar links connected by spherical joints to the tool center point on one side and to the corresponding prismatic joint at another side. Several assembly modes of these robots depends upon the solutions of direct kinematic problem is shown in Fig. 1b and following is the constraint equations:

$$\begin{aligned}(x-1)^2 + (y-\rho_1)^2 + z^2 &= L^2; (x+1)^2 + (y-\rho_2)^2 + z^2 \\ &= L^2; x^2 + y^2 + (z-\rho_3)^2 = L^2\end{aligned}$$

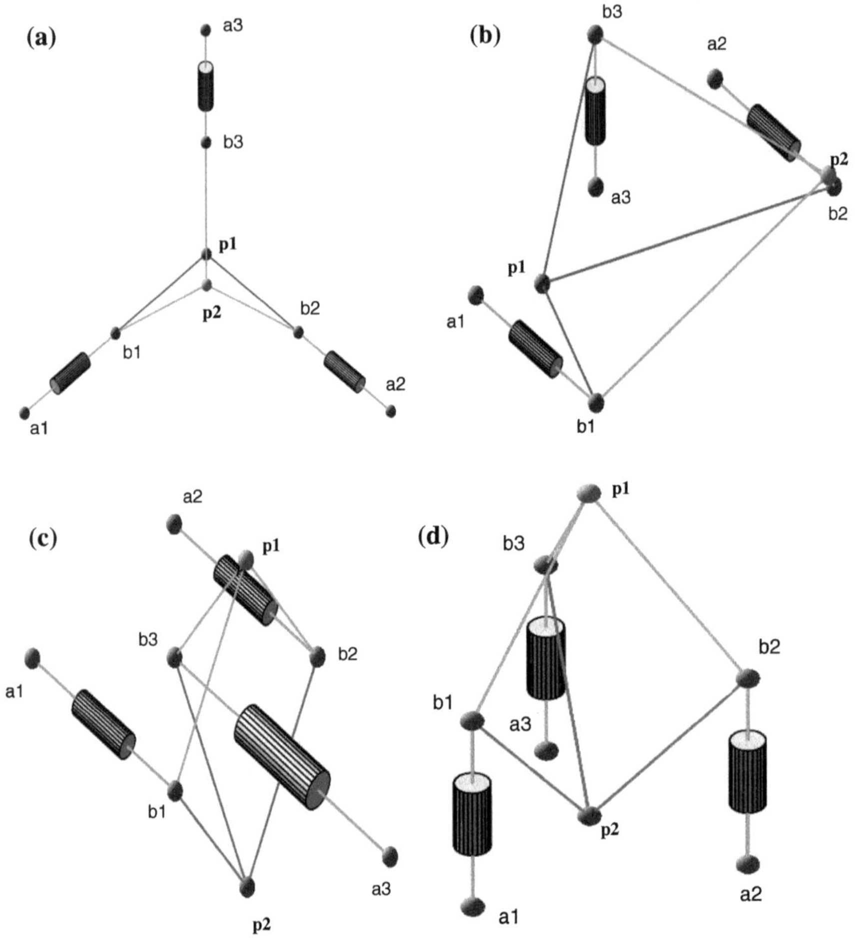

Fig. 1 Configuration plot for Orthoglide (**a**), Hybridglide (**b**), Triaglide (**c**) and UraneSX (**d**) robot

The *Triaglide* manipulator is driven by three actuated prismatic joints, in which all the three actuators parallel to each other and placed in the same plane. The architecture of the Triaglide is shown in Fig. 1c and the constraint equations are defined as:

$$(x-1)^2 + (y-\rho_1)^2 + z^2 = L^2; (x+1)^2 + (y-\rho_2)^2 + z^2 = L^2; x^2 + (y-\rho_3)^2 + z^2 = L^2$$

The *UraneSX* is similar to triaglide, but instead of three actuators in the same plane, it is placed in different planes. The architecture of the Triaglide is shown in Fig. 1d and the constraint equations are as follows:

$$(x-1)^2+y^2+(z-\rho_1)^2=L^2;\,(x+1/2)^2+(y-\sqrt{3}/2)^2+(z-\rho_2)^2=L^2;$$
$$(x+1/2)^2+(y+\sqrt{3}/2)^2+(z-\rho_3)^2=L^2$$

3 Singularity Analysis

Singularities of a robotic manipulator are important feature that essentially influence its capabilities. Mathematically, a singular configuration may be defined a rank deficiency of the Jacobian describing the differential mapping from the joint space to the workspace and vice versa. Differentiating the constraint equations of the robot with respect to time leads to the velocity model: $\mathbf{At}+\mathbf{B}\dot{\mathbf{q}}=\mathbf{0}$ where $\mathbf{A}$ and $\mathbf{B}$ are the parallel and serial Jacobian matrices, respectively, $\mathbf{t}$ is the velocity of $\mathbf{P}$ and $\dot{\mathbf{q}}$ defines the joint velocities. The parallel singularities occur whenever $\det(\mathbf{A})=0$ and the serial singularities occur whenever $\det(\mathbf{B})=0$ [8].

$$\begin{aligned}
\det(\mathbf{A})_o &= -8\rho_1\rho_2\rho_3+8\rho_1\rho_2z+8\rho_1\rho_3y+8\rho_2\rho_3x\\
\det(\mathbf{A})_h &= -8\rho_1\rho_3x+8\rho_2\rho_3x-8\rho_1\rho_3+8\rho_1z-8\rho_2\rho_3+8\rho_2z+16\rho_3y\\
\det(\mathbf{A})_t &= 8\rho_1z+8\rho_2z-16\rho_3z\\
\det(\mathbf{A})_u &= 4\sqrt{3}(3z-\rho_1-\rho_2-\rho_3+\rho_3x+\rho_2x-2\rho_1x)+12\rho_3y-12\rho_2y
\end{aligned}\tag{3}$$

Parallel and serial singularities as well as their projections in workspace and joint space are computed using a Gröbner based elimination method. This usual way for eliminating variables (see [10]) computes (the algebraic closure of) the projection of the parallel singularities in the workspace. In the same way, one can compute (the algebraic closure of) the projection of the parallel singularities in the joint space. Both are then defined as the zero set of some system of algebraic equations and we assume that the considered robots are generic enough so that both are hypersurfaces. $\det(\mathbf{A})_o$, $\det(\mathbf{A})_h$, $\det(\mathbf{A})_t$ and $\det(\mathbf{A})_u$ are the parallel singularities of Orthoglide, Hybridglide, Triaglide and UraneSX, respectively. Starting from the constraint equations and the determinant of the Jacobian matrix, we are able to eliminate the joint values. This elimination strategy is more efficient than a cascading elimination by means of resultants which might introduce many more spurious solutions: singular points that are not projections of singular points. Due to the lack of space, other equations associated with serial singularities are not presented in this paper. Figure 2 represents the projections of parallel singularities in projection space $(\mathbf{x},\mathbf{y},\mathbf{z})$. These surfaces represent the singularity associated with the eight working modes.

In Table 1, a comparative study of five parameters among the family of delta like robot is presented. We have tabulated the main characteristics of the polynomials (In three variables) used for the plots (Implicit surface): their total degree, their number of terms and the maximum bitsize of their coefficients. We have also

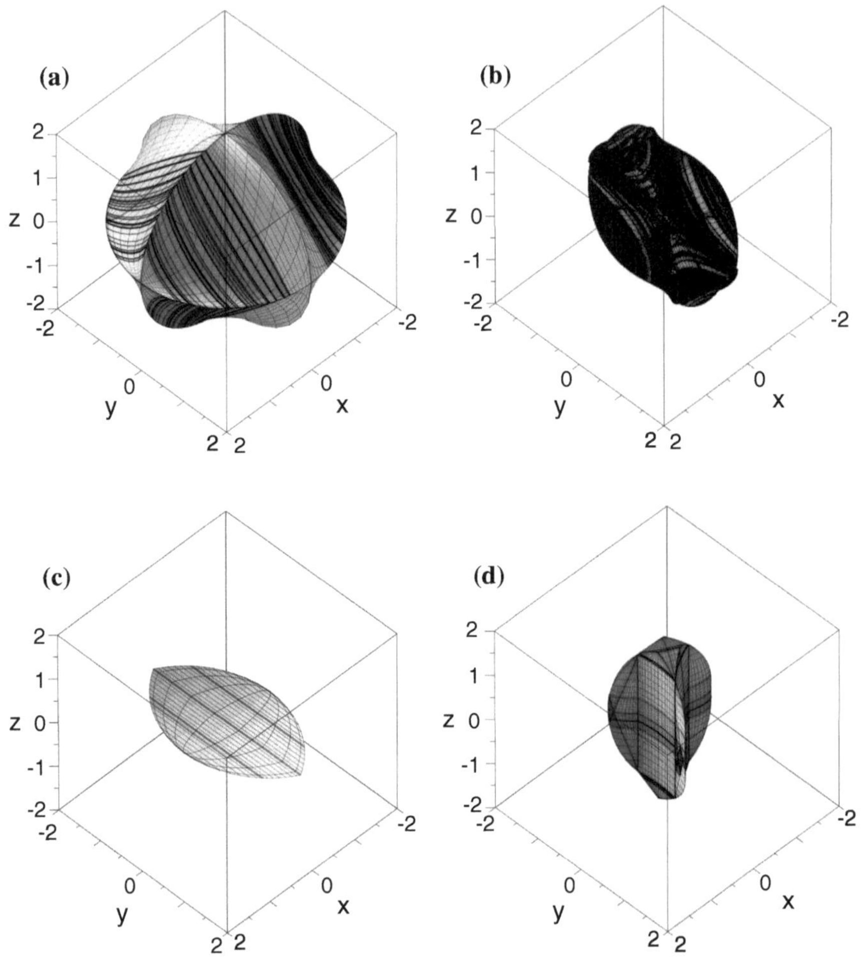

Fig. 2 Projection of parallel singularities of Orthoglide (**a**), Hybridglide (**b**), Triaglide (**c**) and UraneSX (**d**) in workspace. All the computations are done with joint limits

reported the time (In seconds) for plotting the implicit surface which they define and the number of cells computed by the CAD, as well as the number of cells in the final result after gluing those that are adjacent and belongs to the same connected component. Several functions are used which involves the discriminant variety, Gröbner bases and CAD computations, computed in Maple 18 with a Intel(R) Core (TM) i7-3770 CPU @ 3.40 GHz (14 Gb RAM). As can be seen from Table 1, there exists higher values of all the parameters for the Hybridglide, among all manipulators listed, which infers that it has more complex singularities, whereas for the Triaglide all the values are least which intuits the less complicated singularities. For example, the computation times for the Hybridglyde for parallel singularities is high

Table 1 Comparison of the different parameters associated with the singularities for the robots

Manipulators	Types of Singularities	Plotting time (s)	Degrees	No. of terms	Binary Size	No. of cells
Orthoglide	Parallel	0037.827	18[10, 10, 10]	097	015	[02382,0272]
	Serial	0005.133	18[12, 12, 12]	062	012	[00044,0004]
Hybridglide	Parallel	5698.601	20[16,08,12]	119	017	[28012,1208]
	Serial	0007.007	18[12, 12, 12]	281	017	[00158,0027]
Triaglide	Parallel	0010.625	03[00,00,03]	002	002	[00138,0004]
	Serial	0005.079	06[06,06,06]	042	007	[00077,0017]
UraneSX	Parallel	0022.625	06[06,04,00]	015	040	[02795,0070]
	Serial	0018.391	12[12, 12, 12]	252	151	[00392,0142]

compared to the one for the Othoglide, even if the surface has similar characteristics. This is due to the geometry of the surface which is more difficult to decompose in the case of the Hybridglide: the Cylindrical Algebraic Decomposition is described by 1208 cylindrical cells in the case of Hybridglide while it is described by 272 cells for the Orthoglide.

4 Workspace Analysis

The workspace analysis allows to characterize of the workspace regions where the number of real solutions for the inverse kinematics is constant. A CAD algorithm is used to compute the workspace of the robot in the projection space (x, y, z) with some joint constraints taken in account. The three main steps involved in the analysis are [6, 14, 16]

- Computation of a subset of the joint space (resp. workspace) where the number of solutions changes: the *Discriminant Variety*.
- Description of the complementary of the discriminant variety in connected cells: the Generic *Cylindrical Algebraic Decomposition* (CAD).
- Connecting the cells belonging to the same connected component in the counterpart of the discriminant variety: *interval comparisons.*

The different shapes of workspace for the delta-like robots is shown in Fig. 3, where blue, red, yellow and green regions correspond to the one, two, four and eight number of solutions for the IKP. A comparative study is done on the workspace of the family of delta-like manipulator and the results are shown in Fig. 3. All the workspace are plotted in the rectangular box, where $\mathbf{x} \in [-2, 2]$, $\mathbf{y} \in [-2, 6]$ and

$\mathbf{z} \in [-2, 6]$, so that the shapes of these workspace can be compared. From the Fig. 3 it can be intuited that the Triaglide will be good selection, if the task space is more in horizontal plane, whereas the Orthoglide is good for the three dimensional task space.

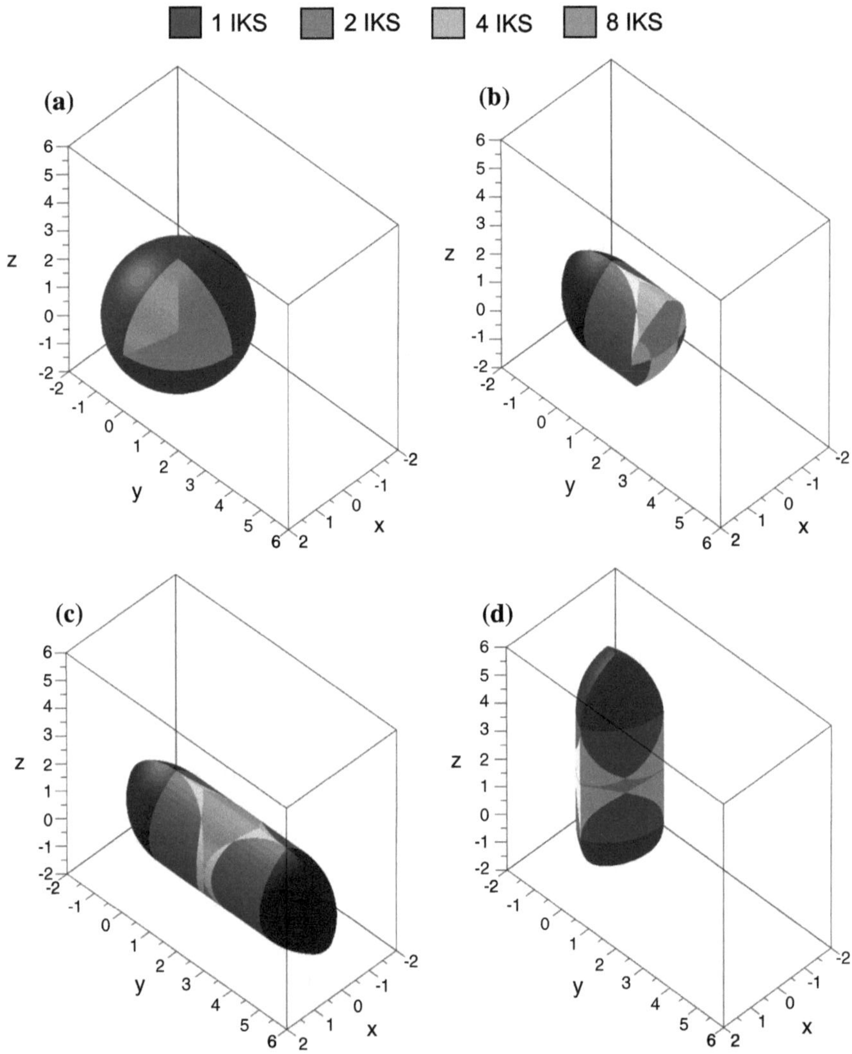

Fig. 3 Workspace with joint constraints for Orthoglide (**a**), Hybridglide (**b**), Triaglide (**c**) and UraneSX (**d**) robot with the number of inverse kinematic solutions. *Blue*, *red*, *yellow* and *green* regions corresponds to the one, two, four and eight number of solutions for the IKP

5 Conclusions

A comparative study on the workspace of different delta-like robots gives the idea about shape of the workspace, which further plays an important role in the selection of the manipulator for the specific task or for the trajectory planning. The main characteristics associated with the singularities are tabulated in Table 1, which also gives some information about the complexity of the singularities, which is an essential factor for the singularity-free path plannings. From these data, it can be observed that the singularities associated with the Hybridglide are complicated, whereas the structure of those associated with the Triaglide is rather simple.

Acknowledgements The work presented in this paper was partially funded by the Erasmus Mundus project "India4EU II".

References

1. Bohigas, O., Manubens, M., Ros, L.: A complete method for workspace boundary determination on general structure manipulators. IEEE Trans. Robot. **2**(8/5), 993–1006 (2012)
2. Bonev, I.A., Gosselin, C.M.: Analytical determination of the workspace of symmetrical spherical parallel mechanisms. IEEE Trans. Robot. **22**(5), 1011–1017 (2006)
3. Bonev, I.A., Ryu, J.: A new approach to orientation workspace analysis of 6-DOF parallel manipulators. Mech. Mach. Theory **36**(1), 15–28 (2001)
4. Castelli, G., Ottaviano, E., Ceccarelli, M.: A fairly general algorithm to evaluate workspace characteristics of serial and parallel manipulators. Mech. Based Des. Struct. Mach. **36**(1), 14–33 (2008)
5. Chablat, D., Jha, R., Rouillier, F., Moroz, G.: Non-singular assembly mode changing trajectories in the workspace for the 3-RPS parallel robot. In: 14th International Symposium on Advances in Robot Kinematics, Ljubljana, Slovenia, pp. 149–159 (2014)
6. Chablat, D., Jha, R., Rouillier, F., Moroz, G.: Workspace and joint space analysis of the 3-RPS parallel robot. In: Proceedings of ASME 2014 International Design Engineering Technical Conferences, Buffalo (2014)
7. Chablat, D., Moroz, G., Wenger, P.: Uniqueness domains and non singular assembly mode changing trajectories. In: Proceedings of IEEE International Conference on Robotics and Automation (2011)
8. Chablat D., Wenger P.: Working modes and aspects in fully-parallel manipulator. In: Proceedings of IEEE International Conference on Robotics and Automation, pp. 1964–1969, May (1998)
9. Chablat, D., Wenger, P., Majou, F., Merlet, J.P.: An interval analysis based study for the design and the comparison of 3-DOF parallel kinematic machines. Int. J. Robot. Res. **23**(6), 615–624 (2004)
10. Cox, D.A., Little, J., O'Shea, D.: Using algebraic geometry, 2nd edn. Springer, Graduate Texts in Mathematics (2004)
11. Gosselin, C.: Determination of the workspace of 6-DOF parallel manipulators. ASME J. Mech. Des. **112**, 331–336 (1990)
12. Hansen, J., Gupta, K.C., Kazerounian, S.M.K.: Generation and evaluation of the workspace of a manipulator. Int. J. Robot. Res. **2**(3), 22–31 (1983)

13. Hebsacker, M., Treib, T., Zirn, O., Honegger, M.: Hexaglide 6 DOF and triaglide 3 DOF parallel manipulators. In: Parallel Kinematic Machines: Theoretical Aspects and Industrial Requirements, pp. 345–355. (1999)
14. Lazard, D., Rouillier, F.: Solving parametric polynomial systems. J. Symbolic Comput. **42**(6), 636–667 (2007)
15. Merlet, J.P.: Geometrical determination of the workspace of a constrained parallel manipulator. In: Proceedings of 3rd International Symposium on Advances in Robot Kinematics, pp. 326–329 (1992)
16. Moroz, G.: Sur la décomposition réelle et algébrique des systèmes dépendant de paramètres. Ph.D. thesis, l'Université de Pierre et Marie Curie, Paris, France. (2008)
17. Ottaviano, E., Husty, M., Ceccarelli, M.: Identification of the workspace boundary of a general 3-R manipulator. ASME J. Mech. Des. **128**(1), 236–242 (2006)
18. Pashkevich, A., Chablat, D., Wenger, P.: Kinematics and workspace analysis of a three-axis parallel manipulator: the Orthoglide. Robotica **24**(1), 39–49 (2006)
19. Siciliano, B., Khatib, O.: Springer handbook of robotics. Springer (2008)
20. Snyman, J.A., Plessis, L.J., Duffy, J.: An optimization approach to the determination of the boundaries of manipulator workspaces. ASME J. Mech. Des. **122**(4), 447–456 (2000)
21. Soni, A., Tsai, Y.: An algorithm for the workspace of a general n-R robot. ASME J. Mech. Trans. Automat. Des. **105**, 52–57 (1983)
22. Zein, M., Wenger, P., Chablat, D.: An exhaustive study of the workspace topologies of all 3R orthogonal manipulators with geometric simplifications. Mech. Mach. Theory **41**(8), 971–986 (2006)

Optimal Trajectory Planning of 3RRR Parallel Robot Using Ant Colony Algorithm

O. Hamdoun, L. El Bakkali and F.Z. Baghli

Abstract By taking 3RRR planar parallel robot the object of this study, this work presents an optimal trajectory planning of 3RRR parallel robot. Indeed, Parallel robot is fundamentally a closed-loop kinematic chain mechanism in which the end-effector is connected to the base by several independent kinematic chains. Due to the potential advantages of parallel robots such as high rigidity, high accuracy and great carrying payload capability, they have fascinated significant attention and interest amongst the researchers in the past decade. After describing the mechanism, a kinematic analysis is reported. Based on the geometry of the manipulator, a workspace analysis is performed and an optimal trajectory is defined using ant colony algorithm in order to avoid singular configurations and optimize the kinetostatic performances of the manipulator.

Keywords Inverse kinematic · Parallel robot · Workspace · Planning of trajectory · Ant colony algorithm

1 Introduction

In recent years, parallel robots have received a big attention for different researches and applications. Parallel manipulators are closed-loop mechanisms in which the separate serial chains (links and joints) are connected to both the fixed base and the moving platform.

This new type of robots can achieve certain tasks that serial robots are not able to perform. Even though serial robots have advantages like simple dynamic and kinematic models, large workspace, but parallel robots have advantages more potential than serial robots as better stiffness, high speed, higher kinematical precision, large load carrying capacity and the most important is the possibility of

O. Hamdoun (✉) · L. El Bakkali · F.Z. Baghli
Modeling and Simulation of Mechanical Systems Laboratory, Abdelmalek Essaadi, Faculty of Sciences, BP.2121, M'hannech, 93002 Tetouan, Morocco
e-mail: wafae.hamdoun@gmail.com

S. Zeghloul et al. (eds.), *Robotics and Mechatronics*,
Mechanisms and Machine Science 37, DOI 10.1007/978-3-319-22368-1_13

fixing actuators on the base, which can be used for fast and precise operations. Parallel robots have also some drawbacks, the major of them are a limited workspace that can be seen as the intersection of the individual workspace of each serial arm and also the complicated singularities [1].

The planar parallel manipulator translate along the x and y-axes, and rotate around the z-axis, they are useful for manipulating an object on a plane. Gosselin and Clement presented a kinematical study of a planar parallel robot, where a moving platform is connected to a fixed base by three links, each leg consisting of two binary links and three parallel revolute joints [2]. Sefrioui and Gosselin [3] give an interesting numerical solution in the inverse and direct kinematics of this kind of planar robot. In opposite to serial robots, parallel robots can accept not just multiple inverse kinematic solutions, although multiple direct kinematic solutions. This property generates more complex kinematic models but allow more flexibility in trajectory planning.

Thus, in this work we can formulated our problem as follow: determine the inverse kinematic problem, given specific points inside the boundary of a desired workspace, establish a workspace analysis of 3RRR planar parallel robot and finally use ant colony algorithm in order to obtain an optimal trajectory planning.

2 Description of 3RRR Parallel Robot

This work presents a geometric description of 3RRR parallel robots which composed by a mobile platform (MP) and three RRR serial chains that join it to a fixed base (Fig. 1). Each RRR chain is a serial chain composed by three rotational R joints.

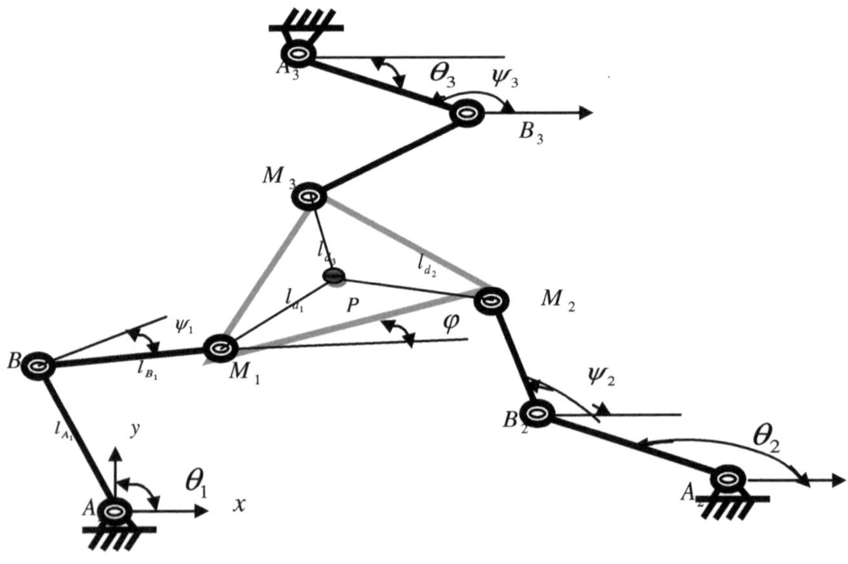

Fig. 1 Geometric description of 3RRR parallel robot

Allow P(x, y) be the end-effect or position in the plane and φ its orientation. Let O be the origin of the fixed reference frame and let A_i, B_i and C_i with $i = 1, 2, 3$ define the rotational articulations of each leg. A_i Points are actuated, so that the actuators are fixed to the base. l_{A_i} is the length of Link A and l_{B_i} is the length of Link B.

3 Inverse Kinematic Model of 3RRR Parallel Robot

The inverse kinematic problem consists of determining θ_1, θ_2 and θ_3 for specified values of x, y and ϕ. Thus, the loop closure equation for the 3-RRR parallel robot is as follows:

$$OP = OA_i + A_iB_i + B_iM_i + M_iP \tag{1}$$

Writing the equations in component form:

$$\begin{aligned} P_x &= l_{Ai}\cos(\theta_i) + l_{Bi}\cos(\psi_i) + l_{di}\cos(\sigma_i + \varphi) + x_{Ai} \\ P_y &= l_{Ai}\sin(\theta_i) + l_{Bi}\sin(\psi_i) + l_{di}\sin(\sigma_i + \varphi) + y_{Ai} \end{aligned} \tag{2}$$

where:

σ_i: is the angle between the lines from the point P to the joint at M_i. It is considered from the line between M_1 and M_2 on the moving platform.

To compute the inverse kinematics, we can rewrite the equation as follows:

$$K_i \sin\theta + F_i \cos\theta = E_i \tag{3}$$

where:

$$\begin{aligned} E_i = {} & P_x^2 + P_y^2 + l_{Ai}^2 + l_{Bi}^2 + l_{di}^2 + x_{Ai}^2 + y_{Ai}^2 - 2P_x l_{di}\cos(\sigma_i + \varphi) - 2P_x x_{Ai} \\ & + 2l_{di}x_{Ai}\cos(\sigma_i + \varphi) - 2P_y l_{di}\sin(\sigma_i + \varphi) - 2P_y y_{Ai} + 2l_{di}y_{Ai}\sin(\sigma_i + \varphi) \end{aligned}$$

$$\begin{aligned} F_i &= -2l_{Ai} + 2l_{Ai}l_{di}\cos(\sigma_i + \varphi) + 2l_{Ai}x_{Ai} \\ K_i &= -\left[-2P_y l_{Ai} + 2l_{Ai}l_{di}\sin(\sigma_i + \varphi) + 2l_{Ai}y_{Ai}\right] \end{aligned}$$

The inverse kinematics solution for Eq. (3) is:

$$\theta_i = A\ \tan 2(K_i, F_i) \pm A\ \tan 2\left(\sqrt{(K_i^2 + F_i^2 - E_i^2)}, E_i\right) \tag{4}$$

where: $i = 1, 2, 3$.

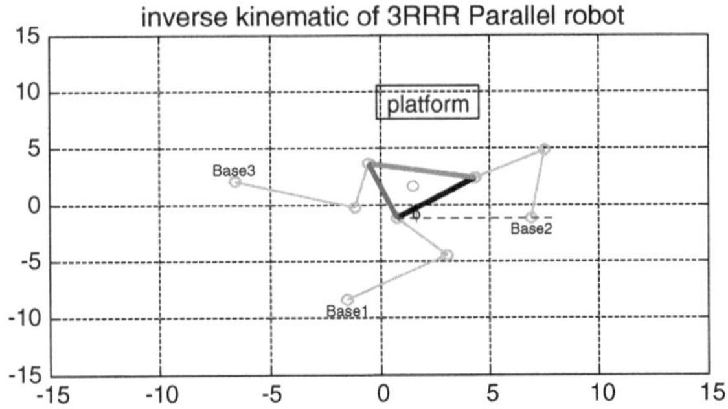

Fig. 2 A configuration of 3RRR parallel robot

4 Numerical Implementation

The equation obtained can simply be solved for any number of legs by simply changing the subscript in equations. For the particular robot there are two possible solutions given by the solutions to equations. When the roots are imaginary (i.e. there is no real solution) in some given poses, it means that the position/orientation of that configuration is outside the reachable workspace (Fig. 2).

5 Workspace Analysis

The mobility region can be established by every leg of this manipulator, it being encircled by the singularity curve which is the closed curve separating the region where the leg has mobility from that in which it does not. However the solution of the inverse kinematic problem is unique since the two branches meet, at the points of the workspace located on this closed curve [4].

The workspace of the 3RRR Parallel Robot is obtained by the intersection of the three mobility regions of the three legs, which is described in the x-y plane.

Thus, the singularity curve (the limit of the workspace) for each leg can be obtained as follow:

$$\frac{l_{A_i}^2 - l_{B_i}^2 + x_{2i}^2 + y_{2i}^2}{2l_{A_1}\sqrt{x_{2i}^2 + y_{2i}^2}} = \pm 1 \tag{5}$$

The workspace is obtained by the intersection of the three previous mobility regions, which are described in the x-y plane by the following equations:

$$(x - x_i)^2 + (y - y_i)^2 = (l_{A_i} \pm l_{Bi})^2, \quad i = 1, 2, 3 \tag{6}$$

where:

$$
\begin{aligned}
x_1 &= l_{d1}\cos(\varphi + \pi/6) \\
y_1 &= l_{d1}\sin(\varphi + \pi/6) \\
x_2 &= 1 - l_{d2}\cos(\varphi - \pi/6) \\
y_2 &= -l_{d2}\sin(\varphi - \pi/6) \\
x_3 &= \frac{1}{2} + l_{d3}\sin(\varphi) \\
y_3 &= \frac{\sqrt{3}}{2} - l_{d3}\cos(\varphi)
\end{aligned}
$$

Each of the annular regions is bounded by two concentric circles whose centers have coordinates (x_i, y_i) for $i = 1, 2, 3$ (Fig. 3).

To obtain the concentric circles, we choose alternatively the plus and minus sign in equations (Fig. 3).

$$
(l_{A_i} + l_{B_i})\, and\, |l_{A_i} - l_{B_2}| \tag{7}
$$

This Fig. 4 presented the workspace of 3RRR parallel robot for $l_{Ai} = 0.355, l_{B_i} = 0.15$ and $l_{di} = 0.125$ when $\varphi = 0°$.

The location of the three regions of mobility depends on the orientation angle φ. If the angle orientation is incremented, the centers of the regions will move around the circles of radii l_{di}. For $\varphi = \pi$, the area of the workspace will be a minimum and the distance between the regions whose intersection defines the workspace of the robot will be a maximum as shown in Fig. 5.

In this figure as shown the workspace is vanished.

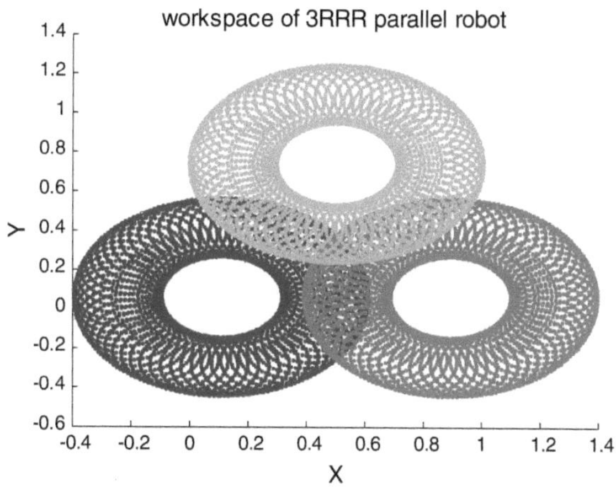

Fig. 3 Intersection between the three regions of mobility of the robot

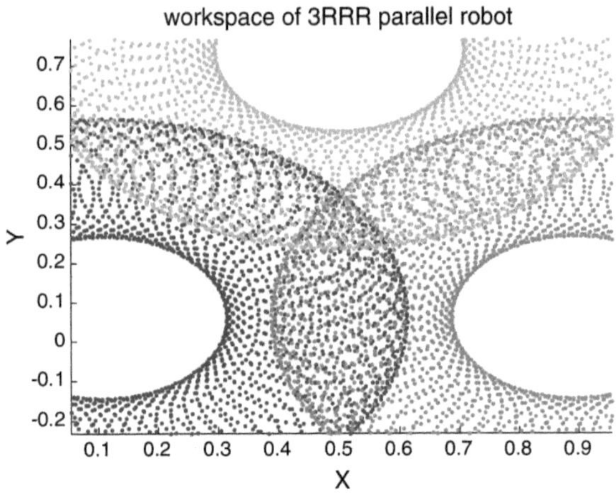

Fig. 4 Workspace for 3RRR parallel robot

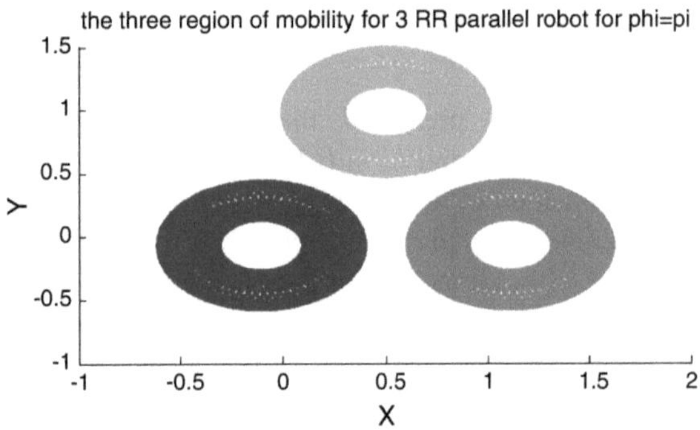

Fig. 5 workspace of robot for $\varphi = \pi$

Thus, to get a non-vanishing workspace the manipulator must verify these conditions:

$$\begin{aligned} &3(l_{A_i} + l_{B_i})^2 \geq (\sqrt{3} l_{di} + 1)^2 \\ &and \\ &3(l_{A_i} - l_{B_i})^2 \leq (\sqrt{3} l_{di} - 1)^2 \end{aligned} \tag{8}$$

6 Optimal Planning Trajectory Using Ant Colony Algorithm

6.1 Ant Colony Algorithm Description

In the ant colony optimization (ACO), a colony of artificial ants cooperates in searching good solutions to difficult optimization problems. Cooperation is a key design component of ACO algorithms [5].

Every ant make a complete tour by choosing the cities according to a probabilistic state rule. Mathematically, the probability with which ant k in city r chooses to move to the city s is:

$$P_k(r,s) = \begin{cases} \frac{\left(\sum_{i=1}^{r}[\gamma^{r-i}\cdot\tau(i,s)]\right)^{\alpha}\cdot[\eta(r,s)]^{\beta}}{\sum_{v\in J_k(r)}\left(\left(\sum_{i=1}^{r}[\gamma^{r-i}\cdot\tau(i,v)]\right)^{\alpha}\cdot[\eta(r,v)]^{\beta}\right)} & \text{if s} \in J_k(\text{r}) \\ 0 & \text{Otherwise} \end{cases} \tag{9}$$

where τ is the pheromone, η is the visibility which is the inverse of the distance $\delta(r,s)$, $JK(r)$ is the situate of cities that remain to be visited by ant r positioned on city r, α and β are two coefficients which build the pheromone information or the visibility information with respect to one another and the parameter $\gamma > 0$ establishes the relative influence of pheromone values corresponding to earlier decisions, earlier places in the permutation.

The good solutions found so far and in the present generation are used to revise the pheromone information. However, before that, some portion of pheromone is evaporated according to:

$$\tau_{rs} = (1-\rho)\tau_{rs} \tag{10}$$

where k is the evaporation rate with $0 \le \rho < 1$ and $(1-\rho)$ is the trail persistence. Let $\tau_{rs}(t)$ be the intensity of trail on $edge(r,s)$ at time t. Each ant at time t chooses the next city, where it will be at time t + 1. Therefore, after each cycle, after each ant has determined a tour, the pheromone trail is updated using the founded solutions according to the following formula:

$$\tau_{rs}(t+n) = \rho\cdot\tau_{rs}(t) + \sum_{k=1}^{m}\Delta\tau_{rs}^{(k)} \tag{11}$$

where $\Delta\tau_{rs}^{(k)}$ is the contribution of the ant k to the pheromone trial between cities r and s.

6.2 Formulated Problem and Results

The problem is finding the shortest path predefined start and final position:

The distance between the two points can be simply calculated in 2D environment as follow:

$$d_i = \sqrt{(x_i - x_{i-1})^2 + (y_i - y_{i-1})^2} \text{ i} = 1,2\ldots\text{n} \tag{12}$$

where n is the number of points on the path, the total length of the path is simply the sum of distances:

$$L = \sum_{i=1}^{n-1} d_i \tag{13}$$

This Fig. 6 presented the convergence curve of the objective function based on the number of iterations.

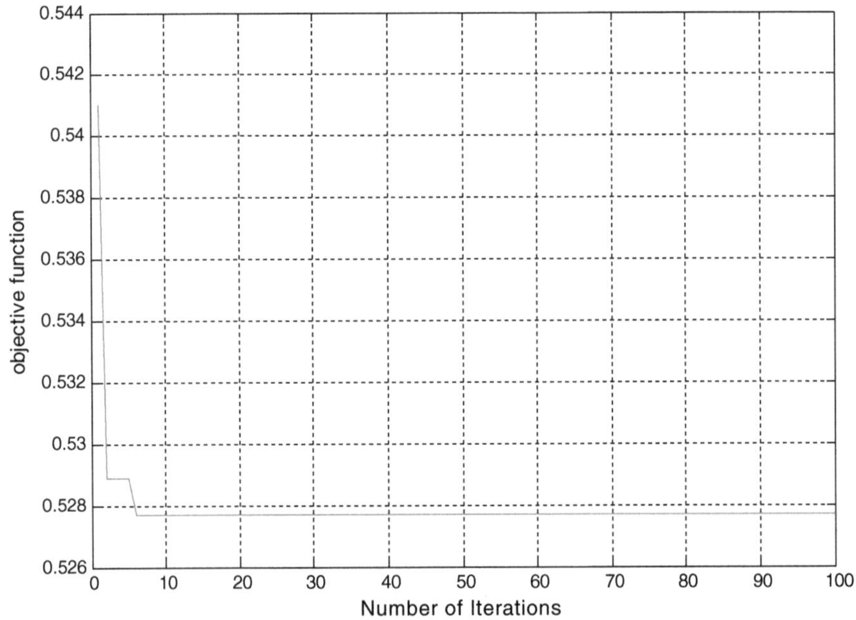

Fig. 6 The diagram of the convergence curve of the objective function

7 Conclusion

In this work we present an approach for determining the inverse kinematics applied on 3-RRR planar parallel, Which have important potential for different applications. Thus the analysis of the workspace of the robot lead us to a great performance and a higher precision in the positioning of the platform and also to an optimal trajectory planning using ant colony algorithm.

References

1. Merlet, J.-P.: Parallel Robots. Kluwer Academic Publishers, Dordrecht (2000)
2. Gosselin, C.M., Lemieux, S., Merlet, J.-P.: A new architecture of planar three-degree-of-freedom parallel manipulator. In: Proceedings, 1996 IEEE International Conference on Robotics and Automation, 1996, vol 4. IEEE (1996)
3. Sefrioui, J., Gosselin, C.M.: On the quadratic nature of the singularity curves of planar three-degree-of-freedom parallel manipulators. Mech. Mach. Theory **30**(4), 533–551 (1995)
4. Huang, M.Z., Thebert, J.-L.: A study of workspace and singularity characteristics for design of 3-DOF planar parallel robots. Int. J. Adv. Manuf. Technol. **51**, 789 (2010)
5. Fan, X., Luo, X., Yi, S., Yang, S., Zhang, H.: Optimal path planning for mobile robots based on intensified ant colony optimization algorithm. In: International Conference on Robotics, Intelligent Systems and Signal Processing, vol. 1, pp. 131–136. IEEE (2003)

Part IV
Medical Robotics I

Force Control Implementation of a Haptic Device for a Medical Use

H. Saafi, M.A. Laribi and S. Zeghloul

Abstract This paper discusses the implementation of the force control of a haptic device. This device is used as a master device for a teleoperation system for minimally invasive surgery. The haptic device applies forces in the user's hand using actuators placed in the active joints. The master device has a spherical parallel architecture. The actuated joint torques are calculated using the jacobian matrix of the master device. Experiments are carried out in order to show the feasibility of the control force feedback. A force sensor was placed on the moving platform to measure the applied force and then to compare with the desired force. The results of the comparison are satisfactory for the desired application.

Keywords Force control · Haptic device · Spherical parallel architecture

1 Introduction

Nowadays, parallel manipulators are widely popular. Thanks to their high load capacity, their stiffness, their low weight and their precision, parallel manipulators are used in many fields such as medicine, where many master devices have been developed with parallel architecture [1, 2]. In previous work [3], new master device was developed to control a surgical robot. This device has a spherical parallel architecture. The master architecture was chosen to have a mechanism able to provide three pure rotations around a fixed point. The geometric parameters of this

H. Saafi (✉)
Department GMSC, Pprime Institute CNRS - University of Poitiers - ENSMA, UPR 3346 Poitiers, France
e-mail: houssem.saafi@univ-poitiers.fr

M.A. Laribi · S. Zeghloul
Institut PPRIME, UPR 3346, University of Poitiers, Poitiers, France
e-mail: med.amine.laribi@univ-poitiers.fr

S. Zeghloul
e-mail: said.zeghloul@univ-poitiers.fr

S. Zeghloul et al. (eds.), *Robotics and Mechatronics*,
Mechanisms and Machine Science 37, DOI 10.1007/978-3-319-22368-1_14

structure were optimized to meet a prescribed workspace. The Forward Kinematic Model (FKM) was solved in [4]. In this paper, the force control implementation is presented. The Jacobian matrix of the haptic device is used to calculate the actuated joint torques needed to apply force in the users hand. Experimental tests were made to validate the developed method. A force sensor is placed on the moving platform to measure the applied force and to compared with the desired torque. The results of these experiments are discussed in this paper.

2 Master Device for a Medical Tele-Operation System

The developed master device is shown in Fig. 1. This device controls the motion of a surgical robot and reproduce the haptic feed-back to the user. The target application is Minimally Invasive Surgery (MIS). In MIS, the instruments enter to the patient body throw tiny incisions. This limits the motion of the instrument to three rotations around the incision and one translation along the instrument axis. For this reason, the spherical parallel architecture was chosen for the master device since it is able to produce the similar motions. The translation motion is not taken into account in this paper because it is uncoupled in the model of the SPM.

The classical SPM has three identical legs, each leg is made of two links and three revolute joints (Fig. 2a). All axes of the revolute joints are intersecting in one common point, called CoR (Center of Rotation). Each link is characterized by the

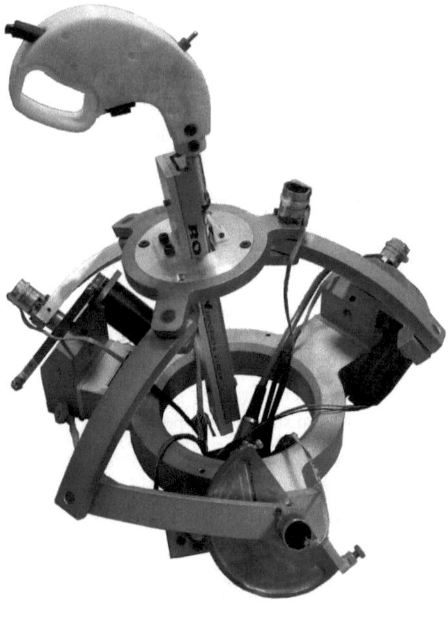

Fig. 1 Master device of a tele-operation system

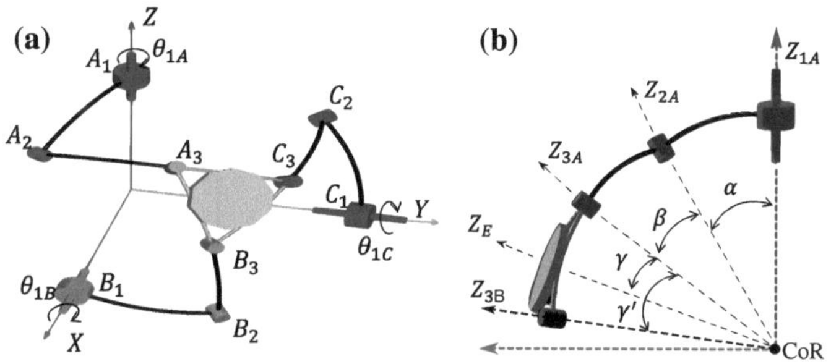

Fig. 2 **a** Classic spherical parallel manipulator. **b** Kinematic of leg A

angle between its two revolute joints, as shown in Fig. 2b. This angle is constant and it represents the dimension of the link. The actuated joint axes are located along an orthogonal frame. The orientation of the SPM is described by the ZXZ Euler angles (ψ, θ and φ). The Jacobian matrix of the SPM can be written as follows:

$$\mathbf{J} = \mathbf{A}^{-1} \cdot \mathbf{B} \tag{1}$$

where,

$$\mathbf{A} = [Z_{2A} \times Z_{3A} \quad Z_{2B} \times Z_{3B} \quad Z_{2C} \times Z_{3C}]^T \tag{2}$$

$$\mathbf{B} = diag(Z_{1A} \cdot Z_{2A} \times Z_{3A}, \quad Z_{1B} \cdot Z_{2B} \times Z_{3B}, \quad Z_{1C} \cdot Z_{2C} \times Z_{3C}) \tag{3}$$

$$\begin{cases} Z_{1K} = R_{0K} \cdot Z \\ Z_{2K} = R_{0K} \cdot R_Z(\theta_{1K}) \cdot R_X(\alpha) \cdot Z \quad for\, K = A, B\, and\, C \\ Z_{3K} = R_{ZXZ} \cdot R_Z(\frac{2\pi}{3}) \cdot R_X(\gamma) \cdot Z \end{cases} \tag{4}$$

$$R_{0A} = \begin{bmatrix} 1 & 0 & 0 \\ 0 & 1 & 0 \\ 0 & 0 & 1 \end{bmatrix} \quad R_{0B} = \begin{bmatrix} 0 & 1 & 0 \\ 0 & 0 & 1 \\ 1 & 0 & 0 \end{bmatrix} \quad R_{0C} = \begin{bmatrix} 0 & 0 & 1 \\ 1 & 0 & 0 \\ 0 & 1 & 0 \end{bmatrix} \tag{5}$$

$$R_{ZXZ} = R_Z(\psi) \cdot R_X(\theta) \cdot R_Z(\varphi) \tag{6}$$

3 Force Control Architecture

The force control architecture is presented in Fig. 3. First, the force applied on the surgical robot is measured using a force sensor. Then, a function determine the actuated torques needed to reproduce the effort to the user. This function solve the Jacobian matrix and calculated the actuated torques using the following equation:

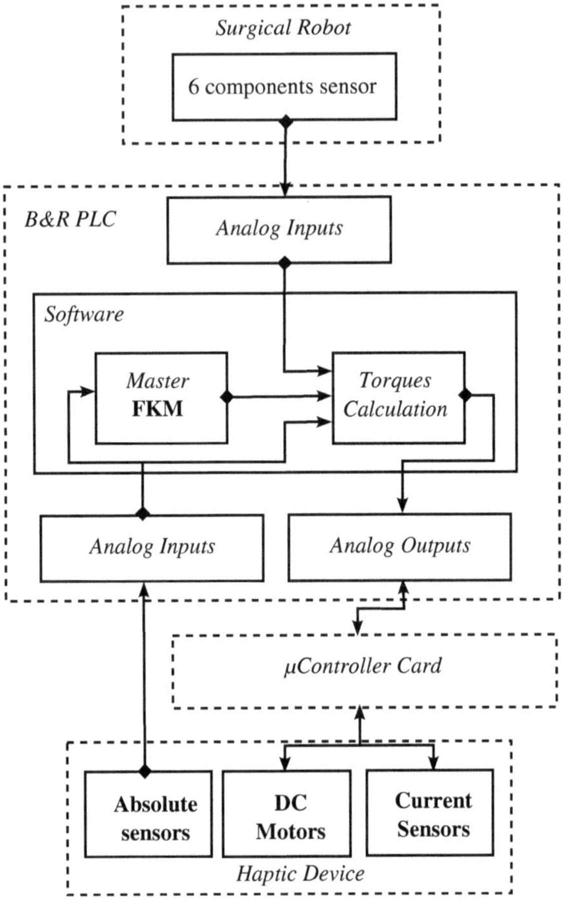

Fig. 3 Material and software architecture of the force control

$$\tau = \mathbf{J}^T \cdot T \tag{7}$$

where τ is the actuated joint torques and T is the reference torques vector.

The jacobian matrix is expressed in the Sect. 2. It depend on the active joint angles and the orientation of the haptic device. The orientation of the master device is calculated using the Forward Kinematic Model (FKM). The FKM and the actuated torques calculation function are implemented in B&R PLC. These models run with period of 1 ms. The actuated joint reference torques are then transferred using an analog signal to *μControllerCard* that control the actuator torques. The torque control scheme is shown in Fig. 4. A PI controller regulates the torque of the actuators. The actuator joint torques are varied using PWM signal and measured using a current sensor.

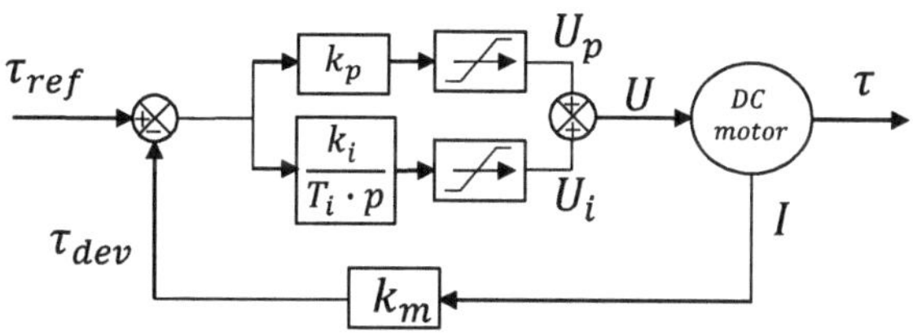

Fig. 4 Actuator torque control loop

In the next section, we present the FKM of the haptic device than we present an experimental test of the force control.

4 Forward Kinematic Model

The FKM calculates the orientation of the end-effector, ψ, θ and φ, for a given position of the active joints (θ_{1A}, θ_{1B}, θ_{1C}). The FKM can be solved using different methods [5–7]. The FKM of the SPM studied by Gosselin et al. Bai et al. [7] proposed a robust procedure to determine the orientation of the end-effector using the input/output equations of spherical four-bar linkages. This approach is used in this paper.

The two considered spherical four-bar mechanisms are described by the following two chains: A2–A3–B3–B2 and A3–A2–C2–C3 (Fig. 5). The corresponding input/output equations can be written as follows:

- For the chain A2–A3–B3–B2 the equation is:

$$L_1(\xi)\cos(\sigma) + M_1(\xi)\sin(\sigma) + N_1(\xi) = 0 \tag{8}$$

Fig. 5 Developed haptic device

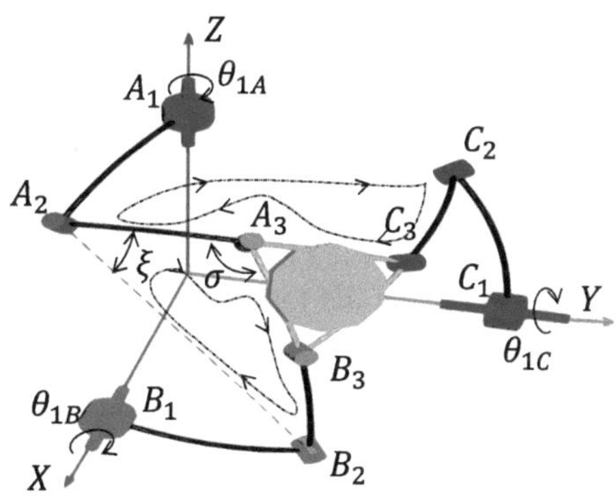

- For the chain A3–A2–C2–C3 the equation is:

$$L_2(\xi)\cos(\sigma) + M_2(\xi)\sin(\sigma) + N_2(\xi) = 0 \quad (9)$$

where, ξ and σ are two angles defined in Fig. 5 and $L_i(\xi), M_i(\xi), N_i(\xi)$ (i = 1, 2) are coefficients that depend on $\cos(\xi)$ and $\sin(\xi)$.

$\cos(\sigma)$ and $\sin(\sigma)$ are obtained using Eqs. (8) and (9) as follows:

$$\begin{cases} \cos(\sigma) = \frac{M_1(\xi)N_2(\xi)-M_2(\xi)N_1(\xi)}{L_1(\xi)M_2(\xi)-L_2(\xi)M_1(\xi)} & (a) \\ \sin(\sigma) = \frac{L_1(\xi)N_2(\xi)-L2(\xi)N_1(\xi)}{L_1(\xi)M_2(\xi)-L_2(\xi)M_1(\xi)} & (b) \end{cases} \quad (10)$$

First, the possible solution of ξ can be found by solving the equation generated by writing (10 − a)^2 + (10 − b)^2 which has only ξ as unknown. This equation is as follows:

$$\begin{aligned} &N_2^2L_1^2 + 2L_1M_2L_1M_1 - 2L_2N_2L_1N_1 \\ &\quad + N_2^2M_1^2 - L_2^2M_1^2 - 2M_2N_2M_1N_1 \\ &\quad - M_2^2L_1^2 + L_2^1N_1^2 + M_2^2N_1^2 = 0 \end{aligned} \quad (11)$$

Equation 11 is a four degree quadratic equation in $\cos(\xi)$ and $\sin(\xi)$. This equation is solved numerically and highly increases the calculation time of the FKM. The obtained solution of ξ is then used to calculate σ using the Eq. (10).

For given values of (θ_{1A},θ_{1B},θ_{1C}), we have at most eight solutions for ξ and σ. Each pair (ξ_i, σ_i), for $1 \le i \le 8$, gives an orientation (ψ, θ, φ) of the end-effector using the forward kinematic of leg A.

The FKM was improved in [4] by using an extra sensor. This sensor reduces the complexity of the FKM.

5 Experimental Test

To validate the force control architecture presented in Fig. 3, experimental tests were made. A force sensor is placed on the moving platform of the haptic device to measure the applied torque (Fig. 6). The force sensor gives the force and moment components F_x, F_y, F_z, M_x, M_y and M_z in sensor frame. Only three component are used to measure the torque applied by the haptic device as follows:

$$\begin{cases} T_{eff.}(x) = r \cdot F_x \\ T_{eff.}(y) = r \cdot F_y \\ T_{eff.}(z) = M_z \end{cases} \quad (12)$$

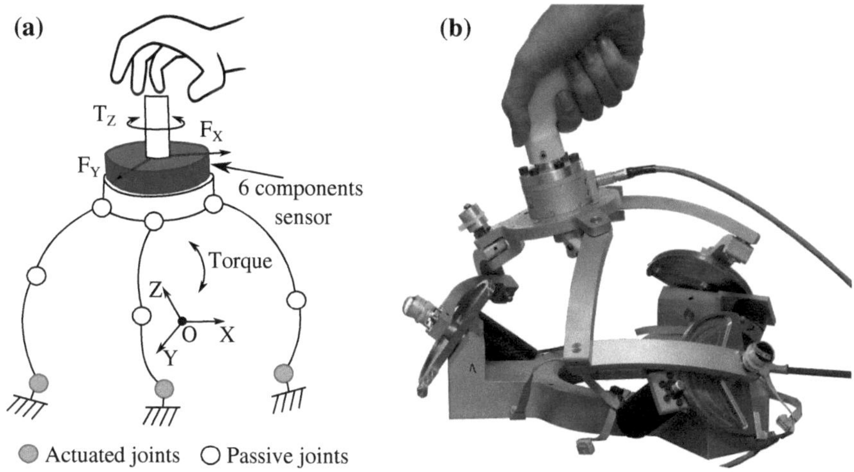

Fig. 6 **a** Illustration of the experimental setup. **b** Experimental setup

$$T_{Cart.} = R_{ZXZ} \cdot T_{eff.} \tag{13}$$

where, r is the distance between the force sensor and the SPM center of motion.

Two experiments were carried out. First, only a self-rotation torque was applied by the haptic device. The desired and the measured torques are shown in Fig. 7. Then, a torque around the Z-axis of the Cartesian space was applied. The comparison between the measured and the desired torques are shown in Fig. 8. Figures 7 and 8 show satisfactory results of the force control.

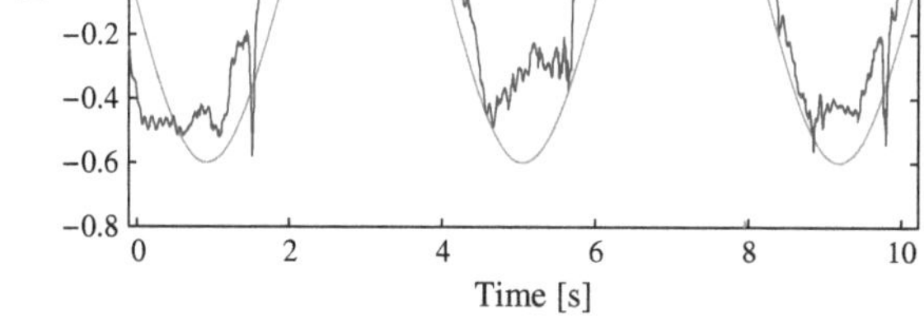

Fig. 7 Comparison between the desired and the measured torques ($T_{eff.}(Z)$)

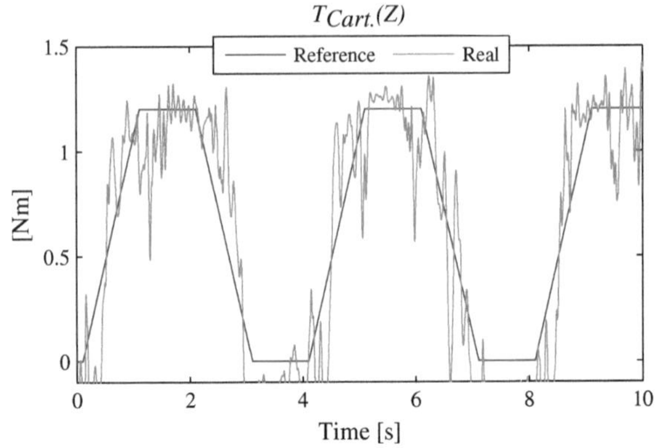

Fig. 8 Comparison between the desired and the measured torques ($T_{Cart.}(Z)$)

6 Conclusions

The implementation of the force control of a haptic device was discussed in this paper. This device is used as a master device for a tele-operation system for minimally invasive surgery. This device controls the motion of a surgical robot and reproduces forces generated by the interaction between the surgical robot and the patient's body. The interaction forces are measured using a force sensor. The actuated joint torques needed to apply the interaction force are calculated using the jacobian matrix of the haptic device. Experiments are carried out and presented in order to show the feasibility of the control force feedback. A force sensor was placed on the moving platform to measure the applied force and then to compare with the desired force. The results of the comparison are satisfactory for the desired application.

Acknowledgments This research is supported by the Poitou-Charentes region 2007–2013 (program project 10 Images and interactivities), in partnership with the European Union (FEDER/ERDF, European Regional Development Fund) and ROBOTEX, the French national network of robotics platforms ($N°$ ANR-10-EQPX-44-01).

References

1. Tobergte, A., et al.: The sigma.7 haptic interface for MiroSurge: A new bi-manual surgical console. In: IEEE/RSJ International Conference on Intelligent Robots and Systems (IROS), 2011, pp. 3023, 3030, 25–30 Sept 2011
2. Ma, A., Payandeh, S.: Analysis and experimentation of a 4-DOF haptic device. In: Haptic Interfaces for Virtual Environment and Teleoperator Systems, Symposium on Haptics, pp. 351, 356, 13–14 March 2008
3. Chaker, A., Mlika, A., Laribi, M.A., Romdhane, L., Zeghoul, S.: Synthesis of a spherical parallel manipulator for a dexterous medical task. Front. Mech. Eng. **7**(2), 150–162 (2012)

4. Saafi, H., Laribi, M. A., Zeghloul, S.: Improvement of the direct kinematic model of a haptic device for medical application in real time using an extra sensor. In: IEEE/RSJ International Conference on Intelligent Robots and Systems (IROS 2014), pp. 1697, 1702, 14–18 Sept. 2014
5. Gosselin, C., Sefrioui, J., Richard, M.J.: On the direct kinematics of spherical three-degree-of-freedom parallel manipulators of general architecture. J. Mech. Des. **116**(2), 594–598 (1994)
6. Celaya, E.: Interval propagation for solving parallel spherical mechanisms. In: Advances in Robot Kinematics, pp. 415–422. Springer (2002)
7. Bai, S., Hansen, M.R., Angeles, J.: A robust forward-displacement analysis of spherical parallel robots. Mech. Mach. Theory **44**(12), 2204–2216 (2009)

Integration of Automated Camera Steering for Robotic Single-Site Surgery

Mohsen Zahiri, Carl A. Nelson, R. Gonzalo Garay-Romero and Dmitry Oleynikov

Abstract Modern surgical approaches favor minimal invasiveness in order to improve recovery time, cosmesis, and post-operative pain. One such technique for abdominal procedures, called robotic laparoendoscopic single-site surgery (R-LESS), uses a single incision to gain access to the abdominal cavity. In this process, as in standard laparoscopic surgery, the surgeon's visualization of the surgical site depends on the performance of an assistant who maneuvers the camera. Variation in skill of the camera assistant at following commands from the surgeon contributes to the difficulty of the procedure. Moreover, giving camera instructions is an additional task for the surgeon and is essentially a distraction from the main task. To solve this problem and provide high-quality visualization to the surgeon, we present a self-guided robotic camera control to replace the camera assistant. The system automatically tracks the robotic tools and automatically manipulates the camera to achieve the best field of view.

Keywords Surgical robotics · Laparoendoscopic single-site surgery · Robotic camera assistant

M. Zahiri · C.A. Nelson (✉)
Department of Mechanical and Materials Engineering, University of Nebraska-Lincoln, Lincoln, USA
e-mail: cnelson5@unl.edu

R. Gonzalo Garay-Romero
Department of Mechanical Engineering, University of Portland, Portland, USA

C.A. Nelson · D. Oleynikov
Department of Surgery, Center for Advanced Surgical Technology, University of Nebraska Medical Center, Omaha, USA

S. Zeghloul et al. (eds.), *Robotics and Mechatronics*,
Mechanisms and Machine Science 37, DOI 10.1007/978-3-319-22368-1_15

1 Introduction

As trends in surgical care are towards less and less invasiveness, the potential role of robotics in surgery is growing in order to maintain high dexterity despite limited access. This can lead to improved outcomes compared to standard manual techniques. For example, laparoscopic single-site surgery (LESS) is a technique which avoids surgical trauma by limiting access to a single incision [7]. Because this complicates tool motion, insertable robotic tools have been developed to restore surgical dexterity. One of the remaining challenges in robotic LESS (R-LESS) is camera guidance, which plays an important role in surgical efficiency and effectiveness. The objective is to build on the concepts of previous work in which various robotic manipulators have been used to guide cameras during surgical procedures [4–6].

A single-site robot for use in R-LESS procedures has recently been developed. The robot is inserted into the insufflated abdominal cavity through the single port. The two robot arms are straightened for insertion and unfolded once fully inserted [3]. The long-term vision for this robotic system is for the surgical arms to be integrated with a vision system, to provide enhanced surgical capabilities as well as telestration, telementoring and telesurgery capabilities.

In this paper, we present a device for integrating an existing steerable laparoscope (Endoeye, Olympus) with the surgical robot. To release the surgeon from the task of controlling the camera, an automated control system was developed based on the surgeon's input to the robot manipulation task. The camera smoothly tracks the field of view which is needed during surgery without any command from the surgeon. In the case of conflict between the automatic tracker and the surgeon's preferred camera view, the surgeon can switch to a manual control mode using joystick input. The design, control approach, and capabilities of this device are explained.

2 Methods

The Endoeye laparoscopic camera (Olympus) is designed as a handheld camera and provides pan/tilt motion via a pair of levers on the handle. The principal objective of the present design is to integrate motorized camera lever control in a form which is easily attachable to the surgical robot. Measurements of the camera were used to create a solid model of the handle shape, including the control levers (Fig. 1). Cross-sections of this model were used as contact points to secure a housing (also shown in Fig. 1) which conforms in shape to the handle, while providing mount points for servomotors.

The levers and servomotor shafts are connected by customized servo arms, which give the operator motorized control of the camera in two degrees of freedom. The no-load angular velocities for the robot arms are 45 rpm. With a rated speed

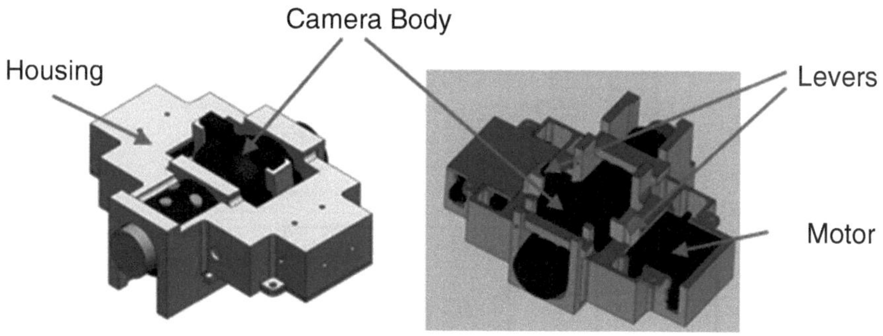

Fig. 1 CAD model of the housing

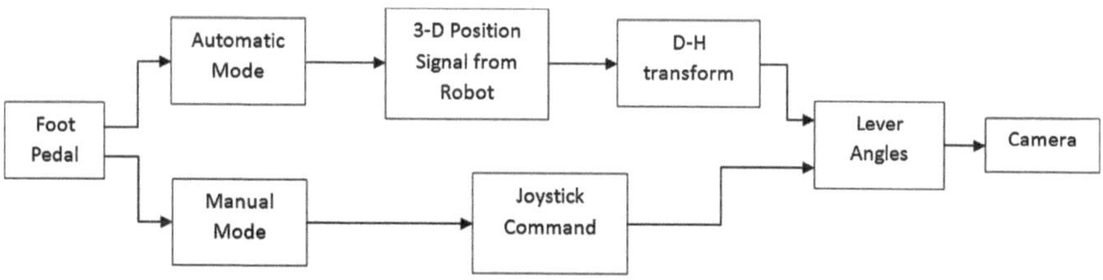

Fig. 2 The control system includes two methods of obtaining target positions, an inverse kinematic transformation, and output to the motor controller

above 50 rpm for the motors driving the camera levers, and an internal amplification of 4× in the camera steering mechanism, the resulting speed of 200 rpm can easily track the robot arms.

A control system was implemented in C# using Visual Studio software, as indicated diagrammatically in Fig. 2. A foot pedal input can be used to toggle between automatic and manual tracking modes. Motor commands are passed from the computer through an Arduino microprocessor via serial communication to the motors, as shown in the physical layout diagram of Fig. 3. These camera control motors are run open-loop based on set-points from the robot.

A user interface was created to implement both manual and automatic camera control modes, with the ability to switch between modes using keyboard input (as a surrogate for the foot switch which would be used in a surgical environment). Figure 4 shows the user interface. The application can detect the Arduino micro-controller's serial port automatically and gives the user a button to open/close the connection, as well as switch between control modes.

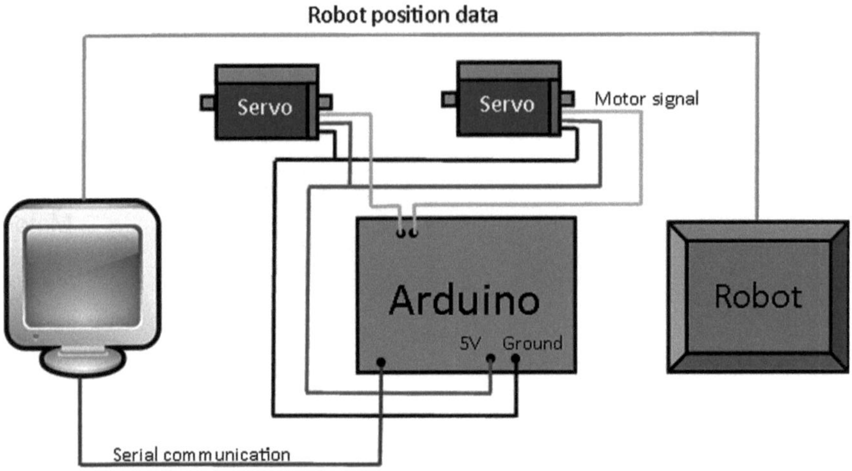

Fig. 3 Electronics layout

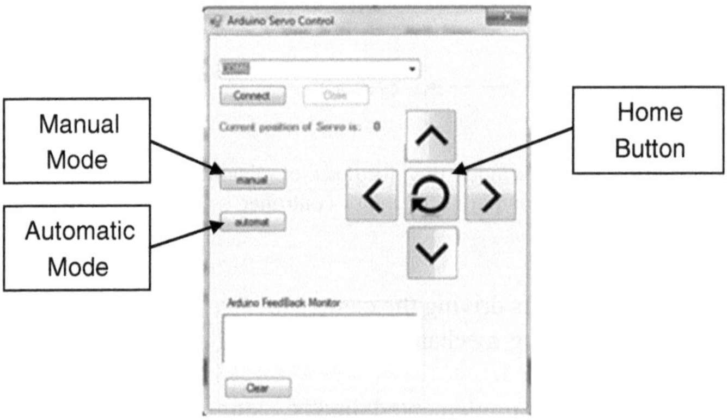

Fig. 4 User interface

2.1 *Automatic Control Mode*

In an automatic tracking mode, the software gets the position of the two robot arms and computes their average, using this as the target camera position. The desired position is sent to the C# code and the respective angular motions are mapped by inverse kinematics. Figure 5 shows the Denavit-Hartenberg (D-H) [1, 2] coordinate frames superimposed on the Endoeye camera. The target position ($x_0y_0z_0$) in the robot's reference frame is transformed to the camera's coordinate system ($x_1y_1z_1$) using a D-H model. The D-H parameters and transformation matrices are given in Table 1 and Eq. (1).

Fig. 5 Denavit-Hartenberg (D-H) coordinate frames for Endoeye camera kinematics

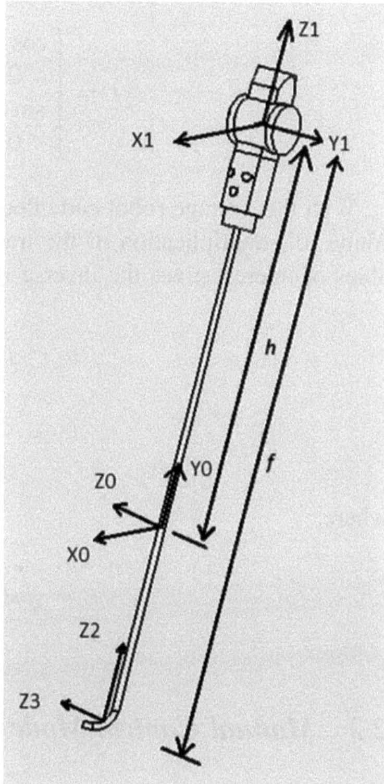

Table 1 D-H transformation parameters for Endoeye camera kinematics

a_{i-1}	α_{i-1}	d_i	θ_i
0	−90°	h	0
0	0	−f	θ_1
0	90°	0	θ_2

$$ {}_0^1T = \begin{bmatrix} 1 & 0 & 0 & 0 \\ 0 & 0 & 1 & h \\ 0 & -1 & 0 & 0 \\ 0 & 0 & 0 & 1 \end{bmatrix} \tag{1} $$

$$ {}_1^2T = \begin{bmatrix} \cos\theta_1 & -\sin\theta_1 & 0 & 0 \\ \sin\theta_1 & \cos\theta_1 & 0 & 0 \\ 0 & 0 & 1 & -f \\ 0 & 0 & 0 & 1 \end{bmatrix} $$

$$
{}_{2}^{3}T = \begin{bmatrix} \cos\theta_2 & -\sin\theta_2 & 0 & 0 \\ 0 & 0 & -1 & 0 \\ \sin\theta_2 & \cos\theta_2 & 0 & 0 \\ 0 & 0 & 0 & 1 \end{bmatrix}
$$

With the average robot end effector position expressed as {x, y, z} in the global frame [0], multiplication of the transformation matrices and isolation of the variables of interest gives the inverse kinematic solution:

$$
\begin{aligned} \theta_2 &= \sin^{-1}\left(\frac{y+f-h}{L}\right) \\ \theta_1 &= -\sin^{-1}\left(\frac{z}{\sqrt{x^2+z^2}}\right) \end{aligned} \tag{2}
$$

where:

$$
L = \sqrt{x^2 + (y+f-h)^2 + z^2} \tag{3}
$$

2.2 *Manual Control Mode*

In the manual mode, the camera gets the commands from a joystick or other input device. Typically the surgeon will use the same joystick or input device which is used to drive the robot arms, and the function of the joystick can be switched from robot to camera and vice versa by the surgeon via a foot switch.

3 Results

A prototype of the laparoscope housing was created using a 3D printing technique. Servomotors were mounted in the housing and connected to the control levers on the laparoscope (Fig. 6). An Arduino microcontroller was programmed to give position commands to the servomotors and interface with the C# program running on a PC.

The experimental validation was divided into two parts: camera control by itself, and integration with the surgical robot. In the first part, the motion of the camera was measured optically using grid paper. The ability of the camera to accurately track the respective motion is shown in Fig. 7. Figure 8 shows the integration of the camera housing with the robot in a benchtop test, which is the second phase of the experimental validation. The camera housing and surgical robot are connected by a small I-beam with a length of 12 inches. The camera followed the tip of the arms by

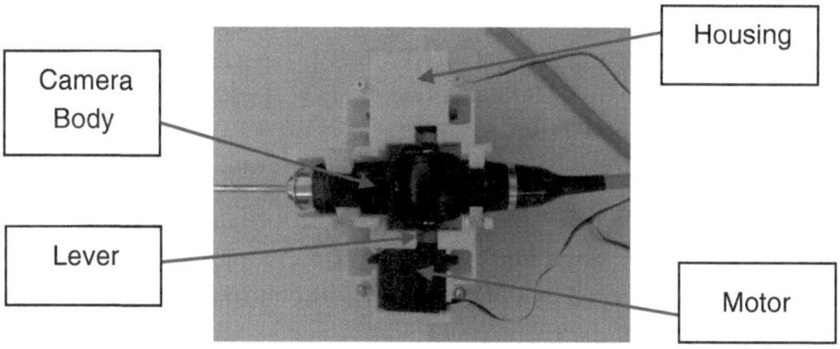

Fig. 6 Prototype of the housing system with camera

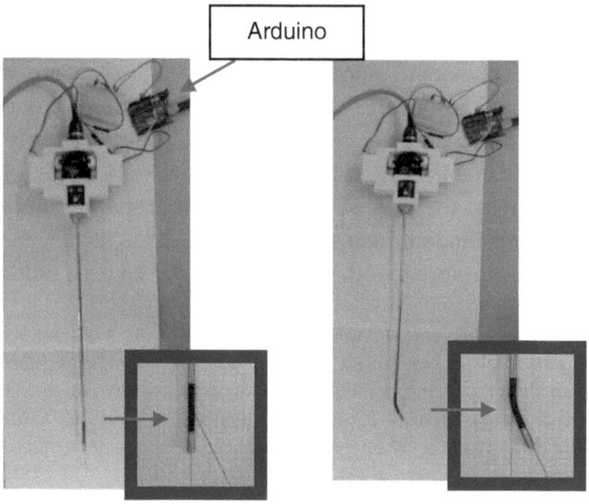

Fig. 7 Benchtop testing outcomes (0° and 30° commanded angles shown)

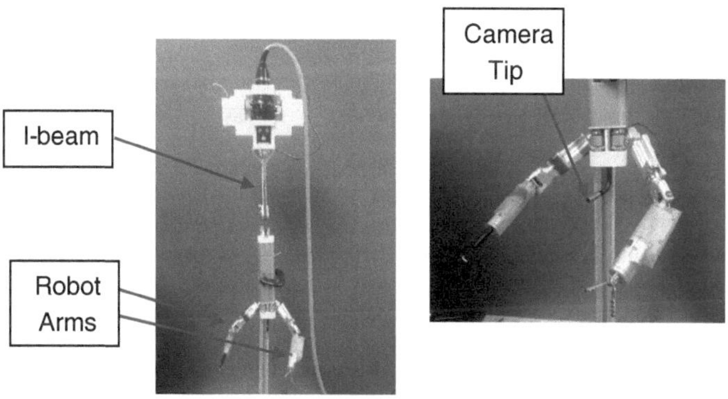

Fig. 8 Integration of Endoeye camera and surgical robot

automatic mode as expected. The desired position was "published" as an input to the C# code and inverse kinematics calculations produced the desired angles of the levers; these control commands were processed through to the motors by the microcontroller.

In vivo experiments in a porcine model further demonstrated the capability of the robot system by tracking tool-tip positions in automatic mode. The preliminary results were satisfactory; however, some minor errors were noted. These errors were associated with imperfect calibration of the initial "home" position of the camera.

4 Conclusions

A motor housing was designed and fabricated to integrate a high-resolution articulating laparoscopic camera (Endoeye) with a bimanual surgical robot. A C# control implementation provides both manual and automatic modes to address visualization needs during surgical procedures. The results of benchtop and in vivo testing demonstrate capability of the system to track the arms of the robot and release the surgeon from the task of controlling the camera. Future work includes further in vivo testing and refinement of the mechanical design as well as the control algorithm, to further smooth and stabilize the camera motion.

Acknowledgments This work was supported by the Office of the Assistant Secretary of Defense for Health Affairs under Award No. W81XWH-14-1-0058. Opinions, interpretations, conclusions and recommendations are those of the author and are not necessarily endorsed by the Department of Defense. The U.S. Army Medical Research Acquisition Activity, 820 Chandler Street, Fort Detrick, MD 21702-5014 is the awarding and administering acquisition office.

References

1. Craig, J.J.: Introduction to Robotics: Mechanics and Control. Prentice Hall, New Jersey (2005)
2. Hartenberg, R.S., Denavit, J.: Kinematic Synthesis of Linkages. McGraw-Hill, New York (1995)
3. Markvicka, E.J.: Design and development of a miniature in vivo surgical robot with distributed motor control for laparoendoscopic single-site surgery. MS thesis, University of Nebraska-Lincoln (2014)
4. Nelson, C.A., Zhang, X., Shah, B.C., Goede, M.R., Oleynikov, D.: Multipurpose surgical robot as a laparoscope assistant. Surg. Endosc. **24**, 1528–1532 (2010)
5. Nelson, C.A., Zhang, X., Guthrie, A.: Automatic laparoscope tracking in minimally invasive surgery using compact robots. In: CARS 2010, Geneva, Switzerland, 23–26 June 2010
6. Omote, K., Feussner, H., Ungeheuer, A., Arbter, K., Wei, G.Q., Siewert, J.R., Hirzinger, G.: Self-guided robotic camera control for laparoscopic surgery compared with human camera control. Am. J Surg. **17**, 321–324 (1999)
7. Tracy, C.R., Raman, J.D., Cadeddu, J.A., Rane, A.: Laparoendoscopic single-site surgery in urology: where have we been and where are we heading? Nat. Clin. Pract. Urol. **5**(10), 561–568 (2008)

Kinematic Models of a New Spherical Parallel Manipulator Used as a Master Device

H. Saafi, M.A. Laribi, M. Arsicault and S. Zeghloul

Abstract The paper discusses the kinematic models of a new spherical parallel manipulator (new SPM). The new SPM is obtained by replacing one leg of a classic 3-RRR SPM. It is used as a master device for a teleoperation system for Minimally Invasive Surgery (MIS). This device controls a surgical robot (slave). The inverse and forward models of the new SPM are studied. Those models are needed to control the motions of the slave robot. A prototype of the new SPM is presented in the end of the paper.

Keywords Spherical parallel architecture · Master device · Forward kinematic model · Inverse kinematic model

1 Introduction

Nowadays, parallel manipulators are widely popular. Thanks to their high load capacity, their stiffness, their low weight and their precision, parallel manipulators are used in many fields such as medicine, where, many master devices have been developed with parallel architecture [1, 2].

In previous works [3], new master device was developed to control a surgical robot. This device has a spherical parallel architecture. The master architecture was

H. Saafi (✉) · M. Arsicault
Department GMSC, Pprime Institute CNRS - University of Poitiers - ENSMA, UPR 3346 Poitiers, France
e-mail: houssem.saafi@univ-poitiers.fr

M. Arsicault
e-mail: marc.arsicault@univ-poitiers.fr

M.A. Laribi · S. Zeghloul
Institut PPRIME, UPR 3346, University of Poitiers, Poitiers, France
e-mail: med.amine.laribi@univ-poitiers.fr

S. Zeghloul
e-mail: said.zeghloul@univ-poitiers.fr

S. Zeghloul et al. (eds.), *Robotics and Mechatronics*,
Mechanisms and Machine Science 37, DOI 10.1007/978-3-319-22368-1_16

chosen to have a mechanism able to provide three pure rotations around a fixed point. The geometric parameters of this structure were optimized to meet a prescribed workspace. However, the main problem of the optimized structure is the presence of parallel singularity in the workspace. To cope with the presence of the singularity, in [4], we proposed to change the architecture of one leg. Then, geometric parameters of the new SPM were optimized in order to have a singular free workspace.

The Forward Kinematic Model (FKM) of the master device is needed to control the surgical robot. The FKM calculates the orientation of the moving platform of the SPM. In this paper, the FKM of the new SPM is presented. The calculation time of the FKM is reduced using an extra sensor. This makes the FKM ready to work in real time.

2 Kinematic of the New Master Device

The new master device is developed to control the motion of a new surgical robot (Fig. 1). The two systems are a part of a tele-operation system for minimally inva-sive surgery (MIS).

In MIS, the instruments enter to the patient body through tiny incisions. This limits the motion of the instrument to three rotations around the incision and one translation along the instrument axis. For this reason, the spherical parallel architecture was chosen for the master device since it is able to produce the similar motions. The architecture of the new master device is shown in Fig. 2.

The new SPM has three legs. The legs B and C are made of two links and three revolute joints (Fig. 3a). All axes of the revolute joints are intersecting in one common point, called CoR (Center of Rotation). The leg A is made of two links, two universal joints and a revolute joint (Fig. 3b). The legs B and C are

Fig. 1 Surgical robot for minemally invasive surgery

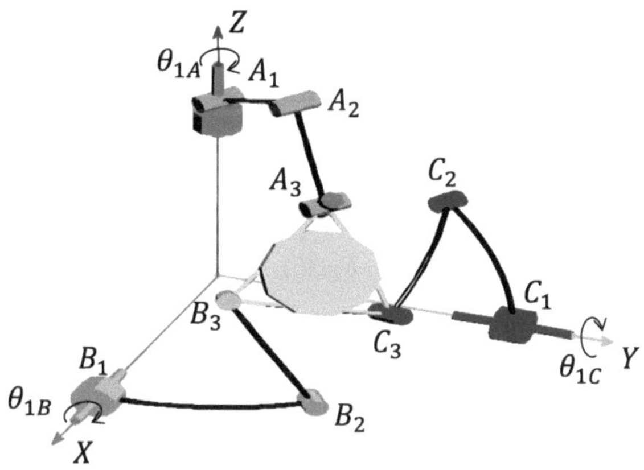

Fig. 2 Master device of a tele-operation system

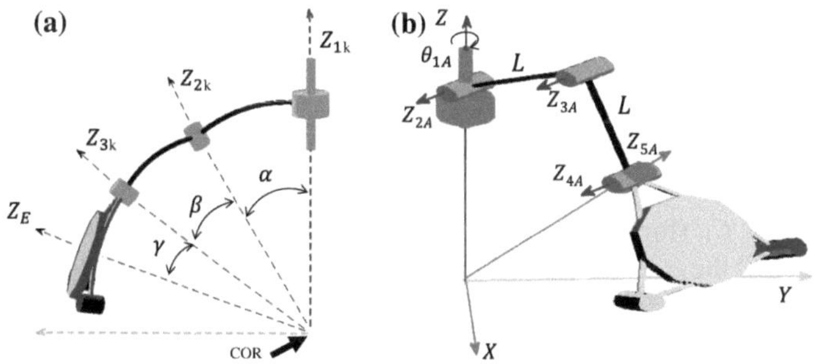

Fig. 3 **a** Kinematic of legs B and C. **b** Kinematic of leg A

defined by the angles α and β and the two links of the leg A are characterized by their length, L.

The actuated joint axes (θ_{1A}, θ_{1B}, and θ_{1C}) are located along an orthogonal frame. The orientation of the SPM is described by the ZXZ Euler angles (ψ, θ, and φ).

The translation motion is not taken into account in this paper because it is uncoupled in the model of the SPM.

3 Inverse Kinematic Model (IKM)

The IKM calculates the active joint angles (θ_{1A}, θ_{1B}, and θ_{1C}) in a function of the orientation (ψ, θ, and φ) of the SPM. Unlike serial robots, the IKM of parallel manipulator is simple. For a given orientation (ψ, θ, and φ) of the SPM, the angles θ_{1B} and θ_{1C} are solved by writing the geometric relation as follows:

$$Z_{2k} \cdot Z_{3k} = \cos(\beta) \quad for\ k = B\ and\ C \tag{1}$$

where,

$$Z_{2K} = R_{0K} \cdot R_Z(\theta_{1K}) \cdot R_X(\alpha) \cdot Z \tag{2}$$

$$Z_{3K} = R_{ZXZ} \cdot R_Z(\frac{2\pi}{3}) \cdot R_X(\gamma) \cdot Z \tag{3}$$

$$R_{0B} = \begin{bmatrix} 0 & 1 & 0 \\ 0 & 0 & 1 \\ 1 & 0 & 0 \end{bmatrix} \qquad R_{0C} = \begin{bmatrix} 0 & 0 & 1 \\ 1 & 0 & 0 \\ 0 & 1 & 0 \end{bmatrix} \tag{4}$$

$$R_{ZXZ} = R_Z(\psi) \cdot R_X(\theta) \cdot R_Z(\varphi) \tag{5}$$

By substituting Z_{2K} and Z_{3K}, we get the following expressions:

$$\begin{cases} A_1 \cos(\theta_{2B}) + B_1 \cos(\theta_{2B}) + C_1 = 0 \\ A_2 \cos(\theta_{2C}) + B_2 \cos(\theta_{2C}) + C_2 = 0 \end{cases} \tag{6}$$

A_k, B_k and C_k (for k = 1 and 2) are variables that depend on the geometric parameters and the orientation of the SPM. These variables define the workspace as follows:

$$\begin{cases} \frac{C_1^2}{A_1^2+B_1^2} \le 1 \\ \frac{C_2^2}{A_2^2+B_2^2} \le 1 \end{cases} \tag{7}$$

Figure 4 shows a simplified representation of the leg A with the end effector. Using the spherical trigonometric relations, the angle θ_{1A} is equal to:

$$\theta_{1A} = -sg(\varphi) a\cos(\frac{\cos(\gamma)\cos(\theta_{2A})\cos(\theta)}{\sin(\theta_{2A})\sin(\theta)}) + \psi \tag{8}$$

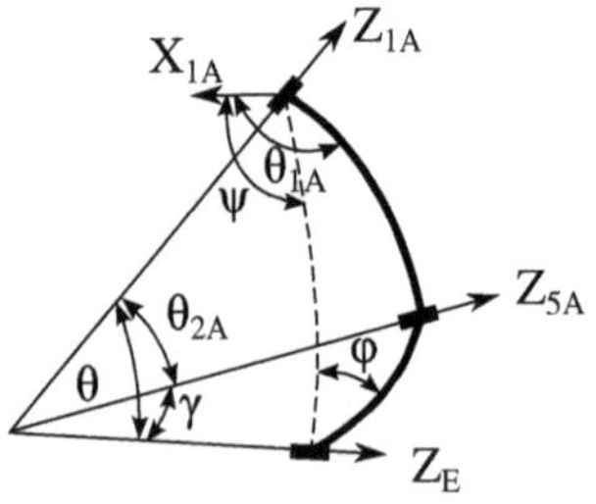

Fig. 4 Simplified representation of the leg A with the end effector

where

$$\theta_{2A} = a\cos(\cos(\gamma)\cos(\theta) + \sin(\varphi)\sin(\gamma)\sin(\theta)) \tag{9}$$

The solutions of the IKM are called working modes. The new SPM has 8 solutions. Only one configuration of the leg A is considered, so, the new SPM has only four working modes presented in Fig. 5.

The working mode presented in Fig. 5a was selected for the master device because it has the larger singular-free area [4]. The dexterity distribution of the new SPM are presented in Fig. 6.

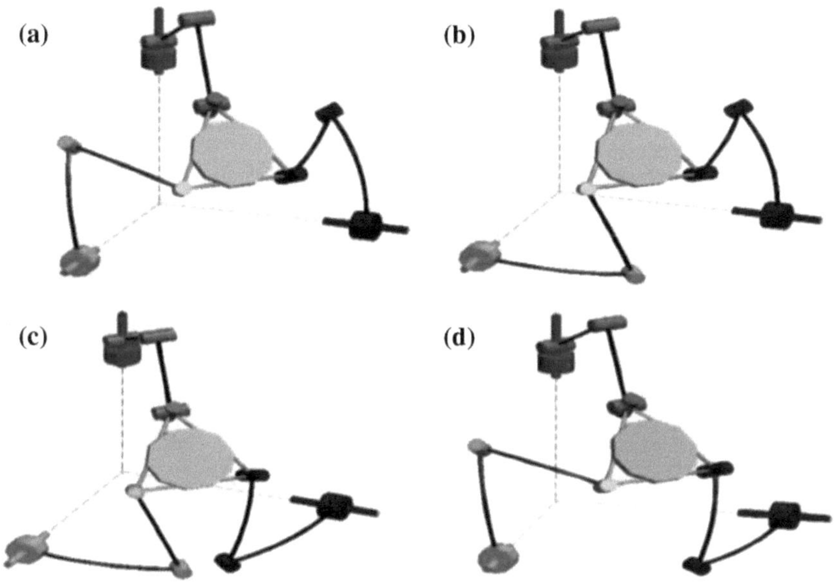

Fig. 5 Working modes of the new SPM

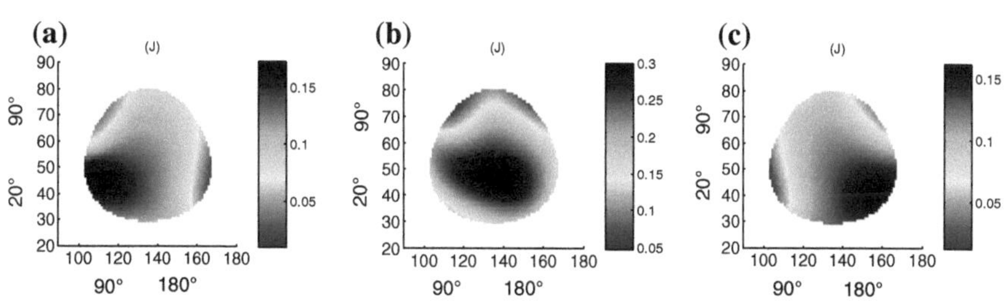

Fig. 6 Dexterity distribution of the new SPM. **a** $\varphi = -50°$. **b** $\varphi = 0°$. **c** $\varphi = 50°$

4 Forward Kinematic Model

The FKM calculates the orientation of the end-effector, ψ, θ and φ, for a given position of the active joints (θ_{1A}, θ_{1B}, θ_{1C}). The FKM can be solved using different methods [5–7]. The FKM is solved in this paper using the input/output equations of spherical four-bar linkages.

For the new SPM, only one four-bar mechanism can be considered (Fig. 7). The input/output equation is as follows:

$$L_1(\xi)\cos(\sigma) + M_1(\xi)\sin(\sigma) + N_1(\xi) = 0 \tag{10}$$

Another equation is needed to solve the system. This equation is obtained by expressing Z_{5A} in function of ξ and σ by using the forward kinematic of leg B as follows:

$$\begin{aligned} Z_{5A} = R_{0B}R_Z(\theta_{1B})R_X(\alpha)R_Z(\xi + \xi')R_X(\beta)R_Z(\sigma - \mu) \\ R_X(\gamma)R_Z(\tfrac{2\pi}{3})R_X(-\gamma)Z_{1A} \end{aligned} \tag{11}$$

Y_{1A} is perpendicular to Z_{5A}, this condition leads to the following equation after the arrangement of ($Y_{1A} \cdot Z_{5A} = 0$):

$$L_2(\xi)\cos(\sigma) + M_2(\xi)\sin(\sigma) + N_2(\xi) = 0 \tag{12}$$

where, ξ and σ are two angles defined in Fig. 7 and $L_i(\xi)$, $M_i(\xi)$, $N_i(\xi)$ (i = 1, 2) are variables that depend on $\cos(\xi)$ and $\sin(\xi)$.

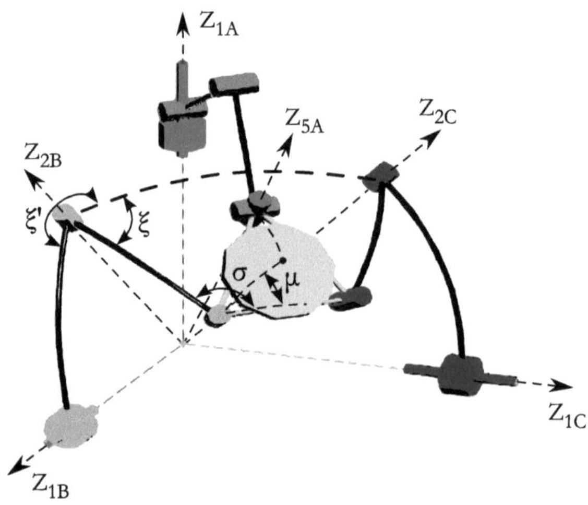

Fig. 7 Considered Four-Bar mechanism

$\cos(\sigma)$ and $\sin(\sigma)$ are obtained using Eqs. (11) and (12) as follows:

$$\begin{cases} \cos(\sigma) & = & \frac{M_1(\xi)N_2(\xi)-M_2(\xi)N_1(\xi)}{L_1(\xi)M_2(\xi)-L_2(\xi)M_1(\xi)} \quad (a) \\ \sin(\sigma) & = & \frac{L_1(\xi)N_2(\xi)-L2(\xi)N_1(\xi)}{L_1(\xi)M_2(\xi)-L_2(\xi)M_1(\xi)} \quad (b) \end{cases} \tag{13}$$

First, the possible solution of ξ can be found by solving the equation generated by writing $(13 - \mathrm{a})^2 + (13 - \mathrm{b})^2$ which has only ξ as unknown. This equation is as follows:

$$\begin{aligned} & N_2^2L_1^2 + 2L_1M_2L_1M_1 - 2L_2N_2L_1N_1 \\ & + N_2^2M_1^2 - L_2^2M_1^2 - 2M_2N_2M_1N_1 \\ & - M_2^2L_1^2 + L_2^1N_1^2 + M_2^2N_1^2 = 0 \end{aligned} \tag{14}$$

Equation 14 is a four degree quadratic equation in $\cos(\xi)$ and $\sin(\xi)$. This equation is solved numerically and highly increases the calculation time of the FKM. The obtained solution of ξ is then used to calculate σ using the Eq. (13).

For given values of $(\theta_{1A},\theta_{1B},\theta_{1C})$, we have at most eight solutions for ξ and σ. Each pair $(\xi_i,\ \sigma_i)$,for $1 \leq i \leq 8$, gives an orientation $(\psi,\ \theta,\ \varphi)$ of the end-effector using the forward kinematic of leg A.

The FKM of the classic SPM was improved in [8] by using an extra sensor. This sensor reduces the complexity of the FKM. This method will be adopted for the new SPM to get a a closed form solution. This work will be presented later.

5 Developed Prototype of the New SPM

A prototype of the new SPM has been developed (Fig. 8). This prototype will be equipped by absolute sensors to solve the FKM of the master device. This new will be used as haptic device. A actuators will be placed on the active joints. The experimental validation of the new SPM will be studied in future works.

Fig. 8 **a** Kinematic of legs B and C. **b** Kinematic of leg A

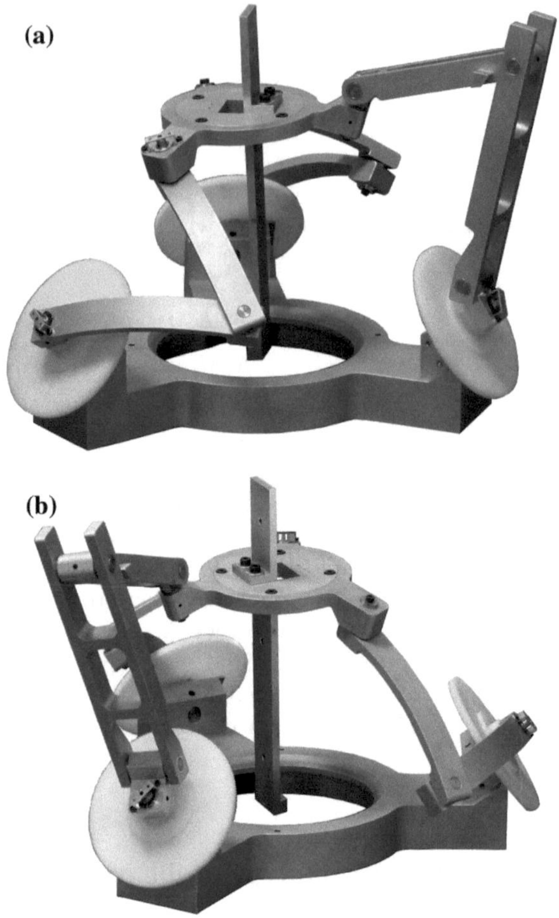

6 Conclusions

The inverse and forward models of a new spherical parallel manipulator were studied in this paper. The inverse model (IKM) was obtained by using the geometric and trigonometric relations of the legs of the new SPM. The IKM is simple and gives a direct solution. The FKM is complex and it was obtained using the equation of the input/output equation of the four-bar mechanism. The new SPM is used as a master device for a teleoperation system for Minimally Invasive Surgery. This device controls a surgical robot (slave). A prototype of the new SPM was presented in the end of the paper.

Acknowledgments This research is supported by the Poitou-Charentes region 2007–2013 (program project 10 Images and interactivities), in partnership with the European Union (FEDER/ERDF, European Regional Development Fund) and ROBOTEX, the French national network of robotics platforms ($N°$ ANR-10-EQPX-44-01).

References

1. Tobergte, A., et al.: The sigma.7 haptic interface for MiroSurge: a new bi-manual surgical console. In: IEEE/RSJ International Conference on Intelligent Robots and Systems (IROS), 2011, pp. 3023, 3030, 25–30 Sept 2011
2. Ma, A., Payandeh, S.: Analysis and experimentation of a 4-DOF haptic device. In: Haptic Interfaces for Virtual Environment and Teleoperator Systems, Symposium On Haptics, pp. 351, 356, 13–14 March 2008
3. Chaker, A., Mlika, A., Laribi, M.A., Romdhane, L., Zeghoul, S.: Synthesis of a spherical parallel manipulator for a dexterous medical task. Front. Mech. Eng. **7**(2), 150–162 (2012)
4. Saafi, H., Laribi, M.A., Arsicault, M., Zeghloul, S.: Optimal design of a new spherical parallel manipulator. In: 23rd International Conference on Robotics in Alpe-Adria-Danube Region (RAAD), pp. 1–6. IEEE (2014)
5. Gosselin, C., Sefrioui, J., Richard, M.J.: On the direct kinematics of spherical three-degree-of-freedom parallel manipulators of general architecture. J. Mech. Des. **116**(2), 594–598 (1994)
6. Celaya, E.: Interval propagation for solving parallel spherical mechanisms. In: Advances in Robot Kinematics, pp. 415–422. Springer (2002)
7. Bai, S., Hansen, M.R., Angeles, J.: A robust forward-displacement analysis of spherical parallel robots. Mech. Mach. Theory **44**(12), 2204–2216 (2009)
8. Saafi, H., Laribi, M. A., Zeghloul, S.: Improvement of the direct kinematic model of a haptic device for medical application in real time using an extra sensor. In: IEEE/RSJ International Conference on Intelligent Robots and Systems (IROS 2014), pp. 1697, 1702, 14–18 Sept 2014

Initial Experiments with the Leap Motion as a User Interface in Robotic Endonasal Surgery

T.A. Travaglini, P.J. Swaney, Kyle D. Weaver and R.J. Webster III

Abstract The Leap Motion controller is a low-cost, optically-based hand tracking system that has recently been introduced on the consumer market. Prior studies have investigated its precision and accuracy, toward evaluating its usefulness as a surgical robot master interface. Yet due to the diversity of potential slave robots and surgical procedures, as well as the dynamic nature of surgery, it is challenging to make general conclusions from published accuracy and precision data. Thus, our goal in this paper is to explore the use of the Leap in the specific scenario of endonasal pituitary surgery. We use it to control a concentric tube continuum robot in a phantom study, and compare user performance using the Leap to previously published results using the Phantom Omni. We find that the users were able to achieve nearly identical average resection percentage and overall surgical duration with the Leap.

Keywords Surgical robot · Concentric tube robot · Continuum robot · User interface · Teleoperation · Endonasal surgery

T.A. Travaglini · P.J. Swaney (✉) · R.J. Webster III
Department of Mechanical Engineering, Vanderbilt University,
Nashville, TN 37212, USA
e-mail: philip.j.swaney@vanderbilt.edu

T.A. Travaglini
e-mail: tate.a.travaglini@vanderbilt.edu

R.J. Webster III
e-mail: robert.webster@vanderbilt.edu

K.D. Weaver
Department of Neurological Surgery, Vanderbilt University Medical Center,
Nashville, TN 37212, USA
e-mail: kyle.weaver@vanderbilt.edu

S. Zeghloul et al. (eds.), *Robotics and Mechatronics*,
Mechanisms and Machine Science 37, DOI 10.1007/978-3-319-22368-1_17

1 Introduction

Many surgical robots are operated using a teleoperation framework in which the user manipulates a master device that records his or her motions and relays them to a patient side (sometimes called a "slave") robot that interacts with the patient. This teleoperation framework is the basis for the widely used da Vinci surgical system (Intuitive Surgical, Inc., USA) and is also found in numerous research systems [8, 25].

Current master devices for surgical robotic systems typically consist of mechanical linkages that include motors and encoders. Toward creating less expensive master interfaces that are not subject to mechanical wear, there has been increasing recent interest in hand tracking as an alternative approach [7, 15, 16, 26, 30]. This trend has been spurred by the recent introduction of low-cost, consumer-based tracking devices such as the Kinect (Microsoft, Inc., USA) and the Leap Motion (Leap Motion, Inc., USA). The Kinect has been used in a variety of hand tracking and teleoperation experiments (see e.g. [9, 10, 20]). It has also been productively used in rehabilitation [27]. But the suitability of optical hand tracking systems like the Kinect or Leap in surgical teleoperation remains an open question.

In this paper we explore the use of the Leap Motion controller in the specific surgical task of pituitary tumor resection. The Leap is an infrared, camera-based system designed to track hands, fingers, and tools. The controller is low cost ($79 as of this writing), has a stated accuracy of 1/100th of a millimeter [17], and is readily available. The research community has found numerous uses for the device in the two and a half years it has been available. For example, Potter et al. used the Leap in a sign language translation application [21]. Igor et al. used the built-in gesture recognition software to control a 3-finger gripper [31]. Bassily et al. used the Leap to control a 6-DOF Jaco robotic arm [3]. Sabir et al. used it to create an interface for visualization of three-dimensional molecular models [23]. There is some evidence that the Leap's accuracy in dynamic settings (e.g. when the user's hand is moving) is substantially less than the manufacturer's 1/100 mm accuracy specification [14, 15, 28]. There is also some indication that touchless interfaces may have a shorter learning curve than touch-based or haptic interfaces [30]. Prior studies have typically focused on evaluating the accuracy and repeatability of Leap measurements in static and dynamic settings [14, 15, 28]. However, it is unclear how these accuracy and precision numbers translate to the feasibility of accomplishing a surgical task, due to the presence of the human surgeon in the loop. Thus, in our current paper we aim to experimentally test the Leap in a specific surgical task, with a specific slave robot. We compare the Leap against the Phantom Omni (formerly Sensable, Inc., USA, now the Geomagic Touch, Geomagic, Inc. USA), which was previously suggested to be a good example haptic device for comparison to the Leap [7]. It is worth noting that the Leap is at least an order of magnitude less expensive than the Phantom Omni at the time of this writing and that the Omni is one of the lower-cost examples of commercial haptic devices, overall.

Our surgical task of interest in this paper is pituitary tumor resections using a concentric tube robot delivered through the nose. This task is useful because we

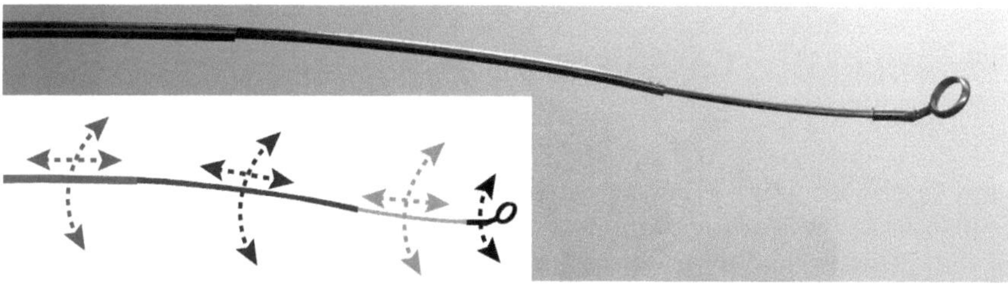

Fig. 1 A concentric tube continuum robot made from three precurved, superelastic nitinol tubes. The robot can bend and elongate by rotating and translating the tubes inside one another. The inset line drawing shows the degrees of freedom. An angled ring curette typically used in hand tools for endonasal tumor removal is attached to the tip of the robot and can be axially rotated

have prior data using the same robot and phantom model, but in which the Phantom Omni was used as the master [24]. There is ample clincal motivation for accomplishing transnasal pituitary resections. Pituitary adenomas account for 15–20 % of primary brain tumors [2]. Endoscopic transnasal techniques are challenging for the surgeon, because of the constrained space inside the nostril, and the challenge of manually manipulating multiple tools in it to accomplish a delicate surgical procedure, motivating robotic assistance. Prior research on robotic systems for transnasal surgery has focused on image-guided drilling [18] and control of surgical endoscopes [19]. We have previously developed a complementary multi-arm robotic system specifically for the tumor resection portion of the surgery, which features needle-sized manipulators that can bend and elongate in a manner conceptually similar to tentacles (see Fig. 1) [5, 13, 22, 24].

2 Experimental Setup

The robot used in our experiments is shown in Fig. 2 and it delivers a concentric tube robot consisting of three telescoping, precurved tubes made from superelastic nitinol (see Fig. 1 and [11, 13, 22] for further information on robots of this type). The robot used in this paper is identical to the one used in [24], and is described in more detail in [5]. It features three tubes with diameters of 2.4, 1.7, and 1.2 mm, each of which can be telescopically extended and axially rotated at its base. The robot also features an axial wrist, which enables the Hardy transsphenoidal curette (P/N SP0007011, Codman) mounted at its tip to rotate as shown in Fig. 3.

We implemented a damped least squares teleoperation approach as described previously [5, 24]. The Leap Software Development Kit (SDK) was used to measure the position of the center of the hand and the roll, pitch, and yaw angles about this point, and these were mapped directly to the tip of the concentric tube robot. When the user wished to move the robot without spinning the curette, the user simply adjusted the pose of his or her hand with fingers spread. When the user

Fig. 2 The actuation unit used to translate and rotate the tubes. Each tube is grasped at its base and may be translated and rotated independently of the others, enabling the concentric tube manipulator to bend and elongate

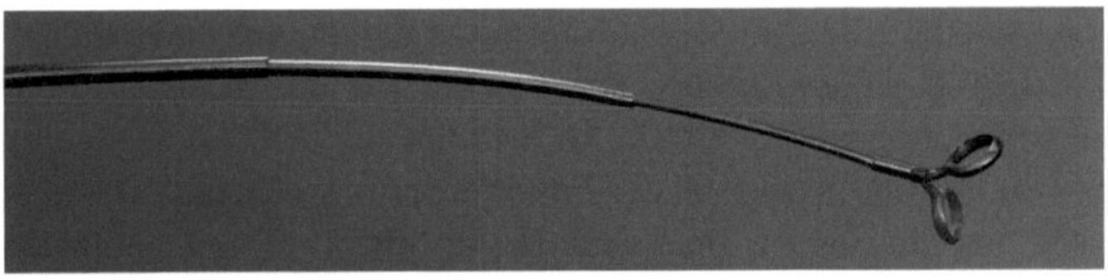

Fig. 3 An illustration of the axial rotation ability of the tip mounted curette. Shown above are two superimposed images of it after a 180 rotation

wished to re-orient the curette, the user adducted his or her fingers, in which case the roll angle (i.e. the angle about the vector aligned with the fingers) was mapped to the axial angle of the curette, causing it to rotate as shown in Fig. 3. The hand movements were scaled down to reduce tremor and to enable precise movements in the small workspace of the pituitary gland. A standard 4 mm, 30, endoscope provided visualization of the surgical site (i.e. the same endoscope used in [24]). The experimental setup was as shown in Fig. 4.

The phantom tumor model was also the same as used in [24]. It was made from a 5:1 ratio of SIM-TEST (Corbin, Inc., USA) to water. This produces a phantom tumor that is similar to the consistency of a typical pituitary tumor [4]. This phantom tumor was placed inside an anatomical skull model (# A20, 3B Scientific, Germany), which had previously been prepared by the surgeon to closely replicate the enlarged sella as commonly found in pituitary tumor patients. The volume of the sella cavity in this model was approximately 6.92 cm^3. During resections, both the endoscope and the skull were fixed in place, and manual suction was used to clean the curette from time to time, but not to directly remove phantom tissue from within the sella.

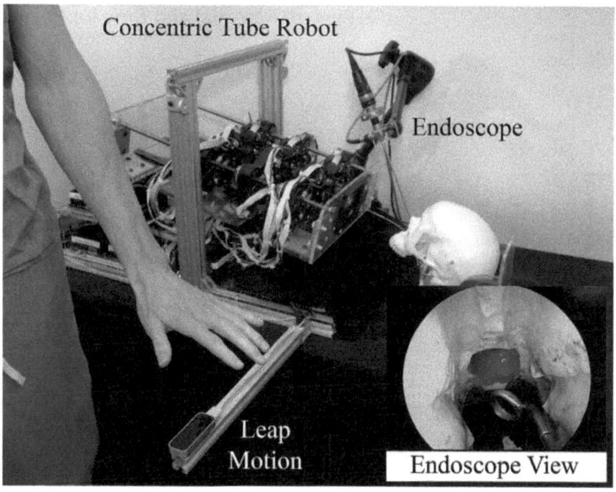

Fig. 4 Experimental setup for phantom tumor resection experiments

3 Experimental Results

Using the setup shown in Fig. 4, co-author Philip Swaney and co-author and neurosurgeon Kyle Weaver performed a total of 16 phantom pituitary tumor resections. It is worth noting that while Philip Swaney was familiar with the robot in general, these were the first resections he had attempted with the robot. We measured the time required to complete the procedure and the quantity of tumor removed. The latter was measured by weighing the skull before introduction of the phantom tumor, weighing it again after introduction of the tumor, and weighing it a final time after resection. Figure 5 presents the results in terms of both time and resection percentage. The average percentage removal was 78.5 $\pm$ 10.9 % and the average time to complete the removal was 12.4 $\pm$ 2.0 min. Figure 6 presents the same metrics, but as previously achieved in [24]. In that work, experienced surgeons performed resections on the same phantom, with the same slave robot, but using the Phantom Omni as the master interface. There, the average percentage removal was 79.8 $\pm$ 5.9 % and the average time to complete the procedure was 12.5 $\pm$ 4.1 min.

4 Discussion and Conclusions

It remains somewhat controversial whether touchless master interfaces will ultimately be suitable for use with surgical robots. At least one group has concluded that they are not yet up to the challenge [15], and at least one other group has concluded that they are promising [7]. In this paper we have studied one particular surgical application in which an existing data set was available for comparison, and found that similar metrics were achieved with both a mechanical linkage and the Leap. This

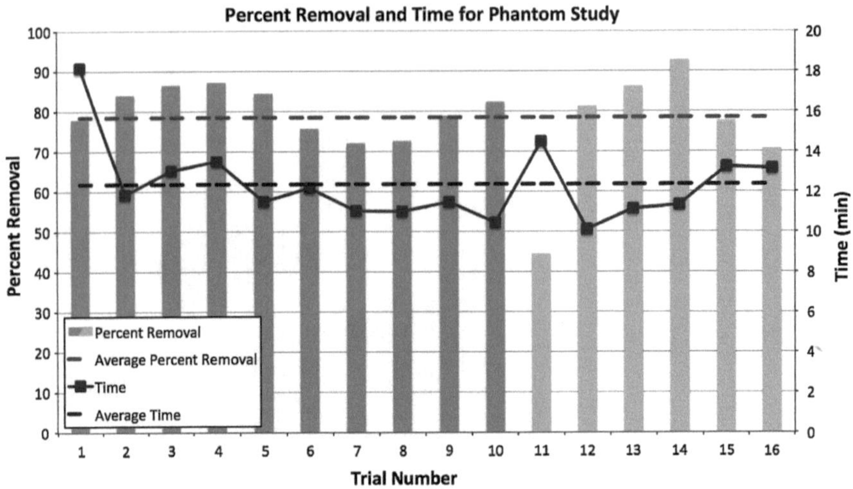

Fig. 5 Percentage removal and time to complete removal are shown for all 16 resections performed with the Leap. The average percentage removal and average time for removal are overlaid on the data. The trials performed by co-author Philip Swaney are shown in blue, and the trials performed by neurosurgeon Kyle Weaver are shown in green. The trials are listed in order of completion—i.e. the first experiment by Weaver had a lower resection percentage, indicating that some learning curve—but apparently not a long one—may be present

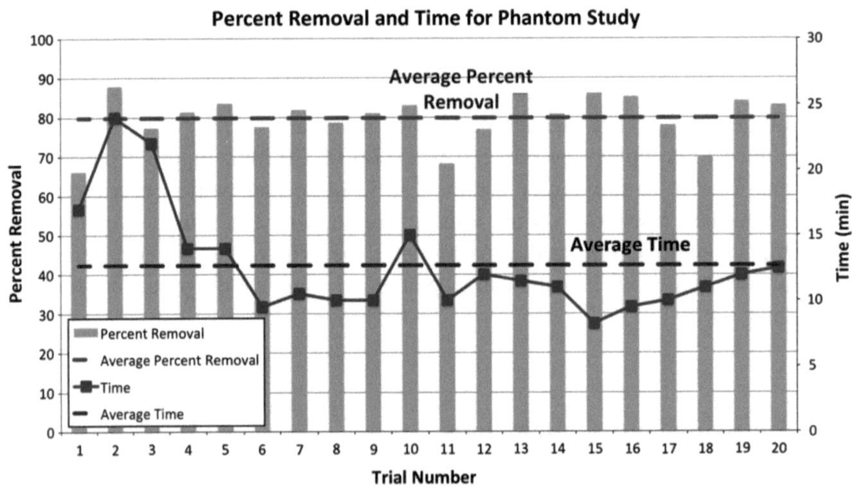

Fig. 6 Percentage removal and time to complete removal are shown for the twenty resections done with the Phantom Omni, and previously presented in [24]. The average percent removal and average time for removal are overlaid on the data

provides an indication that there are likely to be at least some surgeries where a touchless surgical robot interface may be suitable. It also remains to be seen if the loss of haptic feedback with a touchless interface is important. The da Vinci surgical robot has found commercial success without the use of haptic feedback, providing some indication that surgeons can adapt to visual feedback only. The ergonomics of using a touchless interface for surgical procedures is also an open ended research question, as fatigue is an important factor for any lengthy surgical procedure.

As was noted in [24], the 78 % average tumor removal should be considered a successful outcome. This is because pituitary tumors are typically benign and slow growing, and the goal of the surgery is to decompress nearby structures like the optic nerve and corotid artery. Thus, it is not necessary to remove all of the tumor, and 100 % removal is not achieved in many surgeries today. One study identified "definite tumor remnants or at least suspicious findings" in post-operative MRI scans in 42 % of patients [12]. Furthermore, experienced surgeons believe that the results of this experiment are conservative, since the hydrostatic pressure in the brain was not modeled in our phantom. This pressure tends to push tumor material forward where it is easier to resect during a pituitary tumor surgery.

In the future we plan to improve both hardware and software of the system described in this paper. First, we are actively pursuing design of the actuation unit for sterilizability and biocompatibility. One example of a sterilizable actuation unit is given in [6], and we are developing similar concepts for application to a multi-arm system. Furthermore, we are developing miniature wrists to make the manipulators even more dexterous [29]. We are also considering the implementation of virtual fixtures to enforce "no fly zones" around sensitive anatomical structures [1].

In terms of Leap-based control, we are interested in studying learning curves for both novice and expert surgeons using our system. It will also be useful in the future to explore a wider range of surgical procedures and types of slave robot, so that general trends may be observed.

The results of our current study, while by no means a definitive statement on the value of hand tracking as a master interface, do illustrate at least one specific scenario where it appears to work sufficiently well for a human user to accomplish a surgical task. If this holds true in other contexts, in the future one will be able to confidently say that hand tracking is a viable solution for creating effective, lower cost, master interfaces for surgical robots.

Acknowledgments This work was supported in part by the National Science Foundation under Award IIS-1054331 and in part by the National Institutes of Health under Award R01 EB017467.

References

1. Abbott, J.J., Marayong, P., Okamura, A.M.: Haptic virtual fixtures for robot-assisted manipulation. Springer Tracts in Adv. Robot. **28**, 49–64 (2007)
2. ABTA.: American brain tumor association (ABTA) http://abta.org (2011). Accessed on Mar 8 2011

3. Bassily, D., Georgoulas, C., Guettler, J., Linner, T., Bock, T.: Intuitive and adaptive robotic arm manipulation using the leap motion controller. In: ISR/Robotik 41st International Symposium on Robotics, pp. 1–7 (2014)
4. Bekeny, J.R., Swaney, P.J., Webster III, R.J., Russell, P.T., Weaver, K.D.: Forces applied at the skull base during transnasal endoscopic transsphenoidal pituitary tumor excision. J. Neurol. Surg. Part B Skull Base **74**(6), 337 (2013)
5. Burgner, J., Rucker, D., Gilbert, H., Swaney, P., Russell, P., Weaver, K., Webster, R.: A telerobotic system for transnasal surgery. IEEE Tran. Mechatron. **19**(3), 996–1006 (2014)
6. Burgner, J., Swaney, P.J., Lathrop, R.A., Weaver, K.D., Webster III, R.J.: Debulking from within: a robotic steerable cannula for intracerebral hemorrhage evacuation. IEEE Trans. Biomed. Eng. **60**(9), 2567–2575 (2013)
7. Despinoy, F., Sanchez, A., Zemiti, N., Jannin, P., Poignet, P.: Comparative assessment of a novel optical human-machine interface for laparoscopic telesurgery. In: Information Processing in Computer-Assisted Interventions, Lecture Notes in Computer Science, vol. 8498, pp. 21–30. Springer International Publishing, New York (2014)
8. Dogangil, G., Davies, B.L.: Rodriguez y Baena, F.: A review of medical robotics for minimally invasive soft tissue surgery. Proc. Inst. Mech. Eng. [H]: J. Eng. Med. **224**(5), 653–679 (2010)
9. Dragan, A.D., Srinivasa, S.S., Lee, K.C.T.: Teleoperation with intelligent and customizable interfaces. J. Hum. Robot Interact. **2**(2), 33–57 (2013)
10. Du, G., Zhang, P., Mai, J., Li, Z.: Markerless kinect-based hand tracking for robot teleoperation. Int. J. Adv. Rob. Syst. **9**(1), 36–46 (2012)
11. Dupont, P.E., Lock, J., Itkowitz, B., Butler, E.: Design and control of concentric-tube robots. IEEE Trans. Robot. **26**(2), 209–225 (2010)
12. Fahlbusch, R., Ganslandt, O., Buchfelder, M., Schott, W., Nimsky, C.: Intraoperative magnetic resonance imaging during transsphenoidal surgery. J. Neurosurg. **95**(3), 381–390 (2001)
13. Gilbert, H.B., Rucker, D.C., Webster III, R.J.: Concentric tube robots: state of the art and future directions. In: 16th International Symposium on Robotics Research (2013). Springer Tracts in Advanced Robotics (In Press)
14. Guna, J., Jakus, G., Pogačnik, M., Tomažič, S., Sodnik, J.: An analysis of the precision and reliability of the leap motion sensor and its suitability for static and dynamic tracking. Sensors **14**(2), 3702–3720 (2014)
15. Kim, Y., Kim, P., Selle, R., Shademan, A., Krieger, A.: Experimental evaluation of contact-less hand tracking systems for tele-operation of surgical tasks. In: IEEE International Conference on Robotics and Automation, pp. 3502–3509 (2014)
16. Kim, Y., Leonard, S., Shademan, A., Krieger, A., Kim, P.C.: Kinect technology for hand tracking control of surgical robots: technical and surgical skill comparison to current robotic masters. Surg. Endosc. **28**(6), 1993–2000 (2014)
17. Leap Motion, Inc.: Leap Motion | 3D Motion and Gesture Control for PC & Mac (2015). https://www.leapmotion.com/product
18. Matinfar, M., Baird, C., Batouli, A., Clatterbuck, R., Kazanzides, P.: Robot-assisted skull base surgery. In: IEEE/RSJ International Conference on Intelligent Robots and Systems, pp. 865–870 (2007)
19. Nimsky, C., Rachinger, J., Iro, H., Fahlbusch, R.: Adaptation of a hexapod-based robotic system for extended endoscope-assisted transsphenoidal skull base surgery. Minim. Invasive Neurosurg. **47**(1), 41–46 (2004)
20. Oikonomidis, I., Kyriazis, N., Argyros, A.A.: Tracking the articulated motion of two strongly interacting hands. In: IEEE Conference of Computer Vision and Pattern Recognition pp. 1862–1869 (2012)
21. Potter, L.E., Araullo, J., Carter, L.: The leap motion controller: a view on sign language. In: Australian Computer-Human Interaction Conference: Augmentation, Application, Innovation, Collaboration, pp. 175–178. ACM (2013)

22. Rucker, D.C., Jones, B.A., Webster III, R.J.: A geometrically exact model for externally loaded concentric tube continuum robots. IEEE Trans. Robot. **26**(5), 769–780 (2010)
23. Sabir, K., Stolte, C., Tabor, B., O'Donoghue, S.: The molecular control toolkit: controlling 3d molecular graphics via gesture and voice. In: IEEE Symposium on Biological Data Visualization, pp. 49–56 (2013)
24. Swaney, P.J., Gilbert, H.B., Webster III, R.J., Russell III, P.T., Weaver, K.D.: Endonasal skull base tumor removal using concentric tube continuum robots: a phantom study. J. Neurol. Surg. Part B Skull Base **76**(2), 145–149 (2015)
25. Taylor, R.H., Stoianovici, D.: Medical robotics in computer-integrated surgery. IEEE Trans. Robot. Autom. **19**(5), 765–781 (2003)
26. Vargas, H., Vivas, O.: Gesture recognition system for surgical robot's manipulation. In: Symposium on Image, Signal Processing and Artificial Vision, pp. 1–5 (2014)
27. Webster, D., Celik, O.: Systematic review of kinect applications in elderly care and stroke rehabilitation. J. Neuroeng. Rehabil. **11**(108) (2014)
28. Weichert, F., Bachmann, D., Rudak, B., Fisseler, D.: Analysis of the accuracy and robustness of the leap motion controller. Sensors **13**(5), 6380–6393 (2013)
29. York, P.A., Swaney, P.J., Gilbert, H.B., Webster III, R.J.: A wrist for needle-sized surgical robots. In: IEEE International Conference on Robotics and Automation (2015)
30. Zhou, T., Cabrera, M.E., Wachs, J.: Touchless telerobotic surgery—a comparative study. In: IEEE EEE/RSJ International Conference on Intelligent Robots and Systems (2014)
31. Zubrycki, I., Granosik, G.: Using integrated vision systems: three gears and leap motion, to control a 3-finger dexterous gripper. In: Recent Advances in Automation, Robotics and Measuring Techniques, pp. 553–564. Springer (2014)

Part V
Medical Robotics II

Mechatronic Device to Assist Therapies During Hand Fingers Rehabilitation

F. Aguilar-Pereyra and E. Castillo-Castaneda

Abstract This work presents a novel application of a four-bar mechanism to assist therapies during hand fingers rehabilitation. The system is a mechatronic device able to perform fingers flexoextension exercises and can be adapted to different hand sizes. Fingers flexoextension analysis is based on a kinematic finger model from index to little to determine the motion trajectory. A four-bar mechanism is proposed to develop this path and guide the fingertip in rehabilitation therapies. The kinematic analysis of the proposed mechanism is presented and simulated using MATLAB. Finally, a prototype for one finger was designed and built to validate the trajectories.

Keywords Hand rehabilitation · Fingers kinematics · Four-bar mechanism

1 Introduction

The natural flexoextension motion for index, middle, ring and little fingers is developed by the coordinated rotation of their phalanges [3, 10]. The trajectory of each fingertip is a curved line like a logarithmic spiral [5]. Different techniques are used in the rehabilitation therapy for fingers motion [9]; one technique applied by a physiotherapist consists in moving the fingers joints from the neutral position with oscillating movements of low amplitude and velocity. The velocity and amplitude of movements are incremented until develop all the range of motion of the fingers joints of patient. In mechanotherapy, mechanical systems are used for aiding the joint movements in rehabilitation [1].

F. Aguilar-Pereyra · E. Castillo-Castaneda (✉)
CICATA, Instituto Politecnico Nacional, Queretaro, Mexico
e-mail: ecastilloca@ipn.mx; ecast63@yahoo.com

F. Aguilar-Pereyra
e-mail: jaguilarp1200@alumno.ipn.mx

S. Zeghloul et al. (eds.), *Robotics and Mechatronics*,
Mechanisms and Machine Science 37, DOI 10.1007/978-3-319-22368-1_18

2 Kinematic Model of Fingers Position

Different kinematic models of the human hand currently exist [6, 11, 3]. The human hand is the most complex set of jointed bones of human skeleton whose general structure can be divided, for helping its kinematic analysis, into a first set, the carpal bones, and a second set gathering the metacarpals and the phalanges of the five digits [10]. A realistic approach, for developing the fingers flexoextension in rehabilitation therapies, can consider for one finger, index to little, a kinematic chain of four links and five joints with origin at the wrist [3, 7]. However, considering the hand palm is fixed, the metacarpal link (MC) of each finger, except thumb, represents the fixed basis of each kinematic chain, as shown in Fig. 1. In such case, the links are: proximal phalange (PP), medial phalange (MP) and distal phalange (DP) which is the finger end. And the joints are: metacarpophalangeal (MCP) with two d.o.f. that develops flexoextension and abduction-adduction (lateral movements for fingers separation and approximation); proximal interphalangeal (PIP) and distal interphalangeal (DIP), of one d.o.f. each for flexoextension [3, 7].

The Denavit-Hartenberg parameters of fingers (except the thumb) are presented in Table 1, where: LPP, LMP y LDP are respectively the lengths of the proximal, medial and distal phalanges. From those parameters [4, 3], the direct kinematics equation can be obtained, Eq. (1).

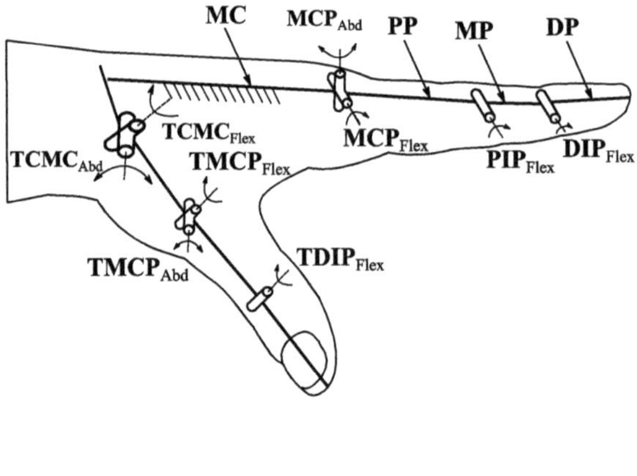

Fig. 1 Kinematic model considering fixed the bones of carpus

Table 1 DH parameters for index, middle, ring and little fingers considering metacarpals as fixed bones

Joint	θ_i	d_i	a_i	α_i
1	$\theta_{MCP\ abd}$	0	0	$\pi/2$
2	$\theta_{MCP\ flex}$	0	LPP	0
3	$\theta_{PIP\ flex}$	0	LMP	0
4	$\theta_{DIP\ flex}$	0	LDP	0

Table 2 Angles of active and passive movements of joints of fingers index to little [4]

Finger\Joint	$\Theta_{MCP\ flex}$ (°)	$\Theta_{PIP\ flex}$ (°)	$\Theta_{DIP\ flex}$ (°)
Index	90	110	80–90
Middle	90	110	80–90
Ring	90	120	80–90
Little	90	135	90

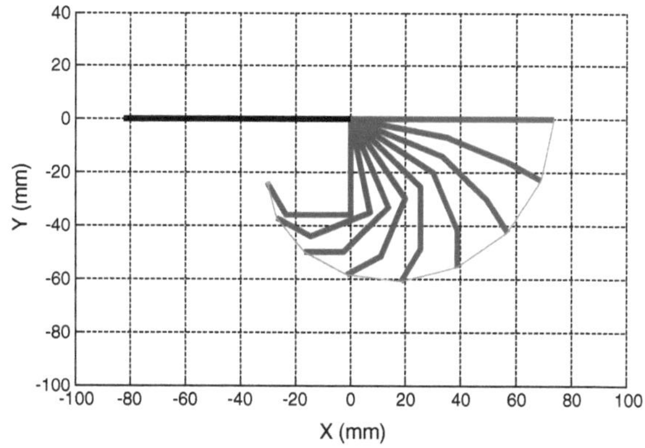

Fig. 2 Natural flexoextension finger path, MCP = PIP = 90°, DIP = 45°

$$
\begin{aligned}
Pi &= {}^{0}_{-1}T(u_i)^{0}_{4}T(\theta_j)i \\
&= {}^{-1}_{0}T(u_i)^{0}_{1}T(\theta_{MCP\,Abd})i^{1}_{2}T(\theta_{MCP\,flex})i^{2}_{3}T(\theta_{PIP\,flex})i^{3}_{4}T(\theta_{DIP\,flex})i
\end{aligned}
\quad (1)
$$

where *Pi* represents a matrix describing position and orientation of the finger tip, *T* (u_i) represents the distance between wrist and the finger reference frame (located at MCP joint), $T(\Theta_j)i$ is a matrix that represents the geometrical transformation between the *i*-finger reference frame and its corresponding fingertip. This matrix is composed by the concatenation of more simple matrixes that represents the contribution of each finger joint [3]. With: *i* = index, middle, ring and little, $j = \Theta_{MCP\ abd}$, $\Theta_{MCP\ flex}$, $\Theta_{PIP\ flex}$ y $\Theta_{DIP\ flex}$. The upper limits of passive flexion range $\Theta_{j\ flex}$ for $2 \leq j \leq 4$, are shown in Table 2. Figure 2 shows the natural flexoextensión path.

Figure 2 presents the flexoextension path of the kinematic chain of three phalanges corresponding to index with a range from 0 to 90° for MCP and PIP and 45° for DIP.

The development of natural flexoextension finger path in rehabilitation mechanism requires the immobilization of the hand palm, it could be done with an ergonomic base and adjustables belts.

3 Four Bar Mechanism for Flexion Fingers Index to Little

3.1 Mechanism Design

The development of a system to assist therapies during physical rehabilitation should consider aspects that determine its design and functionality. Safety is one of the most important factors in mobilization of patient fingers. Hyperextension should be avoided, as well as abrupt movements that cause pain or additional injury to the patient. A second aspect to consider is the effectiveness of finger mobilization; the proposed system must provoke the rotation of three joints that produces finger flexoextension. Finally, efficiency of the system to reach the range of movement for each of finger joints, the velocity and force that are required in rehabilitation exercises. Furthermore, the GUI and its operation should be friendly to encourage the therapists to use it. The coupler link of a four-bar mechanism (Fig. 3) can develop a path very close to the required in flexoextension finger movement when the angle of input link $\boldsymbol{\theta}$, is between 180 and 360° respect to $\boldsymbol{x}$ axis. Therefore, a slider-crank mechanism has been proposed. The curves of flexoextensión index finger and two, slider-crank mechanism, coupler links paths (A and B) are shown in Fig. 4.

Figure 4 shows that A and B paths have the same value in three points. Next step is to determine the length of mechanism links through methods of movement generation of three positions by graphical synthesis [8]. The resulting lengths of mechanism links are: crank length (***OA***) L_1 = 5.20 mm, minimum length of fixed link (***OB*** min) = 21.09 mm, maximum length of fixed link (***OB*** max) = 31.20 mm, coupler link length (***AB***) L_2 = 26.00 mm, coupler link length extension (***AC***) L_3 = 6.00 mm and angle of coupler link extension (respect to ***AB***) $\boldsymbol{\beta}$ = 110°. Figure 4 shows that three paths coincide at the beginning, middle and the end, the path A fits better than B to the finger path.

From mechanism simulations, a CAD modeling of the device was performed in Solidworks®. The Fig. 5 shows the mechanism to mobilize: (a) one finger and (b) four fingers. The range of motion of input link should be limited from 180 to 360 degrees to develop the natural path of finger flexoextension and to avoid hyperextension.

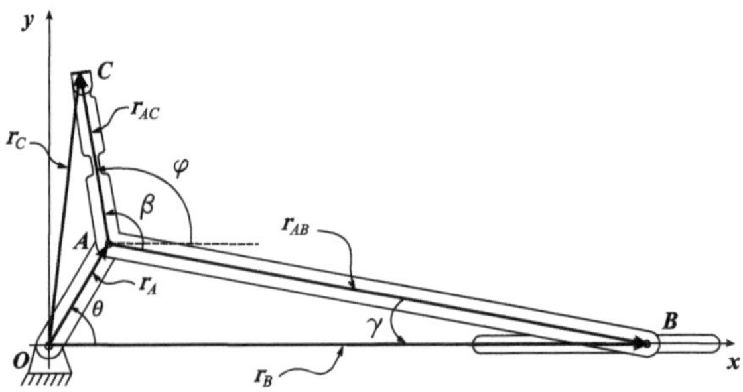

Fig. 3 Four-bar proposed mechanism type slider-crank

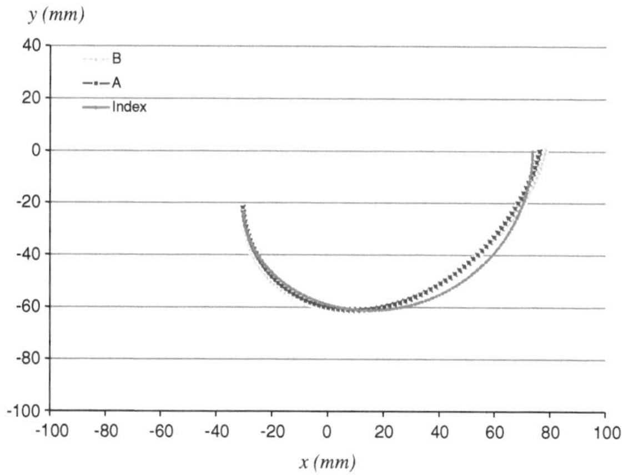

Fig. 4 Flexoextension path of three phalanges and two four-bar mechanisms (*A* and *B*)

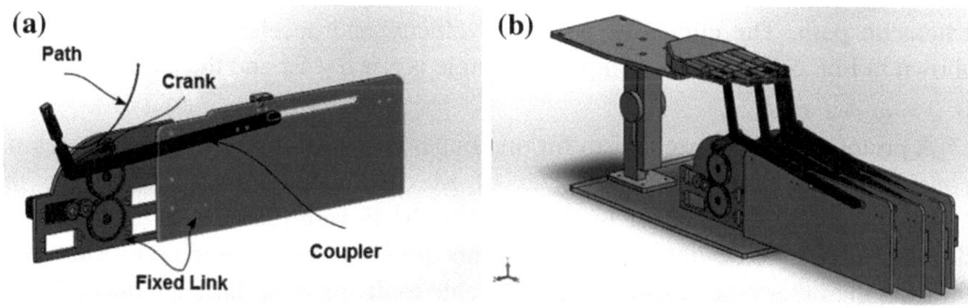

Fig. 5 Mechanism design. **a** One finger mechanism. **b** Four finger mechanism

3.2 Mechanism Kinematic Analysis

The kinematic analysis of the mechanism was preformed through vectorial analysis, it computes position (Eq. 2), velocity (Eq. 3) and acceleration (Eq. 4) of the point of union between fingertip and C point of coupler link. Figure 3 is shows the vectorial diagram of mechanism; vector $\boldsymbol{r}_A$ begins at origin $\boldsymbol{O}$ of reference frame and its ends at the joint with coupler link, point $\boldsymbol{A}$. Vector $\boldsymbol{r}_B$ has starts origin $\boldsymbol{O}$ and ends at $\boldsymbol{B}$, that moves along axis $\boldsymbol{x}$. Vector $\boldsymbol{r}_{AC}$ begins at point $\boldsymbol{A}$ and ends at coupler link extension joint with fingertip, point ***C;*** and vector $\boldsymbol{r}_C$ starts at origin $\boldsymbol{O}$ and ends at ***C***. Finally, vector $\boldsymbol{r}_{AB}$ begins at $\boldsymbol{A}$ and ends at $\boldsymbol{B}$. The coordinates of the point C are computed from the following expressions:

$$\begin{aligned} r_C &= r_A + r_{AC} = (x_A + x_{AC})\hat{i} + (y_A + y_{AC})\hat{j} \\ &= (L_1 \cos\theta + L_3 \cos\varphi)\hat{i} + (L_1 \sin\theta + L_3 \sin\varphi)\hat{j} \end{aligned} \quad (2)$$

$$\begin{aligned} v_C = \dot{r}_C = \dot{r}_A + \dot{r}_{AC} = \dot{x}_C \hat{i} + \dot{x}_{AC} \hat{j} \\ = -\,(L_1 \dot{\theta} \sin\theta + L_3 \dot{\varphi} \sin\varphi)\hat{i} + (L_1 \dot{\theta} \cos\theta + L_3 \dot{\varphi} \cos\varphi)\hat{j} \end{aligned} \quad (3)$$

$$a_C = \ddot{r}_C = \ddot{r}_A + \ddot{r}_{AC} = \ddot{x}_C \hat{i} + \ddot{x}_{AC} \hat{j} \quad (4)$$

$$\begin{aligned} = (-L_1 \ddot{\theta} \sin\theta - L_1 \dot{\theta}^2 \cos\theta - L_3 \ddot{\varphi} \sin\varphi - L_3 \dot{\varphi}^2 \cos\varphi)\hat{i} \\ +(L_1 \ddot{\theta} \cos\theta - L_1 \dot{\theta}^2 \sin\theta + L_3 \ddot{\varphi} \cos\varphi - L_3 \dot{\varphi}^2 \sin\varphi)\hat{j} \end{aligned}$$

where: L_1 is the length of the segment ***OA***, L_3 is the length of the segment ***AC***, $\dot{\theta}$ and $\ddot{\theta}$ are the angular velocity and angular acceleration respectively.

4 Results

The proposed mechanism allows to guide the fingertip through its natural flexo-extension path. The curves of position, velocity and acceleration of *C* point are shown in Fig. 6. The range of input link angle is $\pi < \theta < 2\pi$ and the angular velocity is $\dot{\theta} = \omega = 1$ rad/s.

A prototype of the mechanism for one finger is shown in Fig. 7, it has been built considering the anthropometric sizes of Mexican population [2]. Because the kinematic chain of the finger is redundant in the flexion plane, a larger finger can develop totally the desired path but its joints are going to develop a range less than 90°. On the other case, a shorter finger is able to develop partially the desired path, up an extension limit calculated regarding its length. The mechanism includes a

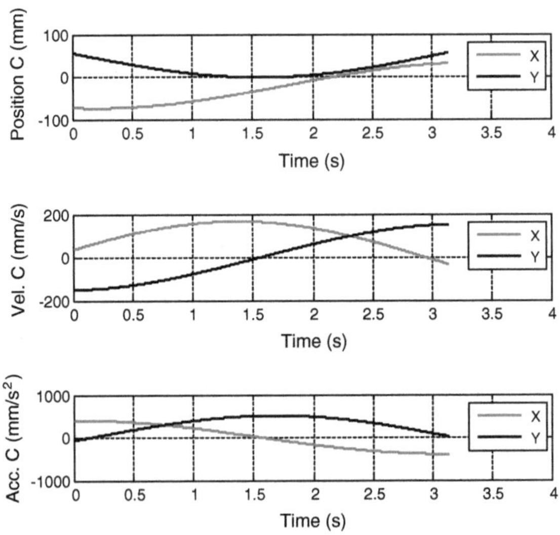

Fig. 6 Results of position, velocity and acceleration of *C* point

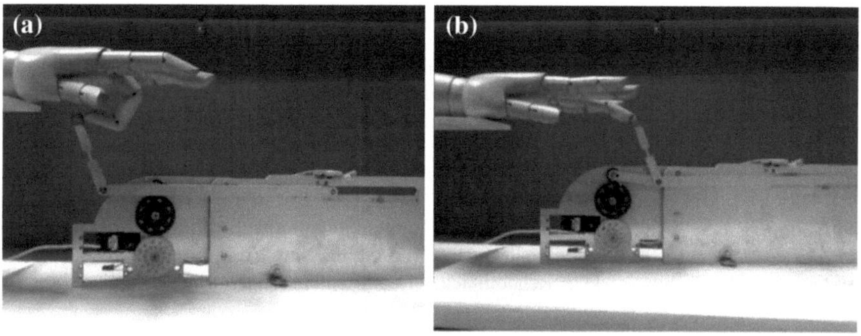

Fig. 7 Mechanism prototype for one finger. **a** Flexion. **b** Extension

position sensor (encoder) in order to know the input link angle θ. A contactless sensor to detect the end of natural finger motion was included as protection against finger hyperextension.

5 Conclusions

The mobilization of joints of different limbs in rehabilitation therapies can be assisted by mechatronic systems. The advantages of external force supply simplifies the therapist work and allow the monitoring and registering of movements in electronic systems for subsequent analysis. Two aspects are relevant to consider: firstly, the mechatronic systems cannot substitute the human therapists in none case, moreover, this kind of systems should be used under strict professional supervision. Secondly, the operation safety of mechatronic systems should be warranted through sensors or mechanic limits to avoid joint mobilization out of natural motion range of limbs.

Acknowledgments This work is financially supported by IPN. The authors also like to acknowledge the help of CRIQ therapists for their advice and discussion about rehabilitation protocols.

References

1. Arcas, M., Galvez, D., Leon, C., Paniagua, S., Pellicer, M.: Manual de Fisioterapia. Modulo i Ebook, MAD-Eduforma (2004)
2. Ávila, R., Prado, L., González, E.: Dimensiones antropométricas de población latinoamericana. Centro de Investigaciones en Ergonomia (2001)
3. Cobos, S., Ferre, M., Uran, S.: Efficient human hand kinematics for manipulation tasks. In: International Conference on Intelligent Robots and Systems, pp. 22–26 (2008)

4. Craig, J.J.: Introduction to Robotics: Mechanics and Control, 3rd edn. Addison-Wesley Longman Publishing Co. Inc., Boston (2004)
5. Kapandji, A.I.: Fisiología articular: esquemas comentados de mecánica humana, 6th edn., p. 349. Editorial Médica Panamericana, Madrid (2006)
6. Kuch, J.J., Huang, T.S.: Human Computer Interaction Via the Human Hand: A Hand Model (1995)
7. Magnenat-Thalmann, N., Zhang, J.J., Feng, D.D. (eds.): Recent Advances in the 3D Physiological Human, pp. 53–71. Springer, London. doi:10.1007/978-1-84882-565-9
8. Norton, R.L.: Design of Machinery: An Introduction to the Synthesis and Analysis of Mechanisms and Machines, 5th edn, p. 167 M. G. Hill, New York (2012)
9. Quesnot, A, Chanussot, J.C.: Rehabilitacion Del Miembro Superior. Médica Panamericana (2010)
10. Tondu, B.: Kinematic modelling of anthropomorphic robot upper limb with human-like hands. In: International Conference on Advanced Robotics, 2009. ICAR 2009
11. Yang, X., Park, J., Jung, K., You, H.: Development and Evaluation of a 25-Degree of Freedom Hand Kinematic Model (2008)

Mechanical Design of a Craniotomy Robotic Manipulator Based on Optimal Kinematic and Force Performance

T. Essomba, C.-T. Wu, S.-T. Lee and C.-H. Kuo

Abstract The present paper introduces the mechanical design of a robotic manipulator for craniotomy application considering its kinematic and force transmission ability. The craniotomy is a surgical technique that allows the surgeon to access the patient's brain by drilling and cutting the skull. The technical requirements of this application are explained. A dedicated mechanical architecture is proposed. This serial spherical arm is analysed in terms of kinematic and force transmission. An optimal mechanism is selected, designed and its actuator are selected to with respect to the identified performances. The global design of the manipulator is analysed and adjusted to reduce the structural deformation.

Keywords Craniotomy · Medical robot · Serial spherical arm · Kinematics · Force analysis

1 Introduction

In traditional neurosurgery, the surgeon needs to operate directly the patient's brain. However, the brain is protected by the solidest bone of the human body, i.e., the skull. In order to provide a physical access the brain, the craniotomy is a surgical

T. Essomba (✉) · C.-T. Wu
Medical Augmented Reality Research Center, Chang Gung Memorial Hospital, Taoyuan, Taiwan
e-mail: Terence.essomba@gmail.com

S.-T. Lee
Department of Neurosurgery, Chang Gung Memorial Hospital, Taoyuan, Taiwan
e-mail: leepei@adm.cgmh.org.tw

C.-H. Kuo
Department of Mechanical Engineering, National Taiwan University of Science and Technology, Taipei, Taiwan
e-mail: chkuo717@mail.ntust.edu.tw

S. Zeghloul et al. (eds.), *Robotics and Mechatronics*,
Mechanisms and Machine Science 37, DOI 10.1007/978-3-319-22368-1_19

operation for temporarily removing a section of the skull. The surgeon must first use a semi-automatic surgical tool to drill several holes called "burr hole" on the skull [1], then he/she use another one to cut and remove an entire section by linking one burr hole to another until a complete piece can be removed, namely "bone flap," from the skull. Once the bone flap is removed, the surgeon can access the brain to perform the brain surgery. The intervention requires an experienced physicians who can provide a high degree of dexterity. But at the same time, the surgeon can rely on robotic technology that provides better stability and suppresses the surgeon's physiological factor such as vibration, fatigue or errors [2, 3].

In 1998, the University of Karlsruhe in Germany, used a PUMA 260 industrial robot to perform a craniotomy procedure [4]. The same research team in collaboration with the University of Heidelberg in 2003 has also raised the RobaCka robot, and successfully conducted human experiments [5] in order to assist physicians for premature malformationns healing. After that, the University of Karlsruhe used another industrial robot (KUKA LWR) and discarded the traditional craniotomy tool by using the method based on CO2 laser craniotomy research [6]. In 2009, Lunghwa University of Science and Technology in Taiwan [7] proposed a method to guide robotic arm using a laser scanning system and to detect the thickness of the skull under the robot trajectory with computer tomography. This method allows guiding of an industrial six-axis robotic arm (Fanuc LR Mate 200iB) for craniotomy purposes.

The above review reflects that robotic systems proposed so far are based on the use of anthropomorphic industrial robots equipped with adapted surgical tools. In 2014, the National Taiwan University of Science and Technology (NTUST) proposed an approach based on a mechanical architecture especially dedicated to the craniotomy requirement [8].

The present paper introduces the design of a new robotic manipulator for craniotomy application. Section 2 is an analysis of the craniotomy requirement including data measurement experimentations. A dedicated mechanical architecture is proposed and its kinematic and force model are defined in Sect. 3. To select appropriate actuation, the optimized mechanism is analyzed in terms of performance in Sect. 4. Section 5 provides the conclusion.

2 Force and Kinematic Requirements

Discussions with involved neurosurgeon, expert in craniotomy, has highlighted several specifications for the robotic system. These have been confirmed by visual observation of real craniotomy in surgical room and described by numerical data through experimentation. The specifications are stated as below:

- The craniotomy task requires 2 rotational DoF and 1 linear DoF aligned with the end effector longitudinal axis;

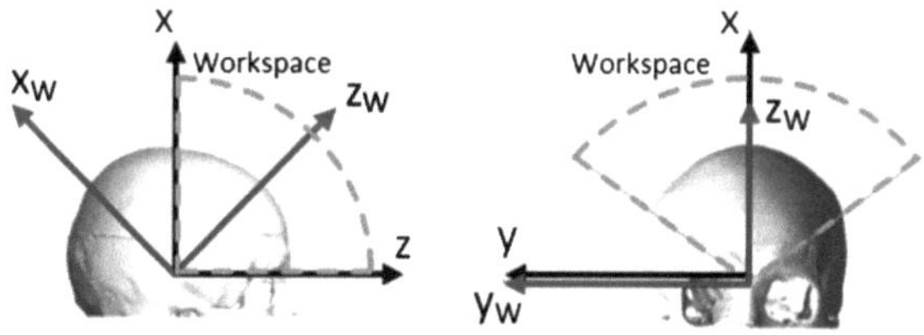

Fig. 1 Rotational workspace of the craniotomy process

- The drill axis has to be perpendicular to the skull surface which is assumed to be a perfect sphere;
- The workspace has been assessed to a cone of 45° apex angle based on the experience of clinic need;
- The center axis of this spherical wrist z_W aligned with the patient's skull sagittal plan (xz) and oriented from 45° from the longitudinal axis (x).

The workspace can be limited to this size because the surgeon always orientates the patient's head so that the operated zone is directed to the top.The resulting workspace is illustrated in Fig. 1.

Experimentations have been performed to measure the force and the motion of the cutting instrument for craniotomy. A regular surgical cutter has been instrumented with a force sensor an optical tracker. Pork bones were used for these experimentation because this animal's tissues are the closest to human's ones. As result, it has been interpreted that this force does not exceed 80 N. The maximum linear velocity is limited to 0.02 m/s. Regarding its low value (0.0001 m/s^2), the linear acceleration will not be considered in following research. The results listed above are used as input data for the optimization of the mechanical architecture and the mechatronic design of the robotic manipulator.

3 Mechanical Architecture Optimization

The literature study reveals that the novelty of these research is usually based on a very accurate control to provide the safety of the patient's brain. The hybrid decoupled Remote Center of Motion (RCM) mechanism of NTUST provides high kinematic performances and very simply control. However, the mechanism was not optimized in term of force transmission at this point of the project.

3.1 Mechanism Definition and Kinematic Model

Regarding the requirements described in Sect. 2, a RRP serial spherical mechanism has been selected. This architecture is famous and widely used for medical application that requires RCM [9]. It is a lot more compact than anthropomorphic robots

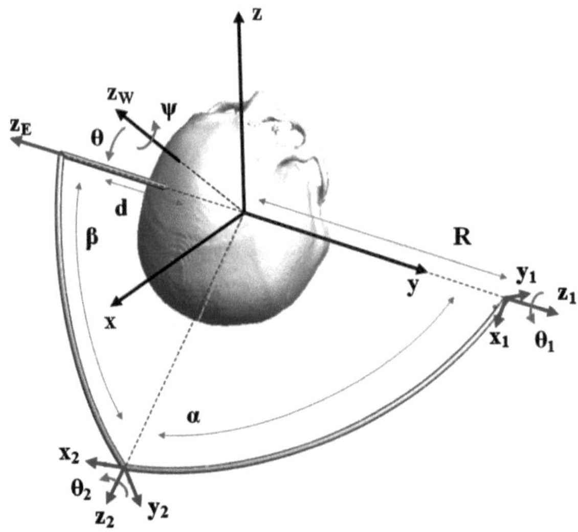

Fig. 2 RRP serial spherical architecture for craniotomy

and its simplicity reduces the number of design variables, which makes its analysis and optimization very simple. Figure 2 illustrates such a mechanism adapted for craniotomy application.

The base of the robot is located to the right side of the patient's skull for practical reasons: the robot is supposed to be attached to a rail on the side of the clinical bed. All axis z_1, z_2 and z_E are aligned with the RCM. Angles α and β are the design variables. They resepectively give the size of the Circular Segment 1 (CS1) and the Circular Segment 2 (CS2). The actuator variables θ_1 and θ_2 respectively orientate the CS1 and CS2. The linear DoF is directed by the axis z_E and is actuated by the actuator variable *d*. The global size of the mechanism is also given by the radius *R* of the CS1 and CS2. It is fixed to 300 mm as a compromise between compactness and free space between the patient and the manipulator.

Using all these variables, the forward kinematic model is written as followed:

$$\begin{aligned} &\theta = \pm\cos^{-1}(C\theta_1 S\alpha C\beta + S\beta(S\theta_1 S\theta_2 - C\theta_1 C\theta_2 C\alpha)) \\ &\psi = ATAN2(-(S\theta_1 S\alpha C\beta - S\beta(C\theta_1 S\theta_2 + S\theta_1 C\theta_2 C\alpha))/S\theta, (C\theta_2 S\alpha S\beta - C\alpha C\beta)/S\theta) \end{aligned} \tag{1}$$

With $C\alpha = \cos\alpha$ and $S\alpha = \sin\alpha$. The inverse kinematic model is determined by the following equations:

$$\begin{aligned} \theta_1 &= \pm\cos^{-1}\left(C\Big/\left(\sqrt{A^2+B^2}\right)\right) + ATAN2(B, A) \\ \theta_2 &= \pm\cos^{-1}((C\alpha C\beta - C\psi S\theta)/CS\beta) \end{aligned} \tag{2}$$

With $A = S\theta_2 S\beta$, $B = S\theta_2 S\beta + C\alpha S\beta C\theta_2$ and $C = S\psi S\theta$.

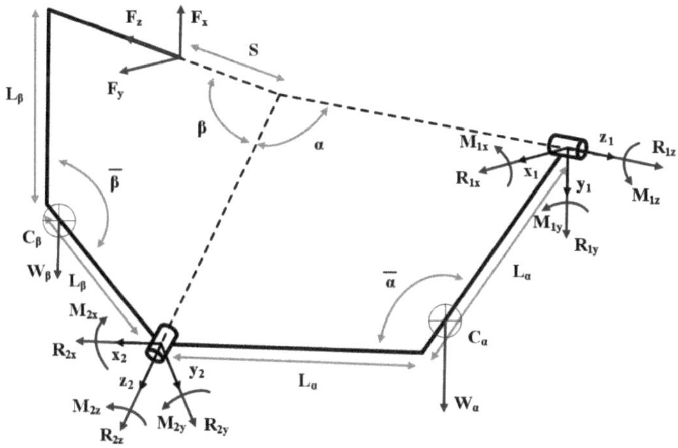

Fig. 3 Force and momentum interaction in the mechanism

To study the force transmission the mechanism, all external forces and momentum applied on each of them are shown in Fig. 3. The objective is to determine the momentum M_{1z} and $M_{2z,}$ which have to be supply by the two actuator torques.

3.2 *Force and Kinematic Optimization*

The objective is to guarantee that the robotic manipulator has enough force to provide drilling and cutting task while maintaining good kinematic performance. The actuators will have to provide enough torque to supply the cutting force that will counter balance the skull reaction force (measured through experiments). Also, the value of this minimum necessary torque has to be as low as possible to reduce the size and weight of mechanical equipment. Based on a genetic algorithm, a series of optimal RRP mechanism have been identified and compared through a Pareto front. Their design variables generate a compromise between force and kinematic indexes.

The serial spherical RR arm which design variables are: $\alpha = 90°$ and $\beta = 50°$ has been selected. This choice is motivated by the kinematic performance generated while maintaining the actuator torque at a low level: 23.76 Nm for the actuator of CS1 and 8.25 Nm for the CS2.

4 Mechatronic Design of the Robotic Manipulator

One optimal RRP mechanism has been selected. However, the weight of the mechanism is not considered at this step the design. The next step is to use all design variable to generate a CAD model of the robotic manipulator. The CAD model is shown on Fig. 4.

The software Solidworks allows providing a very accurate estimation of the structure weight. These new data and the direction of the gravity in all configurations are used as input to compute the two actuator torques in order to consider the weight of the structure. As result, they have computed at 40.36 Nm for the first actuator and 13.83 Nm for the second one. These data will allow selected actuators and reducers to be integrated in the robotic manipulator.

The objective is to select technological solutions to motorize the robotic manipulator to providing the required performances. Another specification of this aspect is the velocity of the manipulator. Experimentation has reveal that the maximum linear velocity is 0.02 m/s. To convert this data into angular velocity for the motorization, the dimension of the CS1 and CS2 are used. For each actuator, the maximum rotational velocity is reached when the linear speed is tangential to the rotation motion as illustrated in Fig. 5.

Using the rotation motion relationship, the angular velocity is estimated at 1.47 rpm for the first actuator and 1.91 rpm for the second one. For the third actuator of linear DoF, it has been estimated that the linear velocity shall be around 10 mm/s and its torque at least 0.076 Nm. The two first gearhead actuators are coupled with Harmonic Drive-type gearheads (HD) to increase the torque and to reduce the velocity. They been selected for their high performance and their light weight. The Table 1 gives all equipment selected to actuate the each joint of the manipulator and their respective generated performances. Torques are given in Newton and speeds in round per minute, except for the third joint which is given in millimeter per second.

Since the manipulator will be subject to high force interaction, its structure has been studied in terms of deformations. Therefore, the design has been adjusted to

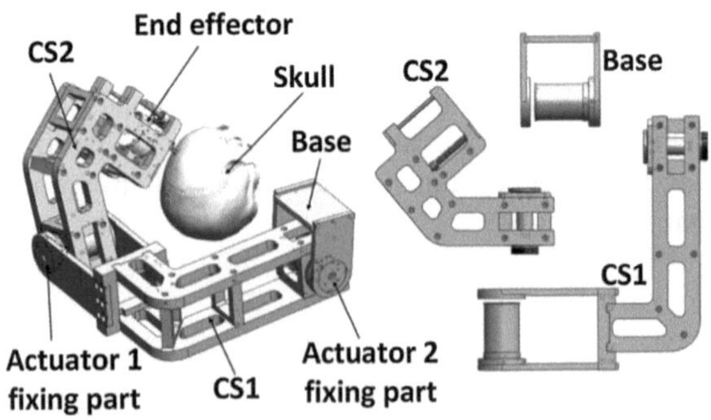

Fig. 4 CAD model of the craniotomy robotic manipulator

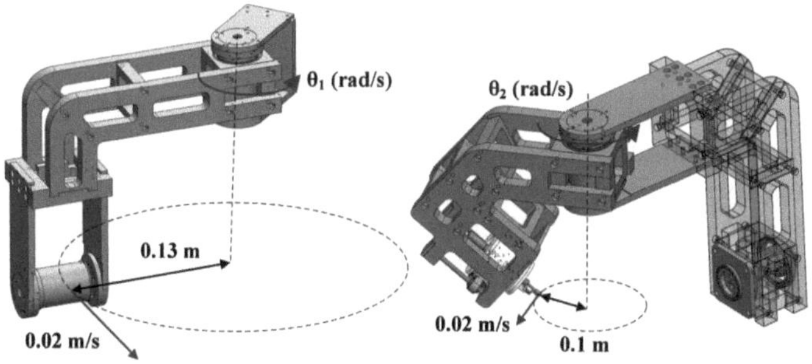

Fig. 5 Estimation of the actuators rotational velocities

Table 1 Actuators integration of the robotic manipulator

Joint	Equipment			Acutal performance		Required performance	
	Motor	Gearhead	HD	Speed	Torque	Speed	Torque
#1	2657-012CXR	26A-16	CSG-17-120	2.65	48.4	1.47	40.36
#2	2342-012CR	17/1-20	CSG-17-120	2.53	25.9	1.91	13.83
#3	1741-012SR	16/7-14	None	15.7	0.094	10	0.076

reduce these deformations. Simulations have been provided for each individual parts. For this study, the assumed skull reaction force has been intentionally increased to 150 N to anticipate potential malutilizations. The Fig. 6 illustrates the maximum deformation computed in these extreme conditions. In such a case, the deformation does not exceed 0.35 mm.

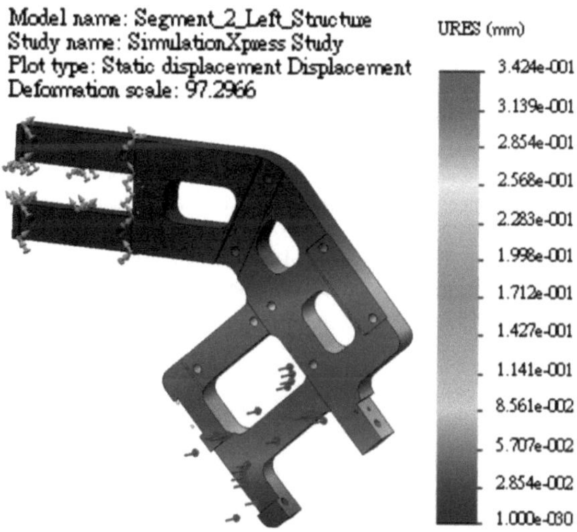

Fig. 6 Estimation of the actuators rotational velocities

5 Conclusions

A RRP serial spherical arm is proposed for a craniotomy robotic system. This mechanical architecture is very simple but adapted for this application, regarding the analyzed user requirements. A mechanism has been optimized in terms of kinematic and force transmission performances to guarantee that the robot generates enough force to drill and cut the skull. The required performance identified through the optimization are used as input data for the selection of manipulator actuator technology. The global design of the structure is refined to reduce the deformation as much as possible.

Acknowledgments This work is financially supported by Chang Gung Memorial Hospital, Taiwan under grant No. CMRPD2C0051-53.

References

1. Hsiao, M.-H., Kuo, C.-H.: A review to the powered drilling devices for craniotomy. In: Design of Medical Devices Conference, Minneapolis, MN, USA, 10–12 Apr 2012
2. Kuo, C.-H., Dai, J.-S., Dasgupta, P.: Kinematic design considerations for minimally invasive surgical robots: an overview. Int. J. Med. Robot. Comput. Assist. Surg. **8**(2), 127–145 (2012)
3. Taylor, R.H., Stoianovici, D.: Medical Robotics in Computer-Integrated Surgery. IEEE Trans. Robot. Autom. **19**(5), 765–781 (2003)
4. Burghart, C., Raczkowsky, J., Rembold, U., Wörn, H.: Robot cell for craniofacial surgery. In: Proceedings of the 24th Annual Conference of the IEEE Industrial Electronics Society, Aachen, Germany, pp. 2506–2511, 31 Aug–4 Sept 1998
5. Korb, W., Engel, D., Boesecke, R., Eggers, G., Kotrikova, B., Marmulla, R., Raczkowsky, J., Wörn, H., Mühling, J., Hassfeld, S.: Development and first patient trial of a surgical robot for complex trajectory milling. Comput. Aided Surg. **8**(5), 247–256 (2003)
6. Mönnich, H., Stein, D., Raczkowsky, J., Wörn, H.: System for laser osteotomy in surgery with the Kuka lightweight robot—first experimental results. In: World Congress on Medical Physics and Biomedical Engineering, Munich, Germany, pp. 179–182, 7–12 Sept 2009
7. Yuan, C.-C.: The study of robot-assisted craniotomy (in Chinese). Master thesis, Graduate School of Engineering and Technology, Lunghwa University of Science and Technology, Taiwan (2009)
8. Kuo, C.-H., Essomba, T., Li, G.-K., Wu, C.-T., Lee, S.-T.: Kinematic Optimization of a Novel Remote Center-of-Motion Mechanism for Robotic Craniotomy. In: The 3rd IFToMM Asian Conference on Mechanism and Machine Science (Asian MMS), 9–10 July, Tianjin, China (2014)
9. Essomba, T., Nouaille, L., Laribi, M.A., Nelson, C.A., Zeghloul, S., Poisson, G.: Spherical wrist dimensional synthesis adapted for tool-guidance medical robots. Mech. Ind. **15**(3), 217–223 (2014)

Dynamic Simulation of a Cable-Based Gait Training Machine

H. Lamine, S. Bennour and L. Romdhane

Abstract In this paper, we present a dynamic analysis of a new type of gait training machine. This latter is based on a conventional body weight support system and a cable driven parallel robot in order to move the lower limbs. This gait training machine can be used to restore walking of neurologically injured patients. The aim of this study is to model and simulate dynamically this machine during a walk cycle. The obtained results will mainly be used in order to size the different actuators of the cable robot.

Keywords Gait rehabilitation · Cable driven robots · Body weight support system · Dynamics

1 Introduction

Restoration of walk motor functions for neurologically injured patients can be achieved through rehabilitation exercises. Conventionally, the training method is based on the Partial Body Weight-supported Treadmill Training System. Indeed, the patient is suspended with a Body Weight Support System (BWS), this latter helps the patient to unload a part of his own weight and thus reducing ground reaction acting on the lower limbs while a therapist manually assists the patient legs

H. Lamine (✉) · S. Bennour · L. Romdhane
Laboratoire de Mécanique de Sousse, Ecole Nationale d'Ingénieurs de Sousse, Université de Sousse, Tunisie, UAE
e-mail: houssein.lamine@gmail.com

S. Bennour
e-mail: sami.bennour@gmail.com

L. Romdhane
e-mail: lotfi.romdhane@gmail.com

L. Romdhane
Mechanical Engineering Department, American University of Sharjah, Sharjah, UAE

S. Zeghloul et al. (eds.), *Robotics and Mechatronics*,
Mechanisms and Machine Science 37, DOI 10.1007/978-3-319-22368-1_20

movement. Therefore training sessions are limited by the performance of the therapists. Motorized training machines overcome this limitation and offer a long duration of training and should lead to a more effective gait recovery [1–3].

These automated machines guarantee a passive guided trajectories of the lower limb. Mainly two methods are used [4], the first one is a basic serial robot attached to the lower limb (e.g., the Lokomat system [5]) and the second is based on programmable foot-plates attached to the down side of the feet, in order to move the lower limbs (e.g., the Lokoiran System [6]). In this work we propose a new design based on a cable driven robot, an orthosis is attached to the leg, which is controlled by a cable system to move the lower limb in the sagittal plane (see Fig. 1). This design has some interesting advantages mainly high dynamic performance, less intrusion and low cost…

Our objective is to present a dynamic modeling of the proposed system and then to determine the reaction forces between the leg and the cable robot's effector.

The paper is organized as follows. In Sect. 2, an overview of the rehabilitation machine and the normal human gait is presented. A dynamic modeling and a simulation are investigated in Sect. 3. Finally, Sect. 4 presents a tension analysis of the cable robot.

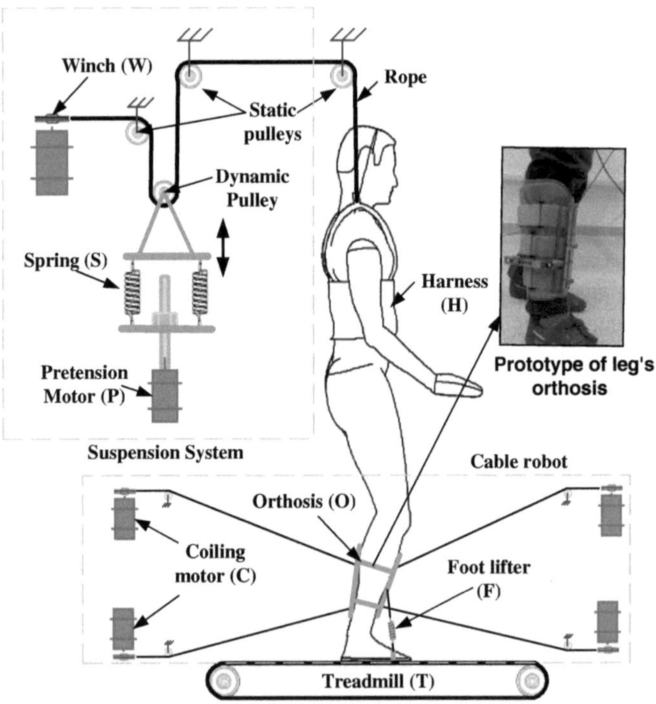

Fig. 1 Mechanical description of the proposed gait training machine

2 Rehabilitation Machine and Human Gait

2.1 Description of the Gait Rehabilitation Machine

The machine shown in Fig. 1, is based on a passive elastic dynamic suspension system [7], a cable robot and a treadmill (T). The patient is wearing a Harness (H), which is linked to the unloading system by a rope and a series of pulleys, the main part of the suspension system is based on a translational system attached to two springs (S). The electric winch (W) helps in lifting the patient from a sitting to a standing position and allows springs to work in a specific range, these latters apply an unloading force on the patient through the moving pulley and the rope. A motor (P) coupled to a screw system serves to adjust the springs pretension and hence the percentage of the unloaded weight.

For the cable robot, an orthosis (O) is placed around the leg and a foot lifter (F) is positioned between the orthosis and the foot to prevent this latter from falling down (similar to the Lokomat system [5]) and a cable set moves the lower limb in the sagittal plane with programmed walk trajectories.

The walk advance movement and the ground reaction are ensured by a treadmill (T).

2.2 Kinematics of a Normal Human Gait

A gait cycle (see Fig. 2) is defined as a sequence of repetitive events. A normal walk cycle is mainly split in two phases: Stance phase and Swing phase. The stance phase is defined as the period when the foot is in contact with the ground and the

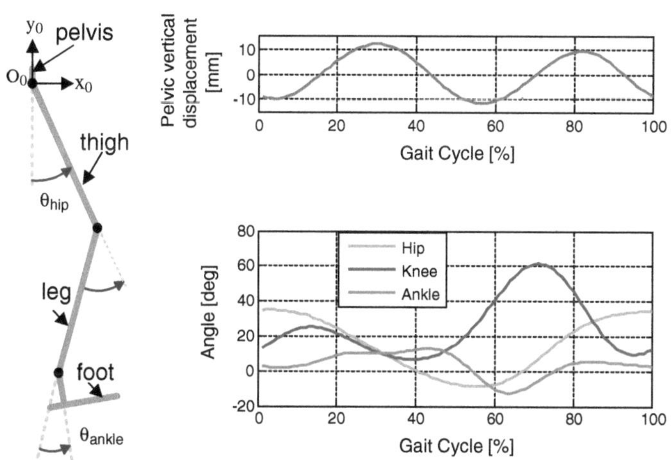

Fig. 2 Lower limb walk kinematics [9]

Swing phase is identified by the time when the foot is flying. The stance phase has two features: single limb stance, if one foot is in contact with the ground, and double support stance when the body weight load is shared between the two limbs. One shall note that, in normal gait, there is a symmetry between the trajectories of the two limbs [8].

The movements of the lower extremity are flexion-extension of the hip, the knee, and the ankle, and the vertical displacement of the pelvis. During a gait cycle, the kinematic data of the lower limb are given in Fig. 2:

3 Dynamic Simulation

3.1 Dynamic Modeling

As mentioned before, lower limb movements are studied in the sagittal plane and the following assumptions are made:

- lower extremity joints are passive and they are free to rotate,
- leg's orthosis is solidary to the leg,
- the upper body has a pure vertical displacement,
- even when the ankle joint is not actuated, the same general trajectory is observed [10].

Dynamic analysis, presented in this study, is based on Newton-Euler equations and free-body diagram formulation. Figure 3 shows the set of loads acting on the lower limb, external loads are: spring upward force F_{spring}, ground reaction R_{ground}, cable robot resultant force and moment (F_{cdr} and M_{cdr}). The joints' transmitted torques are zero since they are passive.

The dynamic simulation data [8] are summarized in Table 1. All dimensions are measured when joint angles are zero, relative to the frame R_0 (see Fig. 2), which is centered at the hip joint with a zero pelvic displacement. The inertia data are computed relative to local frames centered in the centers of mass (CoMs), (see Fig. 3).

3.2 External Wrench Determination

To compute the external wrench applied by the robot cables, we proceed based on a recursive method. First, we write the equilibrium of the upper body, the thigh and the foot and then we solve the equilibrium of the leg.

Moreover, some input data are calculated based on a normalized ground reaction (relative to body weight) measured during a normal gait cycle [9], (see Fig. 4a).

The dynamic equations are as follows:

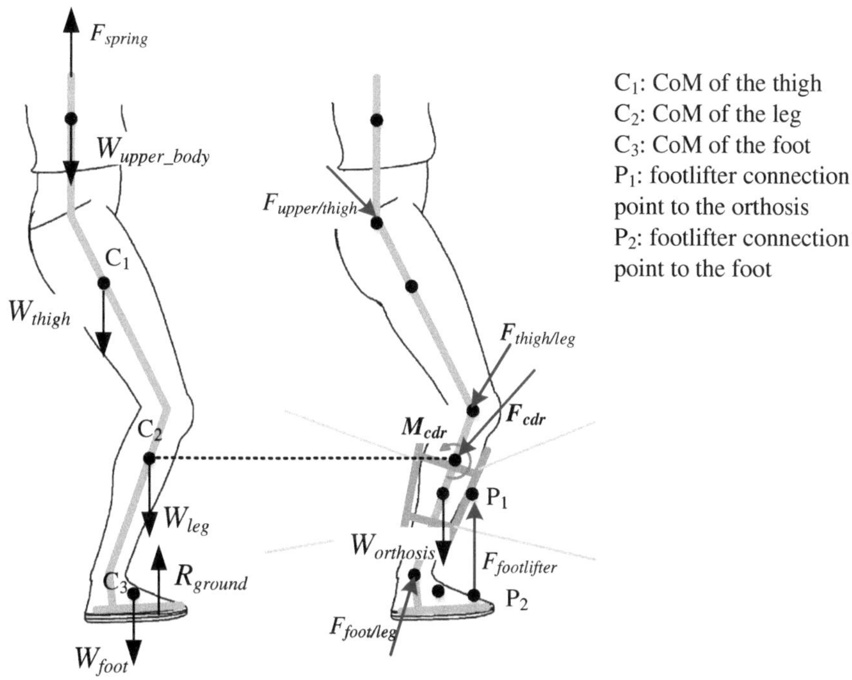

Fig. 3 Loads acting on the lower limb

Table 1 Dynamic simulation data

Mass of the upper body	Thigh (mass; length)	Leg (mass; length)	Mass of foot	Mass of the orthosis	BWS ratio
66.6 kg	10.5 kg; 0.433 m	4.75 kg; 0.435 m	1.45 kg	1.2 kg	50 %

I_{thigh}	I_{leg}	I_{foot}	$I_{orthosis}$	Springs constant	Gait cycle duration
0.1666 kg.m^2	0.0808 kg.m^2	0.0072 kg.m^2	0.0153 kg.m^2	4 kN/m	1.4 s

g	C_1[m × m]	C_2[m × m]	C_3[m × m]	P_1[m × m]	P_2[m × m]
9.81 m/s^2	(0; −0.1878)	(0; −0.6215)	(0.0626; −0.9025)	(0.065; −0.658)	(0.134; −0.928)

- For the upper body, the dynamic equilibrium equation is:

$$F_{spring} + (1/\alpha).F_{thigh/upper} + W_{upper} = m_{upper\_body}.\,a_{spine} \mid Fy_{spring} = K(x_0 + \Delta x)$$
$$x_0 = (BWS.g.m_{body})/K \quad (1)$$

α is a ratio that defines the participation of a limb to support the body weight: α is equal to 1 during single limb stance, zero during swing phase, and during double support time it varies from 0 to 1. This coefficient is calculated based on the

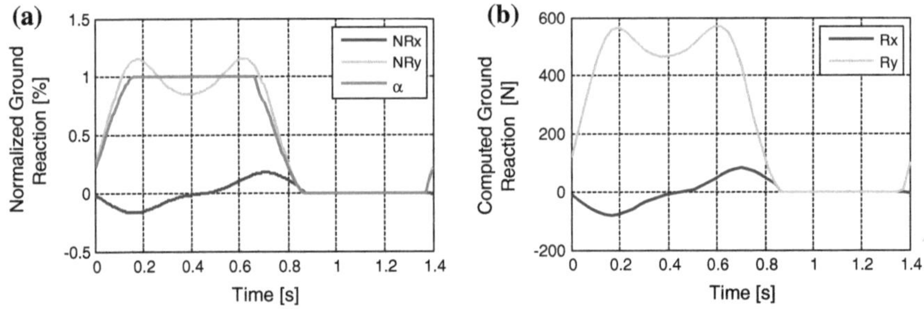

Fig. 4 **a** Normalized ground reaction for one limb. **b** Computed ground reaction [9]

normalized ground reaction component NRy of both feet: $\alpha = $ (NRyright +NRyleft)/NRyright(see Fig. 4a).

The initial spring length x_0 is calculated based on the amount of unloading *BWS*. The output from Eq. (1) is the force $F_{thigh/upper}$.

- For the thigh, the dynamic equilibrium equations are:

$$\begin{aligned} F_{upper/thigh} + F_{leg/thigh} + W_{thigh} &= m_{thigh}.\, a_{thigh} \\ M_{F_{upper/thigh}} + M_{F_{leg/thigh}} + M_{W_{thigh}} &= I_{thigh}.\, \dot{\omega}_{thigh} \end{aligned} \quad (2)$$

The output from Eq. (2) is the force $F_{leg/thigh}$.

- For the foot, we have following equations:

$$\begin{aligned} F_{leg/foot} + F_{footlifter} + R_{ground} + W_{foot} &= m_{foot}.\, a_{foot} \\ M_{F_{leg/foot}} + M_{F_{footlifter}} + M_{R_{ground}} + M_{W_{foot}} &= I_{foot}.\, \dot{\omega}_{foot} \end{aligned} \quad (3)$$

The ground reaction force is obtained from the normalized ground reaction multiplied by the actualized body weight: $m_{body} \times g - F_{spring}$ (see Fig. 4b), i.e. the amount of unloading is subtracted from the body weight. Outputs from Eq. (3) are $F_{leg/foot}$ and $F_{footlifter}$. One shall note that $M_{R_{ground}}$ is neglected since ankle joint is assumed to be passive, also we consider that the sum of $M_{F_{footlifter}}$ and $M_{W_{foot}}$ is null in Eq. (3), therefore $M_{F_{leg/foot}}$ reflects the inertia effect of the foot on the leg.

- And finally for the leg:

$$\begin{aligned} F_{thigh/leg} + F_{foot/leg} + W_{leg} - F_{footlifter} + F_{cdr} &= (m_{leg} + m_{orthosis}).\, a_{leg} \\ M_{F_{thigh/leg}} + M_{F_{foot/leg}} + M_{W_{leg}} + M_{-F_{footlifter}} + M_{cdr} &= (I_{leg} + I_{orthosis}).\, \dot{\omega}_{leg} \end{aligned} \quad (4)$$

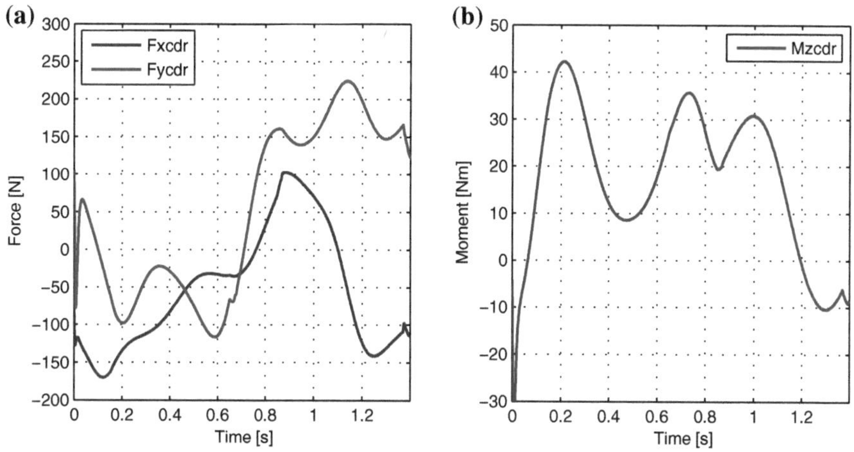

Fig. 5 The external wrench applied by the cables on the leg. **a** Forces. **b** Moment

It is worth mentioning that for Eqs. (3) and (4) the foot lifter force is only active during swing phase since during stance phase the foot is balanced by the ground reaction, moreover this force is computed based on moment equilibrium with foot's weight.

Once the required data ($F_{leg/thigh}$, $F_{leg/foot}$ and $F_{footlifter}$) are obtained from Eqs. (1–3), a simple evaluation of Eq. (4) allows us to calculate the external wrench F_{cdr} and M_{cdr} curves (see Fig. 5).

4 Tension Distribution Analysis

The cable robot is controlled by 4 cables as shown is Fig. 6, location of A and B points are given in Table 2, respectively, with reference to frames R_0 and R_2 (see Fig. 6).

Cable tensions (see Fig. 7) are obtained by solving the feasibility of the equilibrium system of Eq. [11], this is obtained based on a linear programming minimization procedure:

$$\min\left(\sum_{i}^{4} t_i\right) \quad \text{subject to} \quad \text{A.T} = \text{B} \;\; \text{and} \;\; t_i \geq 0 \tag{5}$$

where A is a 3 by 4 structure matrix, T is a 4 dimensional vector that represents the cable tensions and B is a 4 dimensional vector of external wrench (see Fig. 5).

For almost all the gait cycle, cables 2 and 4 are the most active ones, this can be explained by the fact that these two cables can only apply a positive torque and the value of M_{cdr} is positive (see Fig. 5b). Also during the swing phase (from 8.6 to 1.4 s), cables 1 and 2 are more active then cables 3 and 4 since the required external

Fig. 6 The cable robot specifications

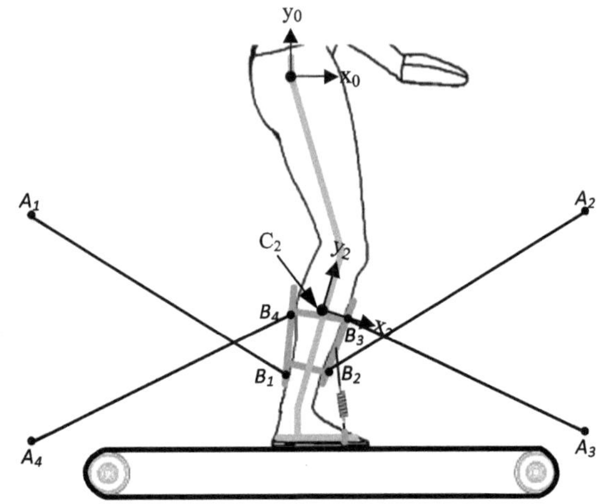

Table 2 Cable driven robot attachment points (x[m]; y [m])

A_1	A_2	A_3	A_4
(−1; −0.2)	(1; −0.2)	(1; −0.9)	(−1; −0.9)
B_1	B_2	B_3	B_4
(−0.06; −0.1359)	(0.06; -0.1359)	(0.07; 0.064)	(−0.07; 0.064)

Fig. 7 The cable tension curves

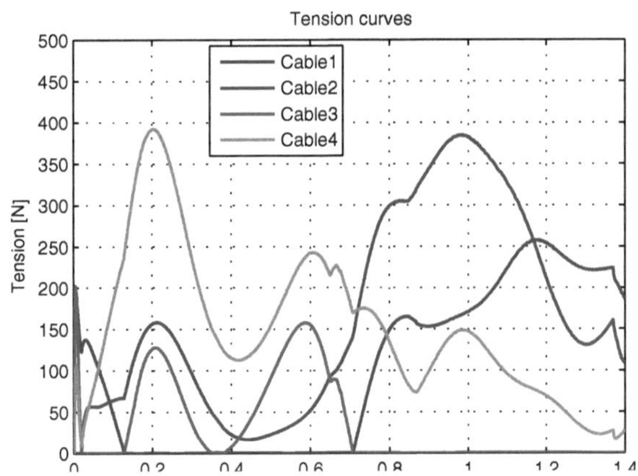

force has a vertical upward direction. The tension distribution could be improved through the choice of attachment points' location and also through the choice of more sophisticated objective functions.

5 Conclusion

A dynamic model and a simulation of a new gait training machine was investigated in this work. Preliminary results concerning the tensions in the cables, were presented. These results will be used to select the appropriate motors for the cable robot. Ongoing work is carried out to present more realistic models to ensure a more robust design.

References

1. Hornby, T.G., Zemon, D.H., Campbell, D.: Robotic-assisted, body-weight-supported treadmill training in individuals following motor incomplete spinal cord injury. Phys. Ther. **85**, 52–66 (2005)
2. Galvez, J.A., Reinkensmeyer, D.J.: Robotics for Gait Training After Spinal Cord Injury. Topics in Spinal Cord Injury Rehabilitation **11**(2), 18–33 (2005)
3. Benito-Penalva, J., Edwards, D.J., Opisso, E., Cortes, M., Lopez-Blazquez, R., Murillo, N., Costa, U., Tormos, J.M., Vidal-Samsó, J., Valls-Solé, J., Medina, J.: Gait training in human spinal cord injury using electromechanical systems: effect of device type and patient characteristics. Arch. Phys. Med. Rehabil. **93**(3), 404–412 (2012)
4. Díaz, I., Gil, J.J., Sánchez, E.: Lower-limb robotic rehabilitation: literature review and challenges. Journal of Robotics **2011**, 1–11 (2011)
5. Colombo, G., Joerg, M., Schreier, R., Dietz, V.: Treadmill training of paraplegic patients using a robotic orthosis. J. Rehabil. Res. Dev. **37**(6), 693–700 (2000)
6. Taherifar, A., Hadian, M.R., Mousavi, M., Rassaf, A., Ghiasi, F.: LOKOIRAN—a novel robot for rehabilitation of spinal cord injury and stroke patients. In: International Conference on Robotics and Mechatronics, ICRoM 2013, pp. 218–223 (2013)
7. Frey, M., Colombo, G., Vaglio, M., Bucher, R., Jörg, M., Riener, R.: A novel mechatronic body weight support system. IEEE Trans. Neural Syst. Rehabil. Eng. **14**(3), 311–321 (2006)
8. Vaughan, C.L., Davis, B.L., O'Connor, J.C.: Dynamics of Human Gait, 2nd edn. Kiboho Publishers, Howard Place, WesternCape 7450, South Africa (1999)
9. Schmitt, C.: Orthèses fonctionnelles à cinématique parallèle et sérielle pour la rééducation des membres inférieurs," Ph.D. dissertation, EPFL, Lausanne, Swiss (2007)
10. Hidler, J., Wisman, W., Neckel, N.: Kinematic trajectories while walking within the Lokomat robotic gait-orthosis. Clin. Biomech. **23**(10), 1251–1259 (2008)
11. Pham, C.B., Yeo, S.H., Yang, G., Chen, I.-M.: Workspace analysis of fully restrained cable-driven manipulators. Rob. Auton. Syst. **57**(9), 901–912 (2009)

An in Vivo Experiment to Assess the Validity of the Log Linearized Hunt-Crossley Model for Contacts of Robots with the Human Abdomen

F. Courreges, M.A. Laribi, M. Arsicault and S. Zeghloul

Abstract A key issue in Human-Robot physical interaction is the real-time perception of contact impedance by the robot. The Hunt-Crossley (HC) model is a popular model of contact force with soft biological tissues as it enjoys accuracy with low-complexity properties and its parameters are physically sound. Because the original HC model is non-linear, the current best known approach of real-time identification consists in identifying the parameters of a log linearized version of the HC model, by means of a Recursive Least Squares (RLS) algorithm. But, the final model used for exploitation in robot control, is the original non-linear HC model with the previously identified parameters. Hence, this approach may be questionable concerning the modeling accuracy and some authors prefer rejecting the HC model. This paper presents for the first time an in vivo experiment to assess the performances of contact models with the human abdomen. In particular we show here through a statistical analysis, that the log linearized HC model should be considered as a contact model on its own and replace the original non-linear HC model for both identification and exploitation.

Keywords Contact modeling · Hunt-Crossley model · Log linearized Hunt-Crossley model · Model selection · Nonparametric goodness-of-fit test · Medical robotics

F. Courreges (✉)
Team Mechatronics, XLIM Institute, CNRS UMR, University of Limoges 7252, Limoges, France
e-mail: fabien.courreges@unilim.fr

M. Arsicault
GMSC Department, PPRIME Institute, CNRS UPR3346 - University of Poitiers, Poitiers, France

M.A. Laribi · S. Zeghloul
Institut PPRIME, UPR 3346, University of Poitiers, Poitiers, France

S. Zeghloul et al. (eds.), *Robotics and Mechatronics*,
Mechanisms and Machine Science 37, DOI 10.1007/978-3-319-22368-1_21

1 Introduction

A key issue in Human-Robot physical interaction is the real-time perception of contact impedance by the robot. The Hunt-Crossley (HC) model [6] is a popular model of contact force with soft biological tissues as it enjoys accuracy [10, 13] with low computational complexity properties [2, 4, 8, 9, 11]. Furthermore, it is also consistent with the physics of impact of viscoeleastic materials through its Coefficient of Restitution [6]. While this model is non-linear, some linear recursive least squares (RLS) approaches of parameters identification, could be proposed [2, 4]; thus enabling a fast and real-time on-line identification. However, because the HC model is non-linear, it is not always preferred [10], probably because it is feared that an RLS based approach may have difficulties to converge or may be inaccurate. In fact in "normal" Human-Robot physical interactions, velocities remain generally low as is the ratio of viscous reaction force over elastic reaction force of soft biological tissues. These conditions thus lead to hindering the identification convergence of the viscous interaction forces, especially with the bootstrapped double stages algorithm of Diolaiti et al. [2] as suggested in [4]. In addition, measurement noise can also contribute to the poor conditioning of the identification problem. Alternatively Haddadi and Hashtrudi-Zaad [4] have proposed to make approximations in the HC model so as to identify a log linearized version of it, with a single stage RLS algorithm. Indeed let us recall the expression of the HC model:

$$\begin{cases} F = kx^n(1 + \mu\dot{x}), & \text{if } x \geq 0 \\ F = 0, & \text{if } x < 0 \end{cases} \tag{1}$$

where, F is the reaction force of the material; x is its indentation; k, μ and n are positive real parameters of the model so that k represents the non-linear stiffness, $\lambda = k.\mu$ represents the non-linear viscous coefficient and n is the dimensionless non-linearity parameter. The exponent n is usually close to 1, and allows taking into account the stiffness increase when the contact area also increases. Remark that the HC model is a generalization of the Hertz model to viscoelastic materials and is in agreement with the intuition that the damping coefficient should vary with the body's relative penetration. Also, importantly, in contrast to the Kelvin-Voigt model the HC force remains continuous upon impact. If we restrict to non-negative indentations x, the natural logarithm of the force yields:

$$\ln(F) = n\ln(x) + \ln(k) + \ln(1 + \mu\dot{x}).$$

Considering that for soft biological tissue like flesh, the coefficient μ remains neatly lower than 1, and because velocities are assumed low in physical Human-Robot Interactions (pHRI), the following first order approximation is proposed:

$$\ln(F) \simeq n\ln(x) + \ln(k) + \mu\dot{x}, \tag{2}$$

thus linearizing $\ln(F)$ in the parameters k, μ and n. However, Haddadi and Hashtrudi-Zaad consider making use of the identified parameters from the linearized model with the non-linear HC model, thus introducing errors in the fit of the model to experimental data. This composite model having the structure of the HC model but with parameters computed from the LLHC model, is denoted HC-LLHC. Furthermore, the safety limit of impact velocity with a robot may reach up to 2 m/s [5], thus the approximation assumptions may not always be satisfied with some particular tissues having substantial viscous effects and which may also be involved in pHRI, such as abdominal tissues. Hence, the aim of this paper is to assess whether the Log Linearized Hunt-Crossley (LLHC) model, obtained by exponentiating Eq. (2) and yielding expression (3), could be considered as an accurate contact model on its own, at least in the fields of collaborative robotics and medical robotics where the human abdomen constitutes the impedance to model and identify.

$$F = ke^{\mu\dot{x}}x^n \tag{3}$$

Our evaluation approach yields the following structure to the article: Sect. 2 describes the protocol to obtain experimental data in vivo with a volunteer subject, so as to obtain stress responses under dynamic loading. To our knowledge, it is the first time that HC type models are statistically evaluated on living human subjects. Section 3, describes how we have processed the data and the chosen statistical goodness of fit test and model selection criterion. Section 4, compares and discusses the modeling fidelity of the HC, LLHC and HC-LLHC models based on the indicator described in Sect. 3.

2 Experimental Protocol

The experimental protocol involved a volunteer male subject (21 years old). An instrumented probe (Fig. 1) was applied on the subject's abdomen and was manually given a cyclical movement by an operator with force loading and unloading cycles. The probe's displacement was performed to be as close as possible to a sinusoidal pattern in a vertical direction by the operator and the subject was lying on a horizontal plane. In this experiment the subject was asked to relax and hold his breathe with empty lungs so as to avoid perturbations from the respiratory motions. In order to measure the relative penetration of the probe in the subjects' abdomen, we used the optical tracking device designed by Vicon™ (Figs. 1 and 2). Several reflective markers were installed on the probe and on the patient' s abdomen (Fig. 1). The probe was also equipped with a force sensor. The cameras used to record the markers position allow a 300 Hz sampling and the force sensor a 900 Hz

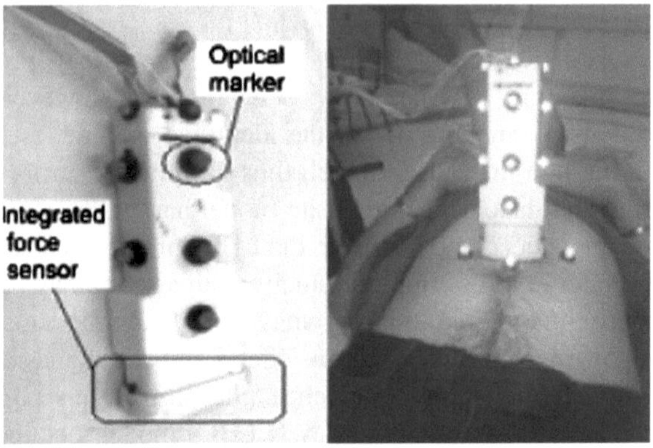

Fig. 1 Probe used in the experiment (*left*) and global experimental setup (*right*), the operator is applying the probe on the subject' s abdomen, there are four markers on the subject abdomen (one is behind the probe)

Fig. 2 Experimental setup: cameras of the Vicon Nexus motion capture system

sampling. The experiment was about 30 s long and the frequency of the loading-unloading cycles was almost constant over this duration. This setup allowed us to record directly the force signal synchronously with the position of the reflective markers. As the reflective markers position did not provide directly the real penetration depth, we needed to reconstruct the probe position. This position was obtained thanks to several geometrical constraints fitted in a least squares sense with the markers position data. These results added to the reflective markers on the subjects' abdomen let us determine the indentation of the abdomen, exerted by the central point of the probe's tip, at each sample time.

3 Experimental Data Analysis Procedure

3.1 Measurements Noise Analysis

To analyze the noise in the probe's end-tip localization, a record of 1116 samples of the probe's markers position was achieved while the probe remained still. As mentioned in the previous section, the end-tip of the probe is reconstructed by a least-squares optimization process subject to some geometrical constraints. From there, the tip position error projected on the probe's axis can be computed. A statistical analysis of these errors showed that they cannot be considered as normally distributed (Shapiro-Wilk test, W = 0.8592, p = 0). Because the probe's end tip position reconstruction is a complicated non-linear process, it is not easy to devise the true parametric form of the underlying noise density. Therefore, we have carried out the experimental data analysis in Sect. 4 by means of a nonparametric approach presented in next subsection.

3.2 Model Quality Evaluation Criteria

We want to assess two properties of the models of force. The first is the adequation or goodness-of-fit to the experimental data, and we need a very accurate test as it is expected that the models considered should perform very closely. The second thing we want to know is which model we should select as better candidate for grabbing the information content of the experimental data. In both cases, we make the standard assumption that if a model perfectly fits the data, then the residuals should be independent and identically distributed (iid) random variables with zero mean.

3.2.1 Goodness-of-Fit Test

For the first problem, the nonparametric likelihood ratio based, goodness of fit statistical test Z_A, from Zhang [14], is used here. To introduce the statistic Z_A, let us denote by X a continuous scalar random variable with distribution function (cumulative density function) $F(x)$. Notice that in our case, X would represent the model residual. This statistic enables to test the null hypothesis

$$H_0 : F(x) = F_0(x), \text{ for all } x \in (-\infty, +\infty)$$

against the general alternative

$$H_1 : F(x) \neq F_0(x), \text{ for some } x \in (-\infty, +\infty)$$

where $F_0(x)$ is a hypothesized and fully known distribution function to be tested. Let us also denote by $F_n(x)$ the empirical sample distribution computed from an iid random sample of size n. The statistic Z_A takes the form:

$$Z_A = \int_{-\infty}^{+\infty} \left\{ \left[(1 - F_n(x))^{-1} \log\left(\frac{F_n(x)}{F_0(x)}\right) + F_n(x)^{-1} \log\left(\frac{1 - F_n(x)}{1 - F_0(x)}\right) \right] \frac{dF_n}{dx}(x) \right\} dx \tag{4}$$

Hence, Z_A can be seen as a discrepancy measure of $F_n(x)$ from a reference distribution $F_0(x)$. Owing to the optimality of the likelihood ratio as established by the Neyman-Pearson lemma, the test based on Z_A enjoys an excellent statistical power (way better than classical nonparametric tests [3, 14] such as Kolmogorov-Smirnov or Cramer-Von-Mises tests). Zhang has proposed in the same vein three different statistics, Z_A, Z_C and Z_K; Z_A and Z_C were shown to be the most powerful and perform very closely. But, Z_A was shown to perform slightly better in the case of composite hypothesis [3]. The importance of this consideration will be understood a bit further. Now we shall determine which reference distribution to use for $F_0(x)$. This distribution should represent the ideal residuals distribution when the model of force perfectly fits the data. But, as discussed earlier this distribution is imperfectly known. Based on the approach of Owen depicted in [12] Chap. 2, we suggest to devise $F_0(x)$ as the maximum likelihood nonparametric empirical distribution satisfying the constraints of an ideal residuals distribution. More precisely, let us assume we managed to grab an n-sample of residuals. The ties free order statistic of the sample is noted $\widetilde{X}_{(1)}, \widetilde{X}_{(2)}, \ldots, \widetilde{X}_{(\widetilde{n})}$, where $\widetilde{n}$ is the number of different values, and $\widetilde{X}_{(i+1)} > \widetilde{X}_{(i)}$ for all $i \in \{1, 2, \ldots \widetilde{n}\}$; furthermore the multiplicity of the ties having the value of rank i, is stored in the variable $m_{(i)}$, so that $\sum_{i=1}^{\widetilde{n}} m_{(i)} = n$. Notice that if there are no ties, then the sequence $\{\widetilde{X}_{(i)}\}$ is exactly the same as $\{X_{(i)}\}$ and $\widetilde{n} = n$. With these notations, from Eq. (4), Z_A simplifies to:

$$Z_A = \sum_{i=1}^{\widetilde{n}} \left[\frac{\log\left\{\frac{F_n(\widetilde{X}_{(i)})}{F_0\left(\widetilde{X}_{(i)}\right)}\right\}}{1 - F_n(\widetilde{X}_{(i)})} + \frac{\log\left\{\frac{1-F_n(\widetilde{X}_{(i)})}{1-F_0\left(\widetilde{X}_{(i)}\right)}\right\}}{F_n(\widetilde{X}_{(i)})} \right] \tag{5}$$

Besides, as suggested by Zhang [14] the empirical distribution, generalized to the case when there are ties in the sample, should be expressed as $F_n(\widetilde{X}_{(i)}) = \frac{\sum_{j=1}^{i} m_{(j)} i - 0.5 m_{(i)}}{n}$. Similarly, we define the distribution $F_0(x)$ as:

$$F_0(x = \widetilde{X}_{(k)}) = \left(\sum_{i=1}^{k-1} p_{(i)}\right) + \frac{p_{(k)}}{2}, \tag{6}$$

where the weights $p_{(i)}$ satisfy the optimal problem:

$$\begin{cases} \max_{\{p_i, i=1,2,\ldots\widetilde{n}\}} \sum_{i=1}^{\widetilde{n}} m_i \log\left(p_{(i)}\right) \\ \text{S.C.} \\ \sum_{i=1}^{\widetilde{n}} p_{(i)} = 1 \\ \sum_{i=1}^{\widetilde{n}} m_{(i)} \widetilde{X}_{(i)} p_{(i)} = 0 \\ p_{(i)} \geq 0, \forall i \in \{1, 2, \ldots, \widetilde{n}\} \end{cases} \quad (7)$$

Now because $F_0(x)$ depends on the sample, the statistical test making use of $F_0(x)$ cannot be considered anymore as a simple hypothesis test, but should be qualified as complex or composite hypothesis test (this is why Z_A is more appropriate than Z_C). Because of this, the distribution of the statistic Z_A cannot be found in pre-computed tables, and should be computed on an ad hoc basis (i.e. for each sample). We have used the Monte Carlo simulation based approach suggested in [7], with 300000 runs.

3.2.2 Model Selection Criterion

The standard and efficient Akaike's Information Criterion (AIC) for model selection [1] is used here. For each model "$\mathcal{M}$" the following reduced AIC can be computed:

$$AIC_{r,\mathcal{M}}(X) = -\log L_{\mathcal{M}}\left(X_{(1)}, X_{(2)}, \ldots X_{(n)}\right)$$

where $L_{\mathcal{M}}\left(X_{(1)}, X_{(2)}, \ldots X_{(n)}\right)$ represents the maximum nonparametric empirical likelihood of the n-sample assuming the model $\mathcal{M}$, so that:

$$L_{\mathcal{M}}\left(X_{(1)}, X_{(2)}, \ldots X_{(n)}\right) = \prod_{i=1}^{\widetilde{n}} p_{(i)}^{m_{(i)}},$$

where $p_{(i)}$ are the weights solution of (7). The AIC is termed "reduced" here, because the original AIC is scaled by a constant factor and has an additive offset depending only on the number of free parameters in the model. However to compare models having the same number of parameters, the scale and offset have no importance.

3.3 Generation of Samples of Residuals

Because the experiments were made in vivo, and the total duration of the experiments were significantly large with respect to the duration of the loading and unloading cycles, the biomechanical parameters to identify may be dynamically affected by physiological phenomena and reactions of the subjects over the whole experiment duration, even when demanding the subjects to hold their breathe. We can nevertheless assume that in a time window including no more than 2 loading-unloading cycles (corresponding to about 110 samples) the biomechanical parameters remain constant. Hence, the whole sample acquired in each experiment has been cut into smaller disjoint sample windows of 110 timely consecutive data points, which represents a time span of about 0.4 s. For each window of 110 samples, the models considered (HC and LLHC) have been fitted to the data, using the total least squares optimization approach. This method of identification is relevant to the practical real-time identification approach which is RLS based ([2, 4]). Because these models are viscoelastic they also need a velocity input. The velocity has been estimated from the reconstructed position by means of a low order and small smoothing window Savitzky-Golay differentiator (1st order polynomial fit with a window length of 7 data points). Only the 35 sample windows having the best statistical homogeneity have been subsequently retained. For each of the retained sample window and each model, the model residuals at each discrete time, have been computed and collected together with the residuals of all sample windows, so as to form one unique pool sample of 3850 residuals. This large size of sample ensures a very good significance of the statistical analysis of the data. The resulting pool sample of residuals has then been processed to compute the previously described criteria in Sect. 3.2. The results are provided and discussed in next section.

4 Processed Experimental Data, Analysis and Discussion

The data processing results are reported in Table 1, below. The "MSSR" is the Mean Sum of Squared Residuals over the whole pool sample; it is added here as an assessment criterion, because the least squares approach of model parameters identification used here, aims at minimizing this criterion. The goodness of fit of a

Table 1 Force models statistical performances

Model type	MSSR	Z_A	p-value (%)	AIC_r	$\exp\left(\frac{AIC_{r,min} - AIC_r}{2}\right)$
HC	0.0285659	0.0580479	85.24079	3.178495679425e + 04	0.99934547
LLHC	0.0285051	0.0536646	85.73203	3.178495548476e + 04	1
HC-LLHC	0.0285804	0.0754974	83.17224	3.178496202057e + 04	0.99673743

model is assessed by its value of Z_A and corresponding p-value. The best model to select should have the lowest AIC_r denoted $\mathrm{AIC}_{r,min}$; and the term $\exp\left(\frac{\mathrm{AIC}_{r,min}-\mathrm{AIC}_{r,\mathcal{M}}}{2}\right)$, represents the relative probability that the model $\mathcal{M}$ minimizes the estimated information loss ([1], Sect. 2.12.4). From Table 1 it is clear that all models perform very closely and are very good at fitting the data. All performance criteria reported here are in agreement to confirm that the HC-LLHC model should be rejected in regards to the two other models contrary to the current practice. Between HC and the LLHC, contrary to what could be thought, it is striking to notice that the LLHC model does not have significant degraded performance compared to the HC model; but on the contrary it performs even slightly better on all indicators. Hence in regards to its advantage for real-time identification and its modeling accuracy, the LLHC model should be definitely considered as a contact model on its own and not only an approximation (of the HC model) and hence should be preferred over the HC and HC-LLHC models.

Acknowledgements This work is supported by the French Research National Agency (ANR) - convention ANR-14-CE27-0016.

References

1. Burnham, K., Anderson, D.R.: Model selection and model inference: a practical information-theoretic approach, 2nd edn. Springer, New-York, USA (2002)
2. Diolaiti, N., Melchiori, C., Stramigioli, S.: Contact impedance estimation for robotic systems. IEEE Trans. Robot **21**(5), 925–935 (2005)
3. Gorbunova, A.A., Lemeshko, B.Y., Lemeshko, S.B., Rogozhnikov, A.P.: Application of nonparametric goodness-of-fit tests for composite hypotheses in case of unknown distributions of statistics. International Workshop—Applied Methods of Statistical Analysis. Simulations and Statistical Inference—AMSA' 2013, pp. 8–24. Novosibirsk, Russia (2013)
4. Haddadi, A., Hashtrudi-Zaad, K.: Real-time identification of hunt-crossley dynamic models of contact environment. IEEE Trans. Robot. **28**(3), 555–566 (2012)
5. Haddadin, S., Albu-Schäffer, A., Hirzinger, G.: On making robots understand safety: embedding injury knowledge into control. Int. J. Robot. Res. **31**, 1578–1602 (2012)
6. Hunt, K., Crossley, F.: Coefficient of restitution interpreted as damping in vibroimpact. ASME J. Appl. Mech. pp. 440–445 (1975)
7. Lemeshko, B.Y., Lemeshko, S.B., Postovalov, R.A.: Real-time studying of statistic distributions of nonparametric goodness-of-fit tests when testing complex hypotheses. In: International Workshop—Applied Methods of Statistical Analysis. Simulations and Statistical Inference, pp. 19–27. Novosibirsk, Russia (2011)
8. Lens, T., Radkhah, K., von Stryk, O.: Simulation of dynamics and realistic contact forces for manipulators and legged robots with high joint elasticity. In: International Conference on Advanced Robotics, pp. 34–41 (2011)
9. Mobasser, F., Hashtrudi-Zaad, K.: Implementation of a rate mode impedance reflecting teleoperation controller on a haptic simulation system. In: International Conference on Robotics and Automation (ICRA), pp. 1974–1979 (2004)
10. Moreira, P., Zemiti, N., Liu, C., Poignet, P.: Viscoelastic model based force control for soft tissue interaction and its application in physiological motion compensation. Comput. Methods Progr. Biomed. **116**, 52–67 (2014)

11. Okamura, A., Verner, L., Yamamoto, T., Gwilliam, J., Griffiths, P.: Surgical robotics, systems applications and visions, Chap. 18, force feedback and sensory substitution for robot-assisted surgery, pp. 419–448. Springer (2011)
12. Owen, A.B.: Empirical Likelihood. Monographs on statistics and applied probability. CHAPMAN & HALL/CRC, Boca Raton (2001)
13. Yamamoto, T., Vagvolgyi, B., Balaji, K., Whitcomb, L., Okamura, A.M.: Tissue property estimation and graphical display for teleoperated robot-assisted surgery. In: IEEE International Conference on Robotics and Automation (ICRA), pp. 12–17. IEEE, Kobe International Conference Center—Japan (2009)
14. Zhang, J.: Powerful goodness-of-fit tests based on the likelihood ratio. J. Royal Stat. Soc.: Series B (Stat. Methodol.) **64**(2), 281–294 (2002)

Part VI
Control and Vision

Real-Time Reconstruction of Heightmaps from Images Taken with an UAV

Jose Gabriel Ramirez-Torres and Ander Larranaga-Cepeda

Abstract In this paper, an approach for a real-time three dimensional reconstruction of a dense point cloud, from images obtained with an unmanned aerial vehicle, is presented. The approach also estimates the model scale, using a known marker as reference, obtaining a fairly precise heightmap of the environment. The proposed image analysis framework is also capable of an off-line 3D reconstruction from monocular images, without known markers in the environment. The proposed framework is an adaptation of the Patch-based Multiview Stereo (PMVS) algorithm, in order to obtain a faster point cloud generation, suitable for real-time environment reconstruction. For the obtained heightmap, the error estimation ranges between 1–1.5 % for manual detection of marker and 4–5 % for automatic marker detection, which we consider is accurate enough for autonomous navigation and path planning.

Keywords Real-time 3D reconstruction · UAV applications · Image analysis · Computer vision · Autonomous robots

1 Introduction

An unmanned aerial vehicle (UAV) can be defined as a powered aerial vehicle that does not carry a human operator, uses aerodynamic forces to provide vehicle lift and can fly autonomously or be remotely piloted [9]. UAVs are classified according

J.G. Ramirez-Torres (✉) · A. Larranaga-Cepeda
Laboratory of Information Technologies, CINVESTAV Tamaulipas, Tamaulipas, Mexico
e-mail: grtorres@tamps.cinvestav.mx

A. Larranaga-Cepeda
e-mail: alarranaga@tamps.cinvestav.mx

S. Zeghloul et al. (eds.), *Robotics and Mechatronics*,
Mechanisms and Machine Science 37, DOI 10.1007/978-3-319-22368-1_22

to the type of wing (fixed or rotary wing) and to its taking off capabilities (vertical take-off or short take-off and landing) [4]. Good performance, low cost, vertical/short take-off and landing allowing access to hazardous areas for humans, low weight and low noise, can be counted among the main advantages of UAVs for different applications. At present, semi-autonomous UAVs are just able to perform very specific automatic tasks, such as marker-based navigation, automatic take-off and landing [20]. In order to expand the actual range of possible applications of UAVs, we must assure a safe and autonomous navigation through unknown environments, which requires a robust mapping ability of the surroundings. Such exploring ability leads to many non trivial problems, such as the UAV localization problem in a GPS denied area or the real-time obstacle evasion when navigating through an unknown environment. In order to solve these problems, it is required to generate a dependable metric representation of the environment, based on the received data from sensors, that is, a map containing distance, height and angle data of the overflown area.

The most common way to describe the orography of the environment is through heightmap. The generation of a heightmap is a process involving the digital representation of the terrain surface from a sequence of sensor readings (rangefinder or visual sensors), in order to obtain a numeric model known as Digital Elevation Model (DEM), Digital Terrain Model (DTM) or Digital Surface Model (DSM) [22, 25]. The main applications of digital terrain models are the terrestrial parameters extraction, generation of three-dimensional maps for cartographic relief depiction, orthorectification of aerial photographies, analysis of GPS data, analysis of terrain surfaces, flight planning, among others.

In litterature, there exist different proposals for digital terrain model generation from GPS-based images obtained using a UAV [2, 3, 20]. The purpose of this paper is to propose a solution based on a commercial UAV equipped with a monocular camera, in order to provide an easy and affordable way for the generation of digital terrain models, particularly in a GPS-denied environment, useful for academic and civilian purposes. For this work, the selected UAV model is an AR.Drone 1.0, manufactured by the French company Parrot; however, the presented development can be easily adapted to any UAV, with a good quality camera for image acquisition.

This work proposes a digital image analysis for the generation of a point cloud, both online and offline, in order to obtain a three dimensional model with scale estimation, as well as the heightmap construction, exploitable for UAV autonomous navigation. The paper is organized as follows: Sect. 2 describes the previous work in three dimensional modeling, heightmap generation and the odometry and navigation algorithms for the AR.Drone UAV; Sect. 3 describes the structure of the proposed approach for the image analysis; Sect. 4 present the obtained experimental results; Sect. 5 provides a data analysis and interpretation; and finally, Sect. 6 summarizes and presents some final conclusions.

2 Previous Work

In this section, some relevant works on image-based 3D reconstruction and heightmaps generation are presented.

2.1 *Image-Based 3D Reconstruction*

Some approaches, addressing the three dimensional reconstruction of the environment from monocular images, have been published. Due to the complexity of this task, many authors suggest the use of multiple sensors for this purpose [19] or stereoscopic cameras [27] in order to obtain depth data from the images and discard the monocular cameras as not suitable for 3D reconstruction [19]. There are also some other approaches focused on Visual SLAM using an Extended Kalman Filter (EKF) for the three dimensional reconstruction of facades, but with scale uncertainty [6, 16]. For the approaches based on structure-from-motion (SfM) with a single camera, the scene can be recovered, but only up to a scale, since the absolute dimensions cannot be computed [21]. To solve the scale uncertainty, the presence of a known reference in the scene must be assured, so the absolute scale can be recovered from the relative dimensions of the reference in it. We propose to employ a marker that could be easily detected by an algorithm for marker detection, such as the method proposed by Garrido-Jurado et al. [8].

The offline 3D reconstruction problem from images has been successfully addressed with algorithms like the Patch-based Multiview Stereo algorithm (PMVS) [7] and the SfM approaches, such as A Contrario Structure from Motion algorithm (ACSfM) [17], capable to provide a dense and highly detailed point cloud. However, the required processing time for image matching and camera pose computing is too high for real-time three dimensional reconstruction and, on the other hand, the scale factor is also unknown.

2.2 *Scale Estimation for Heightmaps*

There exist several proposals when it comes to heightmap generation from aerial images, but most of them are based on GPS data [1, 26] and stereoscopic cameras [2] to get depth data. Weiss's et al. [26] approach creates a three dimensional model in real time after manually measuring the scale factor, however, only environments with available GPS information are considered. Also, several robust feature descriptors have been proposed for defining an object in a three dimensional space [5, 15, 28], but most of them do not consider occlusions and cannot deal properly with several instances of the same object in a scene.

3 Proposed Approach

Simply put, the UAV, equipped with a monocular camera, must overfly an indoor area or a GPS-denied terrain, while the proposed system analyses the images and constructs a heightmap of the environment.

To define the path that the UAV must follow, we use the marker method proposed by Garrido-Jurato et al. [8] for setting navigation checkpoints on the terrain, assigning instructions for the UAV to each marker. The marker provides three additional and useful advantages: first, the matrix of the camera pose can be estimated in real-time, with relatively little computational cost overhead; second, the a priori knowledge of the size of the marker allows us to find out the point cloud's scale factor; and third, the marker also sets the ground level of the scene, that is, the XY plane used as reference for all the estimated heights.

For the point cloud generation, a real-time image analysis, based on the PMVS algorithm [7], is used. The points of interest (POI) are the corners and the edges of the image, obtained by the Harris' algorithm [10] and the difference of gaussians (DoG) [14] respectively. Each POI is modeled as a patch and can appear in more than one image. When a new image is analysed, patches which are epipolar consistent are merged, according to a photometric consistency criterium measured by cross-correlation. The process works as follows: the images are analysed by blocks of three images and a temporary point cloud is obtained, which is used to update the general point cloud. In this way, the point cloud grows as the UAV advances through the environment.

In order to precisely detect the marker within the scene, we use an adaptation of the Tombari's et al. [24] approach for object detection in a three dimensional space, based in the voting procedure of the Hough Transform [11], adapted for 3D shapes. In our approach, there are two point clouds to be compared: the actual modeled environment, called *scene*, and the point cloud describing the marker, called *model*. The total number of points in both clouds are counted and their normal vectors are computed according to Khoshelman's method et al. [12]. Then, a uniform sampling is performed on both clouds in order to reduce them into small squares and approximating every square by its centroid.

Afterward, the reduced clouds are described by SHOT descriptors, as depicted in [23] and, by applying Euclidean distance, matching pairs among descriptors are found. If a matching is found, the BOARD algorithm [18] assigns a local reference frame (LRF) in order to make it rotational and translational invariant. Finally, Hough voting is performed to create an histogram of orientations with all local reference frames. A variable threshold defines the number of found point correspondences, so it can be concluded whether the reference object (*model*) is present within the *scene* or not. If the marker has been found, a search for its corners is performed to measure it. The relative size in the point cloud and the absolute size of the marker allow to solve the scale uncertainly and obtain a real metric 3D model of the environment.

The proposed approach has been implemented in C/C ++ on a Linux OS, structured in two parts: the first for the control of the UAV and the reconstruction of the temporary point cloud, the second for the global point cloud updating and the marker searching. The complete schema of the proposed system is presented in Fig. 1.

As an additional feature, the system is capable of generating a 3D model from the images, without a known marker, in an off-line procedure. This task is accomplished with the ACSfM algorithm [17] which performs image matching based on the SIFT algorithm [13] in order to compute the camera pose matrices for each image. Finally, with the computed pose matrices, the whole set of images is processed with the PMVS algoritm to generate the global point cloud of the environment.

4 Results

To validate the proposed approach, two distinct scenarios, with different sizes of marker were used, (Fig. 2). For both scenarios, the resulting heightmaps are shown in Fig. 6. Four known heights *A*, *B*, *C* and *D* were used as reference for computing

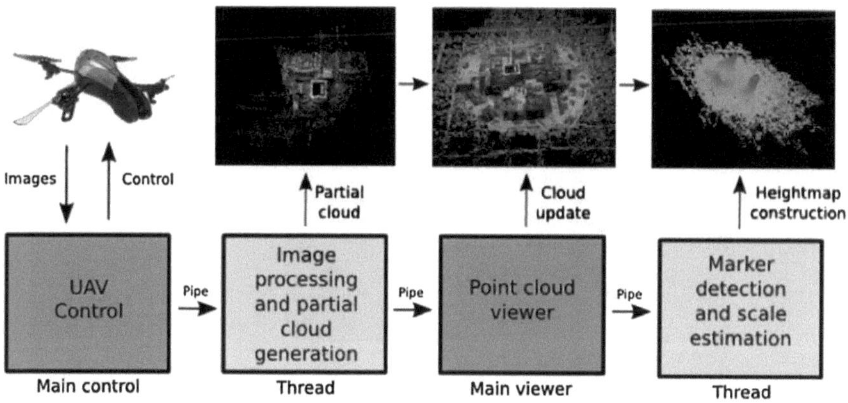

Fig. 1 Proposed approach for real-time 3D heightmap reconstruction

Fig. 2 Scenarios 1 and 2

the estimation error (c.f. Fig. 2). A 14 × 14 cm maker was used in Scenario 1 and a 21 × 21 cm maker was used in Scenario 2 to evaluate the estimation error for different marker sizes.

4.1 Real-Time 3D Modeling

A set containing 72 640 × 480 images was used to obtain the model shown in Fig. 3. The growing process of the global point cloud is shown in Fig. 4. The Fig. 5 shows the relief of the 3D model.

4.2 Marker Detection and Heightmap Generation

Tables 1 and 2 show the estimated height and the error for real heights *A*, *B*, *C* and *D* for each scenario respectively, with different threshold values for Hough voting (Fig. 6).

Fig. 3 Example of the image data set used for the presented results

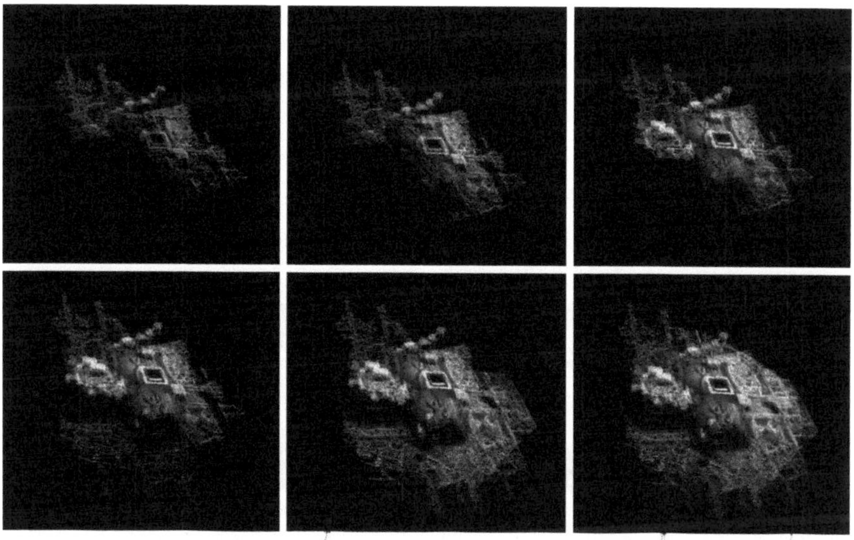

Fig. 4 Real time reconstruction of the point cloud

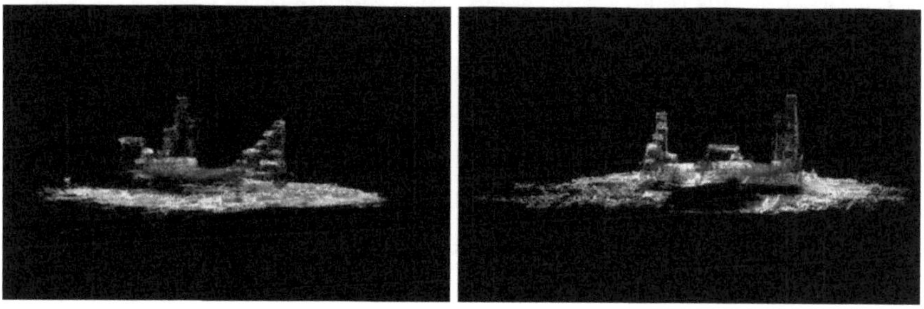

Fig. 5 3D profile of the generated model

Table 1 Height estimation results for Scenario 1, with 14 cm marker

A (0.235 m)	B (0.39 m)	C (0.16 m)	D (−0.14 m)	Avg error (%)	Threshold
0.2308	0.4046	0.1719	−0.1369	3.79	0.01
0.2645	0.4477	0.1952	−0.1523	14.53	0.015
0.3176	0.5474	0.2310	−0.1962	40.01	0.020

Table 2 Height estimation results for Scenario 2, with 21 cm marker

A (0.39 m)	B (0.395 m)	C (0.55 m)	D (0.24 m)	Avg error (%)	Threshold
0.4195	0.4098	0.5645	0.2471	4.22	0.01
0.4261	0.4039	0.5707	0.2457	4.41	0.015
0.4329	0.4143	0.5814	0.2525	6.70	0.020
0.4398	0.4181	0.5777	0.2513	7.09	0.030
0.5171	0.4985	0.7001	0.3071	28.51	0.1

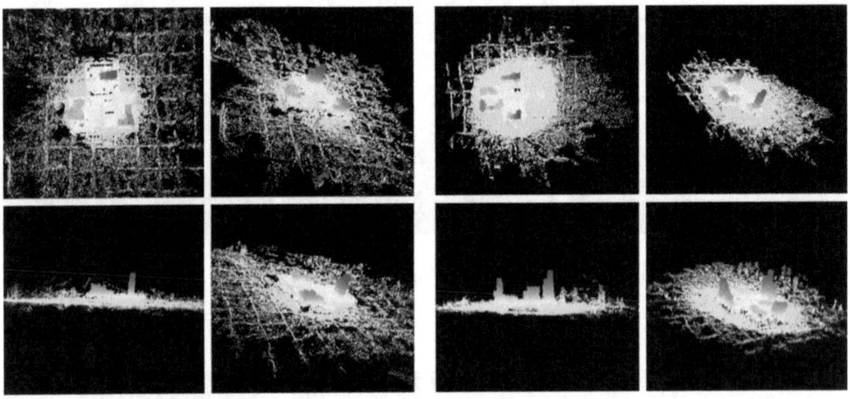

Fig. 6 Generated heightmap for Scenario 1 and 2, respectively

4.3 *Off-line 3D Modeling*

The off-line 3D model generated from the same 72 640 × 480 images is shown in Fig. 7.

5 Discussion

The results reported in Tables 1 and 2 are slightly different because not only the size of the marker, but also the threshold in Hough voting affects the scale estimation process. The image quality and resolution have a significant influence on the camera pose estimation and on the marker detection in the point cloud. This lead us to conclude that scale estimation relies on an accurate marker detection and, at the same time, the precision for the camera pose matrix relies on the visual quality of the marker. For autonomous navigation, 640 × 480 images provide a heightmap in an adequate processing time, enough to accurate fulfill the established requirements of the present work. In these scenarios, a processing time of 1.05 frames per second on average was reached.

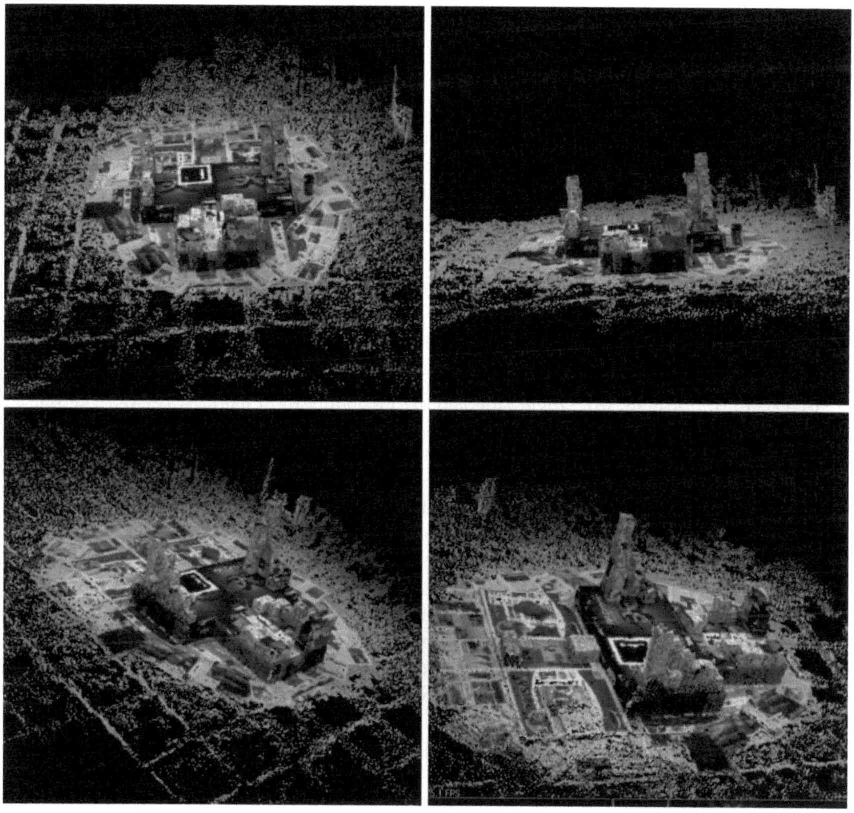

Fig. 7 Different perspectives of the obtained off-line 3D model

The combination of a SfM algorithm for computing camera poses and a patch-based point cloud generation algorithm provided an accurate, dense and highly detailed 3D model in the off-line approach. For this case, the image set is treated after the UAV has overflown the environment.

6 Conclusion

Based on the obtained results, it can be concluded that the generated 3D model is comparable and even better, in terms of visual quality and processing time, when compared to the previous works presented in Sect. 2, for both real-time and off-line approaches, but including the global scale factor estimation. Also, the height estimation errors are suitable and accurate enough for a safe autonomous navigation through the environment. It is planned as future work to test the presented framework with real aerial images to study how the environment and the altitude of the UAV affects the heightmap generation. Processing time reduction, heightmap construction without artificial markers, UAV autonomous indoor navigation and collaborative real-time 3D modeling are also considered.

Acknowledgments The authors would like to acknowledge the support of scholarship fund of CONACyT during A. Larranaga's M.S. degree, and the support of the Cinvestav-LTI for the facilities to perform this work. We would also like to acknowledge the authors of the following libraries: OpenMVG, OpenCV, PMVS2 and Point Cloud Library.

References

1. Barazzetti, L., Scaioni, M.: Orientation and 3D modelling from markerless terrestrial images: combining accuracy with automation. Photogram. Rec. **25**, 2010 (2013)
2. Besnerais, G., Sanfourche, M., Champagnat, F.: Dense height map estimation from oblique aerial image sequences. Comput. Vis. Image Underst. **109**, 204–225 (2008)
3. Call, B.: Obstacle avoidance for unmanned air vehicles. Master's thesis, Brigham Young University (2006)
4. Cetinsoy, E., Dikyar, S., Hancer, C., Oner, K., Sirimoglu, E., Unel, M., Aksit, M.: Design and construction of a novel quad tilt-wing UAV. Mechatronics **22**, 723–745 (2012)
5. Chen, H., Bhanu, B.: 3D free-form object recognition in range images using local surface patches. Pattern Recogn. Lett. **28**, 1252–1262 (2007)
6. Diskin, Y., Asari., V.: Dense point-cloud creation using superresolution for a monocular 3D reconstruction system. In: Proceedings of SPIE 8399, Visual Information Processing XXI, pp. 83,990–N–83,990N–9 (2012)
7. Furukawa, Y., Ponce, J.: Accurate, dense, and robust multi-view stereopsis. Pattern Anal. Mach. Intell. **32**(8), 1362–1376 (2010)
8. Garrido-Jurado, S., Muoz-Salinas, R., Madrid-Cuevas, F., Marn-Jimnez, M.: Automatic generation and detection of highly reliable fiducial markers under occlusion. Pattern Recogn. **47**(6), 2280–2292 (2014)
9. Gertler, J.: US unmanned aerial systems. Congressional Research Service (2012)
10. Harris, C., Stephens, M.: A combined corner and edge detector. In: Proceedings of the 4th Alvey Vision Conference, pp. 147–151 (1988)
11. Hough, P.: Method and means for recognizing complex patterns (1962)
12. Khoshelham, K.: Extending generalized hough transform to detect 3D objects in laser range data. In: Proceedings of ISPRS Workshop on Laser Scanning, pp. 206–210 (2007)
13. Lowe, D.: Distinctive image features from scale-invariant key points. Int. J. Comput. Vis. **60** (2), 91–110 (2004)
14. Marr, D., Hildreth, E.: Theory of edge detection. In: Proceedings of the Royal Society of London, pp. 215–217 (1980)
15. Mian, A., Bennamoun, M., Owens, R.: On the repeatability and quality of keypoints for local feature-based 3D object retrieval from cluttered scenes. Int. J. Comput. Vis. **89**, 348–361 (2010)
16. Motta, C.: Reconstruccin tridimensional de fachadas de edificios empleando imgenes monoculares obtenidas por un vehculo areo no tripulado autnomo. Master's thesis, Centro de Investigacin y de Estudios Avanzados del Instituto Politcnico Nacional (2014)
17. Moulon, P., Monasse, P., Marlet, R.: Adaptive structure from motion with a contrario model estimation. In: ACCV 2012, vol. 7727, pp. 257–270 (2012)
18. Petrelli, A., Stefano, L.D.: On the repeatability of the local reference frame for partial shape matching. In: 13th International Conference on Computer Vision (ICCV), pp. 2244–2251 (2011)

19. Pollefeys, M., Nistér, D., et al.: Detailed real-time urban 3D reconstruction from video. Int. J. Comput. Vis. **78**, 143–167 (2008)
20. Sanfourche, M., Delaune, J., Besnerais, G.L.: Perception for UAV: vision-based navigation and environment modeling. J. AerospaceLab **4**, 1–19 (2012)
21. Scaramuzza, D., Fraundorfer, F., Pollefeys, M., Siegwart, R.: Absolute scale in structure from motion from a single vehicle mounted camera by exploiting nonholonomic constraints. In: 2009 IEEE 12th International Conference on Computer Vision, pp. 1413–1419 (2009)
22. Sirmacek, B., d'Angelo, P., Krauss, T., Reinartz, P.: Enhancing urban digital elevation models using automated computer vision techniques. In: International Conference on Pattern Recognition (2010)
23. Tombari, F., Salti, S., DiStefano, L.: Unique signatures of histograms for local surface description. In: ECCV'10 Proceedings of the 11th European Conference on Computer Vision Conference on Computer Vision, pp. 356–369 (2010)
24. Tombari, F., Stefano, L.D.: Object recognition in 3D scenes with occlusions and clutter by hough voting. In: 4th Pacific-Rim Symposium on Image and Video Technology, pp. 349–355 (2010)
25. Weibel, R., Heller, M.: Digital terrain modelling (1993)
26. Weiss, S., Achtelik, M., Kneip, L., Scaramuzza, D., Siegwart, R.: Intuitive 3D maps for MAV terrain exploration and obstacle avoidance. J. Intell. Robot. Syst. **61**(1–4), 473–493 (2011). Doi:10.1007/s10846-010-9491-y, http://dx.doi.org/10.1007/s10846-010-9491-y
27. Wen-Chung, C., Shu-An, L.: Real-time feature-based 3D map reconstruction for stereo visual guidance and control of mobile robots in indoor environments. In: 2004 IEEE International Conference on Systems, Man and Cybernetics, vol. 6, pp. 5386–5391 (2008)
28. Zhong, Y.: Intrinsic shape signatures: a shape descriptor for 3D object recognition. In: Proceedings of 3DRR Workshop (in conj. with ICCV), pp. 689–696 (2009)

A Human-Machine Interface with Unmanned Aerial Vehicles

D. Soto-Gerrero and J.-G. Ramrez-Torres

Abstract Many UAV applications rely on a ground-based computers to accomplish all processing tasks and the power consumed by such systems are a major roadblock when it comes to portability. Since one of the main goals of this interface is to be ubiquitous, we moved all computational load to a mobile device running Android. On this paper we present an efficient and practical human-machine interface with an unmanned aerial vehicle (UAV). The UAV can respond to both, the movement and body gestures of the user.

Keywords Unmanned Aerial Vehicle (UAV) · Human-robot interaction · Gesture recognition · Particle filter

1 Introduction

The interface proposed and described in this article contributes to what has been observed in many social interactions between unmanned aerial vehicles (UAV) and untrained personnel. This interface does not rely on any type of radio control hardware to function, nor the user is required to know how to maneuver the UAV. Technically, the software developed for this interface interprets what the user does throughout a video stream and translate his actions to control commands. To run the software we used a mobile device, because they already provide a more natural interaction with their touch capabilities and they are overtaking personal computers in common everyday tasks. Furthermore, we avoided distinguishable markers on the user in order to keep the interaction as natural as possible.

D. Soto-Gerrero (✉) · J.-G. Ramrez-Torres
Information Technology Laboratory, Cinvestav Tamaulipas, Tamaulipas, Mexico
e-mail: dsoto@tamps.cinvestav.mx

J.-G. Ramrez-Torres
e-mail: grtorres@tamps.cinvestav.mx

S. Zeghloul et al. (eds.), *Robotics and Mechatronics*,
Mechanisms and Machine Science 37, DOI 10.1007/978-3-319-22368-1_23

A previous study sustain that human-robot interaction research has ignored the social and collocated aspects, which will impact in the future when autonomous flying robots take a step forward in social activities [1]. They attempted to leave behind the remote controlled interaction by using the *Wizard of Oz* technic: a test subject performs gestures to the drone, while an unseen operator controls the UAV wirelessly. They proved that children and grown-ups get emotionally attached to the UAV, thinking of it, as an entity [2].

Among the recent research works on human-machine interaction with UAVs we can cite a sports assistant from the University of Tokyo [3]. A UAV with a camera on board, followed an athlete to provide what is described as *external visual imagery*. Athletes no longer need to picture themselves from the perspective of an external observer to evaluate their skills, strategies and game plans. All that information is delivered by the drone to the athlete through a head mounted display. They used a particle filter for tracking one color in the video stream, which corresponds to the user's jacket color. Although their work is dependent to either a laptop or a PC and that they didn't describe their tracking algorithm's implementation and performance, they successfully tested a UAV as a sports assistant, although the external visual imagery made the athlete feel dizzy.

There are other approaches to achieve interaction between humans and UAVs with significant use of computational power. For example, while achieving interaction with the a UAV, the user's face and gestures can be tracked, although the UAV reference frame has to be initialized before an actual flight can take place [4]; color markers on good lighting conditions can help distinguish all commands to drive a swarm of UAVs [5]. Similar results can be obtained with a Kinect sensor and its corresponding software libraries [6].

2 Human-UAV Interface

On this section we provide a description to our interface's hardware, software and control schemes. Essentially, the UAV reacts to user's commands that it captures through its on board camera; all commands are processed on one mobile device holded by the user. The software is capable of estimating the user's movements and gestures throughout the video stream sent from the UAV (Fig. 1a); it also estimates user's position, relative to the drone, to control the aircraft and make it follow and interpret user's movements and gestures (Fig. 1b).

We used the Parrot's AR-Drone, a WiFi controlled quadcopter, to develop this interface. The communication is accomplished through UDP ports to send and receive: control commands (UDP 5556) for maneuvering the drone, navigation data (UDP 5555) that describe the current drone's state and lastly, the video stream (UDP 5554). As a mobile processing platform, we chose Asus' Transformer Prime, which featured the Tegra 3 processor, the world's first quad core mobile processor from NVIDIA. More on how we take advantage of the tablet's architecture is included in next section.

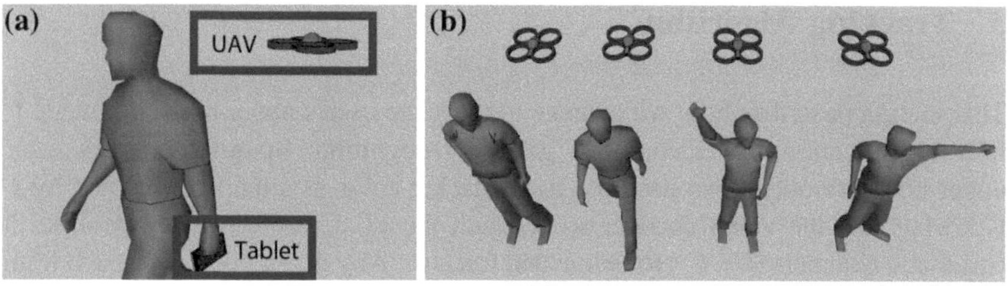

Fig. 1 Interface description. **a** User, followed by the UAV. **b** Four possible scenarios

We built from the ground-up an application named IHRVANT, which is a multi-threaded application that initializes, attends and closes the communication with the UAV. First four threads attend GUI, control commands, navigation data and video stream reception, respectively (see Fig. 2). The video reception thread is far more complex, because it manages all image processing threads, that represent the tracking algorithm and gesture recognition functionality.

The video data stream is sent from the drone to the mobile device coded with Universal Variable Length Code with the following features: YC_bC_r color space subsampled 4:2:0, 8 × 8 DCT, resolution QVGA. Since we based our digital image processing (DIP) techniques in the YC_bC_r color space, the decoder feeds directly all color channels to DIP threads only if they are idle. With this strategy the video stream is displayed flawlessly while DIP threads work on the latest accepted frame. All parameters can be managed with controls available on IHRVANT's interface.

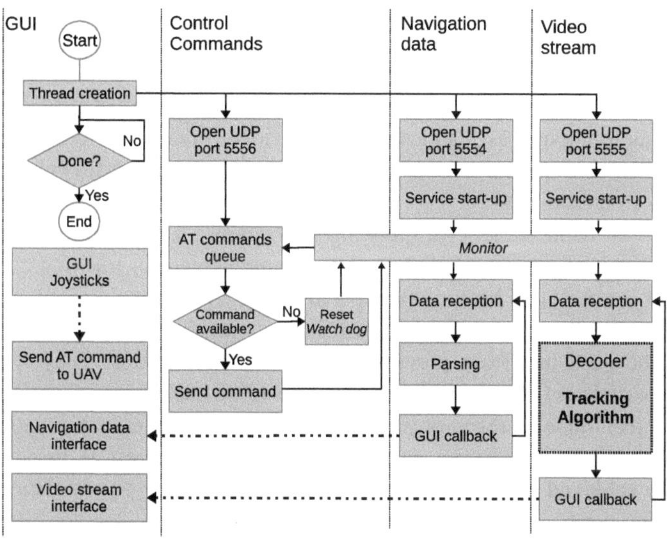

Fig. 2 Thread diagram for IHRVANT. Video stream reception thread decodes and process all received frames

3 Tracking Algorithm

This section describes how we manage to track the user's upper body silhouette for later classification and accomplish gesture recognition. In order to track user's upper body silhouette, we used the user's jacket color as a marker, denoted by $\mathbf{c}^*$. Knowing that the video decoder works with the YC_bC_r color space, we used the chromatic channels (C_b, C_r) to define and track $\mathbf{c}^*$. Any given pixel from each frame is to be segmented if the euclidian distance $d(\cdot)$ between its corresponding chromatic data $\mathbf{c} = (c_b, c_r)$ to a reference $\mathbf{c}^* = (c_b^*, c_r^*)$ is smaller than a threshold T_c (See Eq. 1).

$$\mathbf{C_s}(x, y) = \begin{cases} 1 \text{ if } \mathrm{d}(\mathbf{c}(\mathrm{x,y}),\mathbf{c}^*) < \mathrm{T_c} \\ 0 \text{ if } \mathrm{d}(\mathbf{c}(\mathrm{x,y}),\mathbf{c}^*) > \mathrm{T_c} \end{cases} \quad (1)$$

Indeed, the size of each frame improved the segmentation performance in the mobile platform. The luminance and chromatic channels are subsampled (Sect. 2), which means the resolution for each channel are: Y channel is QVGA, while the C_b and C_r channels are both QQVGA.

Additionaly, the luminance channel provided important contrast information, gray tones segmented from the Y channel are meant to be within one fixed interval $[y^* - 15, y^* + 15]$, where y^* represents the user's jacket gray tone (Eq. 2). To adapt to luminosity changes, y^* is continuously updated with a rate of 1 Hz by IHRVANT.

$$\mathbf{Y_s}(x, y) = \begin{cases} 1 \text{ if } (\mathrm{y}^* - 15) \leq \mathbf{Y}(\mathrm{x,y}) \leq (\mathrm{y}^* + 15) \\ 0 \text{ Oththerwise} \end{cases} \quad (2)$$

Last strategies mentioned for color and gray segmentation would easily lead to non desirable *false-positives*. In other words, not all segmented pixels will necessarily belong to the true user's location on the video frame. To overcome this, while adding resilience to partial occlusion, we fusioned the results from color segmented areas with a particle filter.

There is a wide variety of tracking algorithms, among them, probabilistic algorithms such as particle filters have been implemented for tracking color objects [7, 8] and some have been implemented using AR-Drone's front camera [3]. Particle filters estimate a variable's distribution function using a set of weighted samples (the particles) from previous variable's observations. They work with a weighted particle set $\mathrm{X_t}$ of size I, sampling based on each particle's weight is what makes this technique so relevant [9]. In this particular case, our observation process corresponds to the coordinates in which our area of interest is located on each frame. We achieved this by making all the particles track one specific color histogram using the C_b and C_r channels from the video stream. A color histogram can be seen as a bidimensional table in which all pixels are categorized depending on its corresponding chroma values and n discrete intervals for each chroma channel, giving a total of n^2 bins.

The first important aspect of our implementation is how we defined each particle as a square region W_i with side length l, centered in coordinates (x_t, y_t) at time t. The state vector for each particle is defined as $\mathbf{x_t} = (x_t, y_t, x_{t-1}, y_{t-1})^T$. Secondly, the state transition distribution from which we sample variable's a priori hypothetical state $\tilde{\mathbf{x}}_{\mathbf{t+1}}$ does not require all previous observations. We assumed that the object moves in uniform motion frame by frame, hence we only required the last two observations $\mathbf{x_{t-1}}$ and $\mathbf{x_t}$; this is called a second order dynamic model (see Eq. 3, $\mathbf{v_t} \sim N(0, \sigma)$) [10].

$$\tilde{\mathbf{x}}_{\mathbf{t+1}} = \begin{pmatrix} 2 & 0 & -1 & 0 \\ 0 & 2 & 0 & -1 \\ 1 & 0 & 0 & 0 \\ 0 & 1 & 0 & 0 \end{pmatrix} \mathbf{x_t} + \begin{pmatrix} 1 & 0 \\ 0 & 1 \\ 0 & 0 \\ 0 & 0 \end{pmatrix} \mathbf{v_t} \tag{3}$$

The third relevant aspect is how we weighted all particles; we computed the similarity between the normalized color histogram $\mathbf{q_t}$, taken from W_i, and a normalized reference color histogram $(\mathbf{q}^*)$ with the Bhattacharyya similarity coefficient (Eq. 4). As noted by Pérez et al. [11], we also observed a consistent exponential behavior for the squared distance $\mathbf{D^2}$, hence we used Eq. 5 for weighting all particles ($\lambda = 20$). Later resampling with draw probability $\propto p(\mathbf{x_{t+1}} | \tilde{\mathbf{x}}_{\mathbf{t+1}}, \mathbf{q}_{t+1})$ will suffice to make all particles look as if they were bees looking for honey, all trying to match the reference color histogram $\mathbf{q}^*$.

$$D[\mathbf{q}^*, \mathbf{q_t}(x_t, y_t)] = \left[1 - \sum_{n=1}^{N} \sqrt{\mathbf{q}^*(n)\mathbf{q_t}(x_t, y_t)} \right]^{\frac{1}{2}} \tag{4}$$

$$p(\mathbf{x_{t+1}} | \tilde{\mathbf{x}}_{\mathbf{t+1}}, \tilde{\mathbf{q}}_{t+1}) \propto \exp(-\lambda D^2[\mathbf{q}^*, \tilde{\mathbf{q}}_{\mathbf{t+1}}]) \tag{5}$$

Finally, to fusion the color ($\mathbf{C_s}$) and gray tone segmented ($\mathbf{Y_s}$) areas and particle set $\mathbf{X_t}$, we used morphological operations. We considered color segmented pixels useful only if 3 or more Y pixels are lit, those are marked with green in Fig. 3c After one dilation with structural element diamond (3 × 3), we overlayed all particles for applying the *bush-fire* algorithm using the particles as obstacles and the free space as the color segmented areas [12] (Fig. 3e). We diminished all areas not reached by fire (Fig. 3f) and by doing so, we eliminated all *false-positives* (shown in Fig. 3e in green). Figure 4c shows how IHRVANT displays fusion results in blue, segmented area's centroid (B_c) and its boundary box (B_{bb}) in yellow. During normal operation, the area in blue corresponds to the user's position in the video frame and represents his upper body silhouette, which we can classify to accomplish body gestures interpretation.

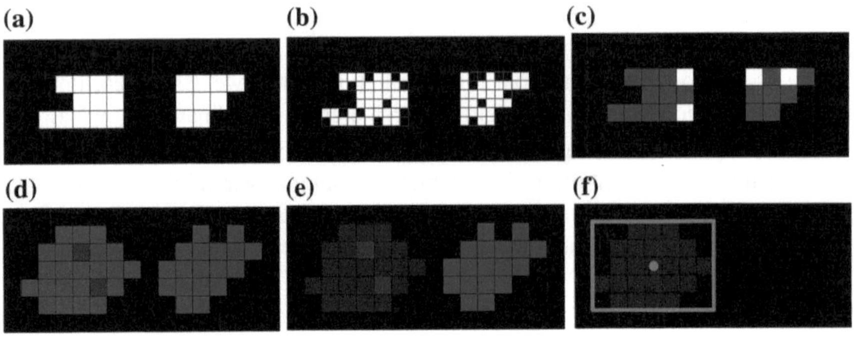

Fig. 3 Fusion of all DIP techniques. **a** Color segmentated area. **b** Gray tone segmented area. **c** Fusion 3a and 3b. **d** Dilation and particles. **e** After brushfire. **f** Area of interest

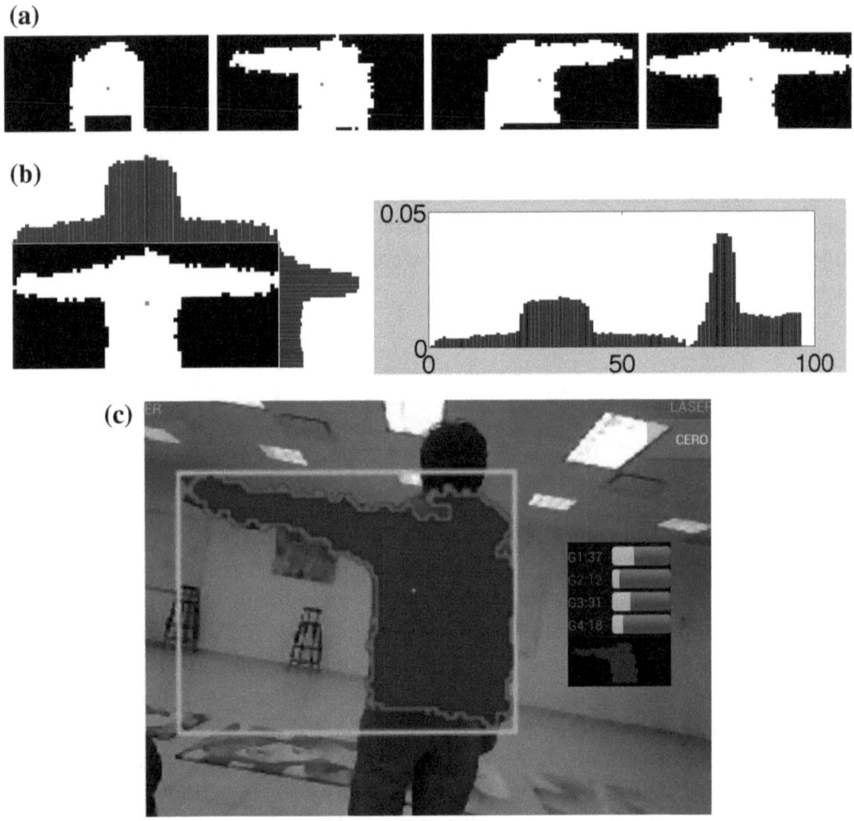

Fig. 4 Histograms for classification. **a** All four gestures proved with IHRVANT. **b** *Vertical* and *horizontal* histograms for gesture 4 and its complete normalized descriptor. **c** Silhouette stabilization and successful classification while the drone is rotated around the roll axis

4 Gesture Recognition

For classifying the user's upper body silhouette, we use a normalized histogram-based descriptor, which we reorientate based on the camera's *roll* angle estimated by the AR-Drone; by absorbing minor misleading rotations caused by normal flight conditions, we greatly improved our descriptor's certainty. To compute our descriptor, the region of interest B_{bb}, is scaled and reorientated by the *roll* angle into one fixed-size bitmap B_s (See Fig. 4c). B'_s width and height (w_s and h_s, respectively) ought to follow human proportions. On our tests, a proportion equal to $w_s : 2.5h_s$ suffices to depict human's upper body proportions when arms are horizontal or close to the body. We first scales boundary box's height (h_{bb}) to match B'_s (height scale factor $s = h_{bb}/h_s$), while translating B_c to the center of B_s; then, we rotate the scaled area according to the UAV's *roll* state variable, using as pivot the center of B_s. In Fig. 4c we show their corresponding B_s bitmap after scaling, translating and rotating is done.

Our descriptor is based on normalized histograms, from B_s we obtain its vertical and horizontal histograms, which in turn can be unified into one normalized histogram **h** (see Fig. 4b). The classifier is trained with the average of 31 samples of **h**, taken 1.2 s apart from each other while the user holds his pose in front of the UAV's camera; the averaged histogram $\mathbf{h}^*$ becomes the descriptor for the gesture trained. With two or more trained gestures, the classifier starts measuring the square error between any given descriptor against all available gestures using Eq. 6, in other words, with two or more trained gestures, all B_s images are classified at the same rate at which the segmentation process delivers its results, which is 14 times per second. The gesture for which the error is minimum will be considered as the best match. All errors computed are shown by IHRVANT as a bar graph, the winner's label is shown in red to the user (see Fig. 4c).

$$e = \sum_{i=1}^{N} (h_i^* - h_i)^2 \tag{6}$$

IHRVANT supports up to four different gestures and it triggers one valid gesture detection event if the upper-body silhouette is classified continuously for one and a half seconds i.e. the descriptor of the gesture has been classified 21 consecutive times as either one of the available gestures. Figure 4a shows all four tested gestures.

5 Control

In Fig. 5 we show AR-Drone's flight altitude relative to the ground (g) and the three fixed axis that are used to describe any aircraft's pose on air. For each of these references there is one control variable we can modify to govern the AR-Drone.

Fig. 5 Control scheme

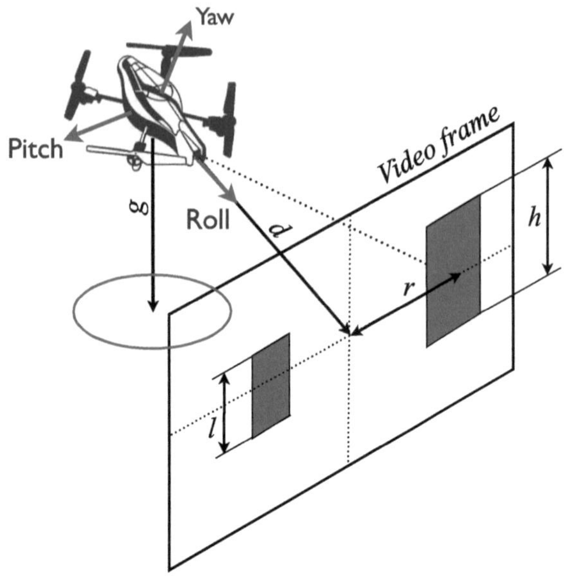

If we modify the variable related to the *roll* axis, we make the AR-Drone go sideways, the *pitch* variable will make it go forward or backwards, the *yaw* variable will make it spin and the altitude variable will make it go up and down.

For each variable just mentioned, we used a closed loop PID control scheme. We used the estimated flight altitude we were receiving from the drone as input to the altitude's controller. From the tracking algorithm (Sect. 3) we obtain the user's segmented area, boundary box dimensions and its centroid, from which we can estimate the user's position relative to the drone. In Fig. 5 a gray rectangle describes the possible user's position with a gray area and the distance from its centroid to the video frame's center (r), this result is used as input to the controller for the *yaw* axis. Figure 5 shows two gray areas with a different height with $l < h$, they illustrate that the closer the user is to the drone, the bigger h becomes. The distance d serves as input to the PID controller for UAV's *pitch* axis.

6 Results

The reader is welcome to watch a video on YouTube named "IHRVANT",[1] it shows how the UAV follows the user and reacts to body gestures; it also includes one link to another video showing how we used the tablet's touch capabilities to initialize all segmentation processes. Since our tracking algorithm is working along

[1] Look for this video in the uploaded videos section available on one of the author's YouTube channel: http://www.youtube.com/danielsoto888.

side the video decoder, we reached more than sufficient performance. Consider that the UAV sends the video at 18 FPS; color and gray tone segmentation: 14 FPS, Particle filter ($I = 200, l = 7, n = 4$) 7 FPS, after the fusion of all results, we reached an overall 14 FPS. The control loop for all four flight variables (Sect. 5) is updated for every computed frame at the same rate the segmentation process delivers one frame (14 Hz). In gesture classification, trying with *roll* angles in the [0°, 15°] range, we measured our classifier performance statistically. By measuring the proportion of actual positives correctly identified, we reached a sensitivity of 89 %. In the other hand, by measuring the proportion of actual negatives identified as such, we reached a specificity of 90 %. It is worth mentioning that we didn't use architecture's specific instructions to achieve this performance, including NEON instructions nor GPU capabilities.

7 Conclusion

We presented a feasible implementation of a portable Human Machine Interface with an UAV. Capable of successfully tracking the user indoors throughout the video stream without many light restrictions and a dynamically changing background; we also showed how to accomplish a vision based control system in most recent portable devices, capable of autonomously drive the UAV without the need of high performance computing. The AR-drone is not able to carry any significant additional payload and that made impossible for us to add exteroceptive sensors onboard the drone, that has become part of our future job. Our main effort is with the idea that sooner or later, robots will take a step forward into our everyday social interaction.

Acknowledgments The authors would like to thank to Cinvestav and CONACYT for providing the means to carry on with our research.

References

1. Ng, W.S., Sharlin, E.: Collocated interaction with flying robots. In: RO-MAN, 2011 IEEE, pp. 143–149, 31 July–3 Aug 2011
2. Duncan, B.A., Murphy, R.R., Shell, D., Hopper, A.G.: A midsummer night's dream: social proof in HRI. In: 2010 5th ACM/IEEE International Conference on Human-Robot Interaction (HRI), pp. 91–92, March 2010
3. Higuchi, K., Shimada, T., Rekimoto, J.: Flying sports assistant: external visual imagery representation for sports training. In: Proceedings of the 2nd Augmented Human International Conference, AH'11, pp. 7:1–7:4, New York, NY, USA, 2011. ACM
4. Monajjemi, V.M., Wawerla, J., Vaughan, R., Mori, G.: HRI in the sky: creating and commanding teams of UAVs with a vision-mediated gestural interface. In: 2013 IEEE/RSJ International Conference on Intelligent Robots and Systems (IROS), pp. 617–623, Nov 2013

5. Nagi, J., Giusti, A., Gambardella, L.M., Di Caro, G.A.: Human-swarm interaction using spatial gestures. In: 2014 IEEE/RSJ International Conference on Intelligent Robots and Systems (IROS 2014), pp. 3834–3841, Sept 2014
6. Pena Palacio, J.A., Londono Arevalo, S.: Remote control flying RC helicopter through natural user interface. In: 2014 9th Iberian Conference on Information Systems and Technologies (CISTI), pp. 1–6, June 2014
7. Chen, Y., Yu, S., Fan, J., Chen, W., Li, H.: An improved color-based particle filter for object tracking. In: Second International Conference on Genetic and Evolutionary Computing, 2008, WGEC'08, pp. 360–363, Sept 2008
8. Zhang, T., Fei, S., Li, X., Lu, H.: An improved particle filter for tracking color object. In: 2008 International Conference on Intelligent Computation Technology and Automation (ICICTA), vol. 2, pp. 109–113, Oct 2008
9. Thrun, S., Burgard, W., Fox, D.: Probabilistic Robotics (Intelligent Robotics and Autonomous Agents). The MIT Press, Cambridge (2005)
10. Satoh, Y., Okatani, T., Deguchi, K.: A color-based tracking by kalman particle filter. In: Proceedings of the 17th International Conference on Pattern Recognition, 2004, ICPR 2004, vol. 3, pp. 502–505, Aug 2004
11. Pérez, P., Hue, C., Vermaak, J., Gangnet, M.: Color-based probabilistic tracking. In: Proceedings of ECCV, pp. 661–675 (2002)
12. Choset, H., Lynch, K.M., Hutchinson, S., Kantor, G.A., Burgard, W., Kavraki, L.E., Thrun, S.: Principles of Robot Motion: Theory, Algorithms, and Implementations. MIT Press, Cambridge (2005)

Design and Simulation of Robot Manipulator Position Control System Based on Adaptive Fuzzy PID Controller

F.Z. Baghli and L. El Bakkali

Abstract The problem of manipulator control is a highly complex problem of controlling a system wich is multi-input multi-output, non linear and time variant. In this paper we present a new approach for a robot manipulator with two degrees of freedom based on the intelligent adaptive fuzzy PID, Parameters which guide the functioning of Proportional Integral Derivative controller are dynamically adjusted with the assistance of fuzzy control to ensure the position robot control strategy, the proposed control scheme is based on nonlinear dynamic model derived using Lagrange-Euler formulation. Our robot manipulator's simulated in Matlab Simulink environment, the results obtained present the efficiency and the robustness of the proposed control with good performances compared with the classical PID.

Keywords Robot manipulator · Fuzzy logic · PID · Adaptive fuzzy PID

1 Introduction

Research on the dynamic modeling and control of the arms manipulators has received increased attention since the last years due to their advantages, this system is a mechanical system multi-articulated, in which each articulation is driven individually by an electric actuator is the most robot used in industry, this system need an efficient control strategy.

A robot manipulator is a high-speed process that is highly nonlinear, dynamically coupled and often it is not adequate to use linear servo control, if accurate performance in high bandwidth operations is desired. Many efforts have been made in developing control scheme to achieve the precise tracking control of robot manipulators [1–3].

F.Z. Baghli (✉) · L. El Bakkali
Modeling and Simulation of Mechanical Systems Laboratory, Faculty of Sciences, University Abdelmalek Essaadi, BP. 2121 M'hannech, 93002 Tetouan, Morocco
e-mail: baghli.fatimazahra@gmail.com

S. Zeghloul et al. (eds.), *Robotics and Mechatronics*,
Mechanisms and Machine Science 37, DOI 10.1007/978-3-319-22368-1_24

The traditional PID controller with simple structure and stable performance is widely used. But it is difficult to meet the high precision and fast response, moreover the parameters tuning of classical PID controller is so complex [4]. Therefore fuzzy algorithm is introduced. Fuzzy control is a particular type of intelligent control, has a great potential since it is able to compensate for the uncertain nonlinear dynamics using the programming capability of human control behavior. The main features of fuzzy control is that a control knowledge base is available within the controller and control actions are generated by applying existing conditions or data to the knowledge base, making us of inference mechanism [2, 5]. Also, the knowledge base and inference mechanism can handle no crisp, incomplete information; the knowledge itself will improve and evolve through learning and past experience [2].

Fuzzy logic control does not require a conventional model of the process, whereas most conventional techniques require either an analytical model or an experimental model. Fuzzy logic control is particularly suitable for complex and ill-defined process in which analytical modeling is difficult due to the fact that the process is not completely known and experimental model identification is not feasible because the required inputs and output of the process may not be measurable [5].

In this work after the system modeling, simulation and control robot manipulator using two articulations for motion using MatLab/Simulink software were carried, when the proposed Adaptive Fuzzy PID controlled is used to improve the articulation robot stability. Two types of control PID and Fuzzy PID were studied; analysis and comparative studies were made.

The reminder paper was structured as follow: the robot modeling is presented in second part of this paper, in the third part of this paper Intelligent Fuzzy PID is detailed, the results discussion are presented in the last part of this paper and finally conclusion was given.

2 Intelligent Fuzzy PID Controller

Fuzzy controllers have been widely applied to industrial process. Especially, fuzzy controllers are effective techniques when either the mathematical model of the system is nonlinear or not the mathematical model exists. In this paper, the fuzzy control system adjusts the parameter of the PID control by the fuzzy rule. Dependent on the state of the system. The adaptive PID realized is no more a linear regulator according to this principle. In most of these studies, the Fuzzy controller used to drive the PID is defined by the authors from a series of experiments [6–9]. The frame of the fuzzy adaptive PID controller is illustrated in Fig. 1.

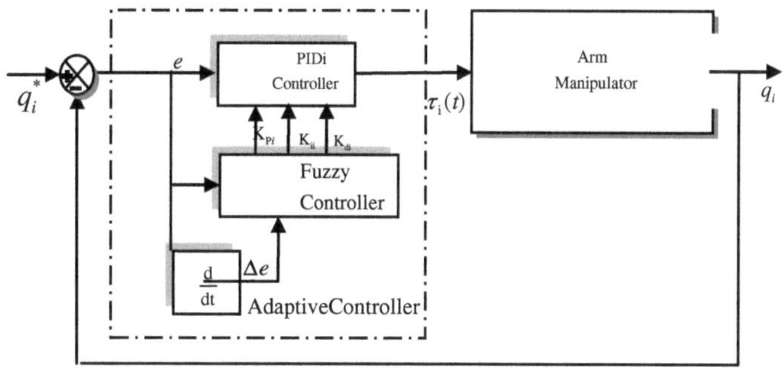

Fig. 1 Arm manipulator robot Fuzzy PID control

2.1 Fuzzy Inference System

The fuzzy inference system has been considered the min-max method (Mamdani), where the implication has been assumed to min and the aggregation has been considered to max. In addition, the Defuzzification method has been considered to the centroid method.

2.2 Input and Output Variables

The position control of the robot arm manipulator requires two fuzzy PID controllers (fuzzy PID1 applied to the first joint and fuzzy PID2 applied to the second joint). Where two inputs and three outputs have been considered for the fuzzy PID1, same thing for fuzzy PID2. The two inputs are *e* (position error), *de* (velocity error) for the first and second controller. The output variables are the proportional gain K_{pi}, the integral gain K_{ii}, the derivative gain K_{di} (i = 1, 2).

2.3 Membership Function

In the proposed fuzzy PID control (1 or 2), a seven-term set {negative big (NB), negative middle (NM), negative small (NS), zero (ZE), positive small (PS), positive middle (PM) positive big (PB)} is applied to defining input and output linguistic variable.

The system structure of Fuzzy PID controller are shown in Fig. 2.

The membership function for outputs variables illustrated in Fig. 3 where the domain of gain K_{pi}, K_{ii}, K_{di} are set by {−0.3 −0.2 −0.1 0 0.1 0.2 0.3};{−0.06 −0.04 −0.02 0 0.06 0.04 0.02};{−3 −2 −1 0 1 2 3}.

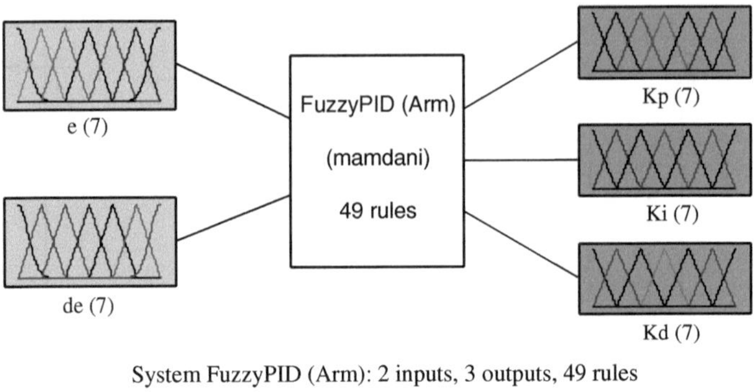

Fig. 2 Structure of Fuzzy PID controller

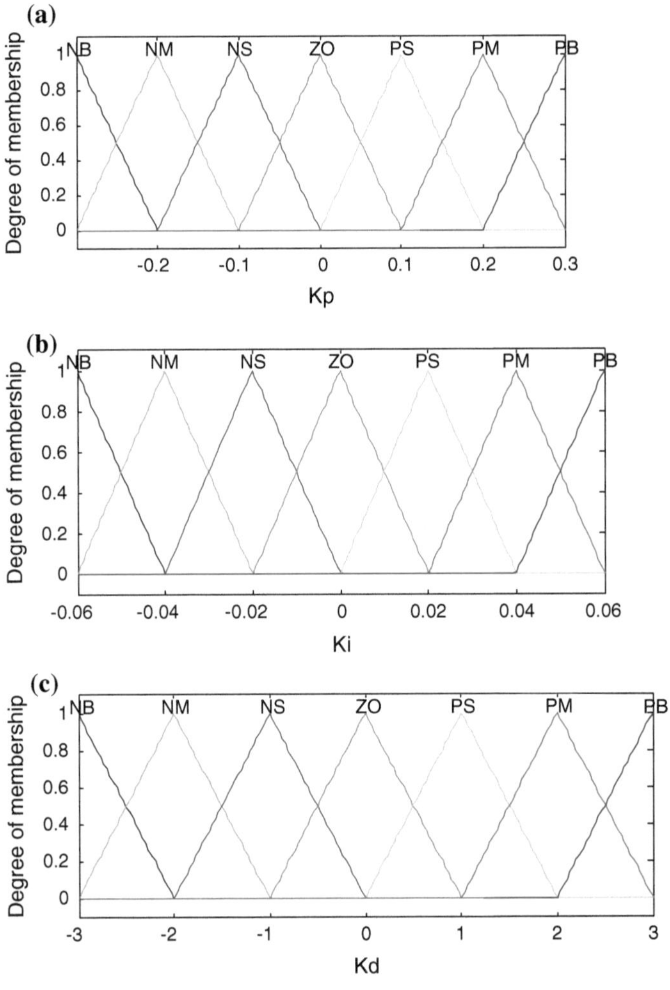

Fig. 3 The membership function for K_{pi}, K_{ii}, K_{di}

2.4 Fuzzy Rules-Base

Building fuzzy logic rule is the key step in the improvement of system performance, which directly affects the quality control system. In this paper, fuzzy control rules are designed with these tune principles as follows:

While error is large, control variable should try to reduce error.

While error is small, the stability of system should be considered, besides eliminating the error (Tables 1, 2, 3 and 4).

Table 1 Effects of increasing a parameter independently

Parameter	Rise time	Overshoot	Settling time	Steady state error
K_{pi} ↑	Decrease	Increase	Small change	Decrease
K_{ii} ↑	Decrease	Increase	Increase	Eliminate
K_{di} ↑	*Minor change*	*Decrease*	*Decrease*	–

Table 2 The fuzzy rule base for K_{pi}

K_{pi}		De						
		NB	NM	NS	ZO	PS	PM	PB
e	NB	PB	PB	PM	PM	PS	ZO	ZO
	NM	PB	PB	PM	PS	PS	ZO	ZO
	NS	PM	PM	PM	PM	ZO	ZO	NS
	ZO	PM	PM	PS	ZO	NS	NS	NM
	PS	PS	PS	ZO	NS	NS	NM	NM
	PM	PS	ZO	NS	NM	NM	NM	NB
	PB	ZO	ZO	NM	NM	NM	NB	NB

Table 3 The fuzzy rule base for K_{ii}

K_{ii}		De						
		NB	NM	NS	ZO	PS	PM	PB
e	NB	NB	NB	NM	NM	NS	ZO	ZO
	NM	NB	NB	NM	NS	NS	ZO	ZO
	NS	NB	NM	NS	NS	ZO	PS	PS
	ZO	NM	NM	NS	ZO	PS	PM	PM
	PS	NM	NS	ZO	PS	PS	PM	PB
	PM	ZO	ZO	PS	PS	PM	PB	PB
	PB	ZO	ZO	PS	PM	PM	PB	PB

Table 4 The fuzzy rule base for K_{di}

K_{di}		De						
		NB	NM	NS	ZO	PS	PM	PB
e	NB	PS	NS	NB	NB	NB	NM	PS
	NM	PS	NS	NB	NB	NB	NM	PS
	NS	ZO	NS	NM	NM	NS	NS	ZO
	ZO	ZO	NS	NS	NS	NS	NS	ZO
	PS	ZO	ZO	ZO	ZO	ZO	ZO	ZO
	PM	PB	NS	PS	PS	PS	PS	PB
	PB	PB	PM	PM	PM	PS	PS	PB

These tune principles we permit to create the fuzzy rule-base for K_{pi}, K_{ii}, K_{di}, These rules can be written in the format of IF-THEN as follows (there are total 49 rules):

1. If (e is NB) and (de is NB), then (K_{pi} is PB) (K_{ii} is NB) K_{di} (is PS);
2. ....

49. If (e is PB) and (de is PB), then (K_{pi} in NB) (K_{ii} is PB) (K_{di} is PB).

3 Simulation

SISO control based on classical PID model and intelligent control based on Fuzzy PID model were tested to sinus response trajectory. This simulation applied to two degrees of freedom robot arm was implemented in Matlab/Simulink. Position error are compared in these controllers.

Figures (4, 5, 6, and 7) are shown error performance, by comparing position error for the first and second link. The FLC PID Control ensure the robot manipulator's stability by maintaining the error position equal to zero ($\varepsilon_1 = \varepsilon_2 = 0$) so PID (i)'s error position is respectively higher than FLC PID (i) where (i = 1, 2).

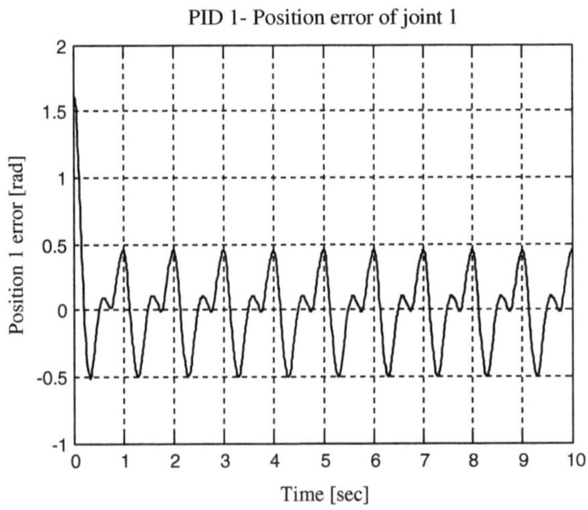

Fig. 4 PID 1 for the first link position error

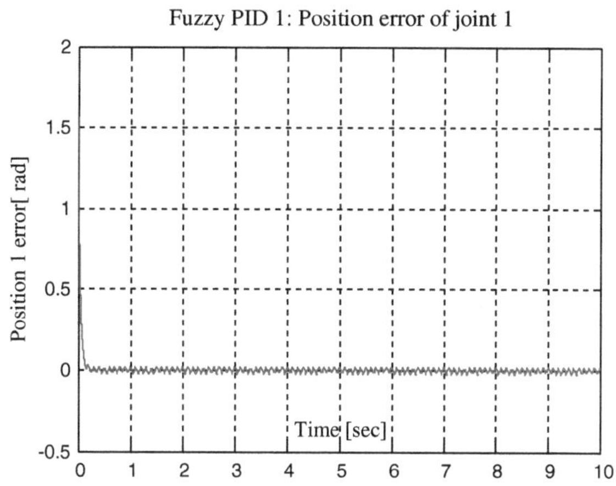

Fig. 5 FLC PID 1 for the first link position error

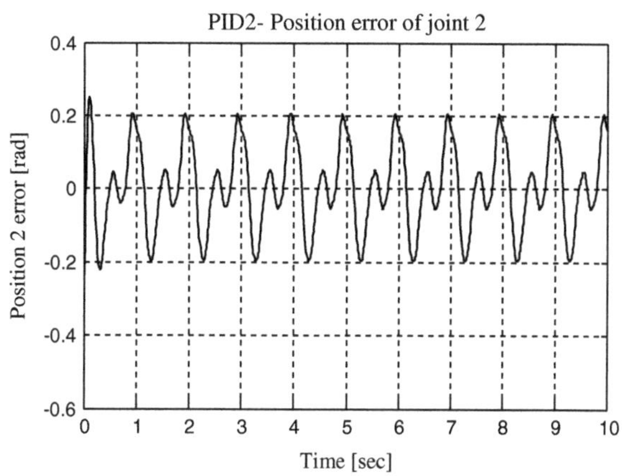

Fig. 6 PID 2 for the second link position error

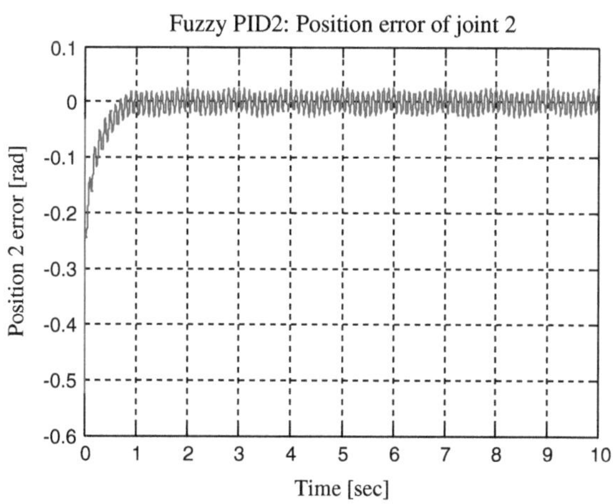

Fig. 7 FLC PID 2 for the first link position error

4 Conclusion

In this present work an arm manipulator robot using two degree of freedom was controlled using two types of controls strategies, SISO control based on classical PID model and intelligent adaptive Fuzzy PID, this last one present maximum control structure of our control model and give more and more efficiency for the robot model with more position stability and good dynamical performances with no overshoot so industrials would take into account the efficiency of the developing control model for the futures two freedom robot design considerations.

References

1. Asada, H., Slotine, J.J.: Robot Analysis and Control. Wiley, New York (1986)
2. Sumathi, P.: Precise tracking control of robot manipulator using fuzzy logic, In: DARH2005 Conference, Session 4.1
3. Lewis, F.L., Abdallah, C.T., Dawson, D.M.: Control of Robot Anipulators. Macmillan, New York (1993)
4. Yu, B., Zhu, H., Xue, C.: Research on adaptive fuzzy PID synchronous control strategy of double-motor I.J. Intell. Syst. Appl. **5**, 28–33 (2011)
5. Passino, K.M., yurkovich, S.: Fuzzy logic. Addison Wesley longman California (1998)
6. Ata, A.A.: Optimal trajectory planning of manipulator: review. J. Eng. Sci. Technol. **2**(1), 32–54 (2007)
7. Ang, K.H., Chang, G., Li, Y.: PID control system analysis, design and technology. IEEE Trans Control Syst. Technol. **13**(4), 559–577 (2005)
8. Astrom, K.J., Hagglund, T.: PID controllers: theory, design, and tuning, (2nd edn.). Instrument society of America (1995)
9. Wang, J.S., Zhang, Y., Wang, W.: Optimal design of PI/PD controller for non-Minimum phase system. Trans. Inst. Meas. Control **28**(1), 27–35 (2006)
10. Bingul, Z.: A new PID tuning technique using differential evolution for unstable and integrating processes with time delay, In: ICONIP, Proceedings Lecture Notes in Computer Science vol. 3316, pp. 254-260 (2004)
11. Allaoua, B., Laoufi, A., Gasbaoui, B.: Multi-drive paper system control based on multi-input multi-output PID controller. Leonardo J. Sci. (16), 59–70 (2010)
12. Baghli, F.Z., El bakkali, L., Lakhal, Y., Nasri, A., Gasbaoui, B.: The efficiency of the inference system knowledge strategy for the position control of a robot manipulator with two degree of freedom. Int. J. Res. Eng. Technol. **2**(7) (2013)

Generating the Optimum Dynamic Trajectory of a Hybrid Cable-Serial Robot

M. Ismail, S. Lahouar and L. Romdhane

Abstract In this work, a geometric path planner and a dynamic controller are designed for a hybrid cable serial robot. The shortest path between two positions in presence of obstacles, is found using a geometric path planner then a dynamic joint space controller finds the fastest way to reach the goal while ensuring that cables tensions stay positive. The goal of this controller is to maintain the tensions in the cables in a bounded interval between a minimum and a maximum value. By limiting the maximum acceleration of the different links, the controller ensures that the tensions stay within an allowable range. However, these constraints limit the maximum speed reachable by the robot, and the controller tries to find the fastest solution, which minimizes the total elapsed time while respecting the constraints. In conclusion, the trajectory planner finds the shortest path and the dynamic controller finds the minimum total elapsed time for this path.

Keywords Hybrid serial cable robot · Total elapsed time · Optimization

M. Ismail · L. Romdhane
LMS, ENISO, Université de Sousse, Sousse, Tunisia
e-mail: mourad.smail@gmail.com

L. Romdhane
e-mail: lotfi.romdhane@gmail.com

S. Lahouar (✉)
LGM, ENIM, Université de Monastir, Monastir, Tunisia
e-mail: samirlahouar@gmail.com

L. Romdhane
American University of Sharjah, Sharjah, UAE

S. Zeghloul et al. (eds.), *Robotics and Mechatronics*,
Mechanisms and Machine Science 37, DOI 10.1007/978-3-319-22368-1_25

1 Introduction

Recently, several hybrid cable-serial robots can be found in the literature, see for example [1–4]. The multi-body (serial part) motion is provided through cables pulling on the end-effector, instead of using rotary actuators mounted in the joints. Removing the actuators from the joints yields a robot with lower moving inertia. This property makes hybrid cable robots more attractive in terms of dynamic capability, achieving complex motions. One promising application of these robots is medical rehabilitation where the use of the serial linkage provides many desirable advantages compared to traditional cable robot like, improved accuracy, moment resistance at the end effector, reduced out of pane compliance. These robots are relatively new and controlling them should take into consideration dynamics and cable tensions. There are few works taking into account these aspects.

In this article we consider dynamics, cable tensions, obstacle avoidance and time minimization all in an integrated method.

The robot used in this work is a hybrid (serial/cable), planar, under constrained and fully actuated (see Fig. 1). It has two degrees of freedom, and it moves in a vertical plane. The robot includes a serial support composed of two passive links. It acts, due to its weight. The end-effector (point M) is manipulated by two cables attached to the tensioning actuators.

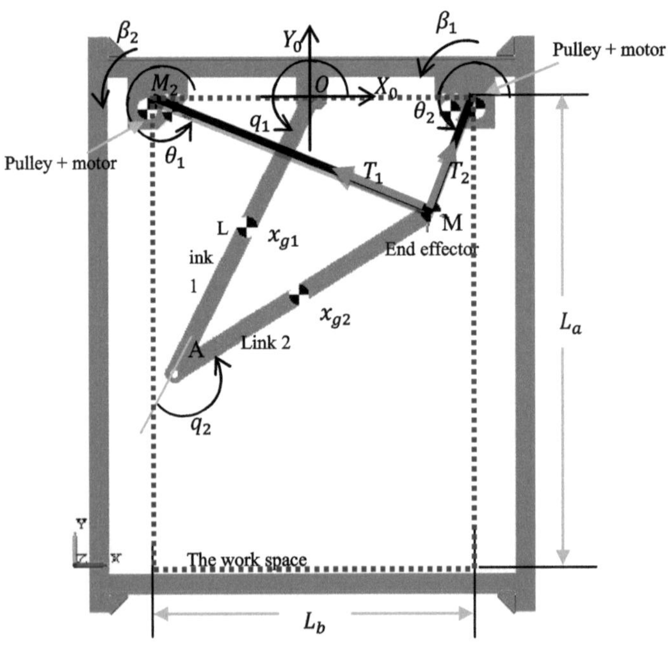

Fig. 1 The studied hybrid serial cable robot

This hybrid structure provides unique features like, high payload to weight ratio, a relatively large workspace, a reduced out of plane compliance, a moment resistance at the end effector, reduced tracking errors generally induced by cable elasticity.

The geometric parameters of the system are defined in Fig. 1. The serial manipulator is characterized by the following parameters:

Serial robot	Link 1	Link 2
Length	$OA = l_1$	$AM = l_2$
Mass	m_1	m_2
Inertia	J_1	J_2
Joint parameters	$\boldsymbol{q} = \begin{bmatrix} q_1 \\ q_2 \end{bmatrix}$	

End-effector position	$\boldsymbol{x} = \begin{bmatrix} x_1 \\ x_2 \end{bmatrix}$
Link1Center of gravity	$\boldsymbol{x}_{g1} = \begin{bmatrix} x_{g11} \\ x_{g12} \end{bmatrix}$
Link2Center of gravity	$\boldsymbol{x}_{g2} = \begin{bmatrix} x_{g21} \\ x_{g22} \end{bmatrix}$
Inertia of the pulleys and motors 1 and 2	$\boldsymbol{j} = \begin{bmatrix} j_1 \\ j_2 \end{bmatrix}$
Pulleys viscous friction	$\boldsymbol{c}$
Pulley radius	r
mass of the end effector	m
Inertia of the end effector	j

Cable robot	Cable 1	Cable 2
Length	$M_1M = L_1$	$M_2M = L_2$
Tension	T_1	T_2
Orientation	θ_1	θ_2
Actuator angles	β_1	β_2
Location	$OM_1 = \frac{L_b}{2}$	$OM_2 = \frac{L_b}{2}$

The paper is organized as follows: Sect. 2 introduces the method used to generate collision free dynamically feasible trajectories. Section 3 contains the obtained results. Some concluding remarks are presented in Sect. 4.

2 Collision Free Dynamically Feasible Trajectory Planning Algorithm

Interaction between different parts of the algorithm is explained in Fig. 2. Given the geometry, the obstacles, the initial position and the final position a trajectory is generated using a path planner.

The controller computes inverse kinematics and torques to be applied to the robot.

These torques are applied to the robot's dynamic model via a numerical solver that gives the final trajectory. An algorithm verifies that cable tensions are within a feasible region. This algorithm is used in order to minimize the total elapsed time.

2.1 The Shortest Collision-Free Trajectory

Collision free trajectory is generated using a path planning algorithm dedicated for cable robots [5]. This algorithm is based on the switch between two modes depth mode and width mode as shown in Fig. 3.

As described in Fig. 3, only the collisions with the end effector are considered. In fact it is not practical to check all the collisions with the cable and the serial manipulator because the reachable workspace region will be greatly reduced. Furthermore, one can loop throw all the positions occupied by the serial support and the cables for each pose and still be able to use this same technique.

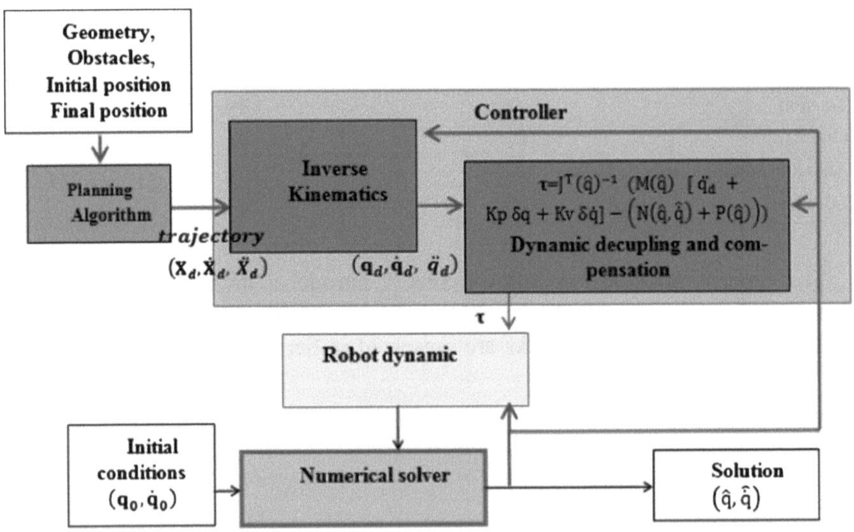

Fig. 2 The controller scheme

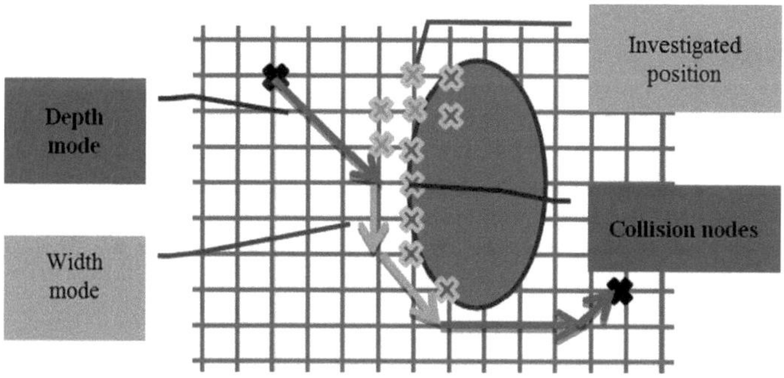

Fig. 3 Collision free trajectory

To achieve smooth trajectory planning witch is needed to unsure velocity and acceleration derivability, Spline interpolation is used.

2.2 *Robot Dynamics*

Using the Lagrange formulation, the equation of motion of the robot can be written as follows:

$$M(q)\,\ddot{q} = N(q,\dot{q}) + P(q) + J^{T}(q)S(\theta)T \tag{1}$$

where M is the generalized mass (2 × 2) matrix of the robot, N is the (2 × 1) vector containing the centrifugal and Coriolis forces, P is the (2 × 1) gravitational force vector, J is the Jacobian matrix of the robot and $J^{T}(q)S(\theta)T$ is the projection of cable tensions T in the frame R0.

2.3 *Total Elapsed Time*

For each cable the chosen velocity during the trajectory has a trapezoidal profile as shown in Fig. 4.

In the case of a trapezoidal speed profile minimizing total elapsed time while keeping cables tensions in a bounded interval can be formulated as:

objective function:

$$\mathrm{Min}_{as,ai,Vlim}\, t_{tran}(as, ai, Vlim) = \frac{D_X}{Vlim} + \left(\frac{Vlim}{2}\right) * \left(\frac{1}{as} + \frac{1}{ai}\right) \tag{2}$$

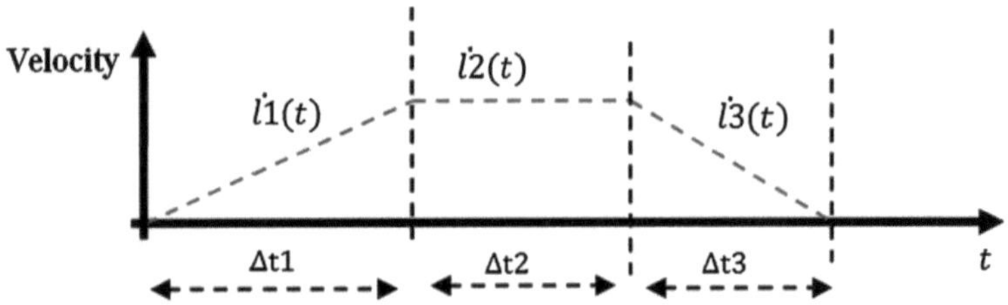

Fig. 4 Velocity profile

Subject to:

$$
\begin{gathered}
-\int S(\theta)^{-1} \cdot J^{T}(q)^{-1}(M(q)\ddot{q} - N(q,\dot{q}) - P(q))dq \leq 0 \\
\int (S(\theta)^{-1} J^{T}(q)^{-1}(M(q)\ddot{q} - N(q,\dot{q}) - P(q)) - Tmax)dq \leq 0 \qquad (3) \\
-Vlim \leq 0, -as \leq 0, -ai \leq 0
\end{gathered}
$$

$\forall q, q, \ddot{q}$ during all the trajectory.

This approach is not limited to a trapezoidal profile. It can be applied to any given trajectory or velocity profile.

2.4 Verification Algorithm

The algorithm of Fig. 5 gives the sum error during the track of a given trajectory using the values Vlim, as and ai.

The optimization scheme is shown in Fig. 6.

3 Results

Given a task in Fig. 7, the end effector of the robot has to move from the upper position A to the lower position B while avoiding the obstacles. We consider in this figure that the obstacles are in the plane of the effector but not in the plane of the serial links. Thus there is no collision between the obstacles and the serial links.

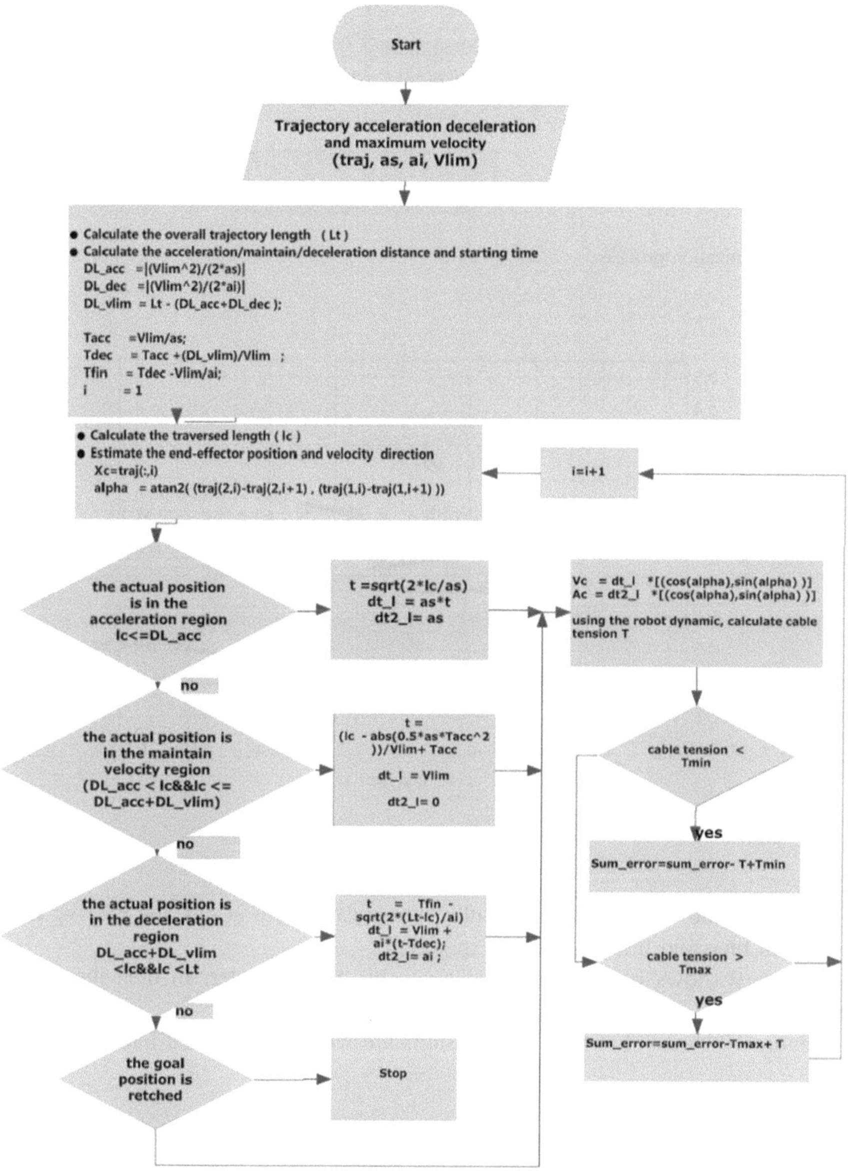

Fig. 5 Verification algorithm

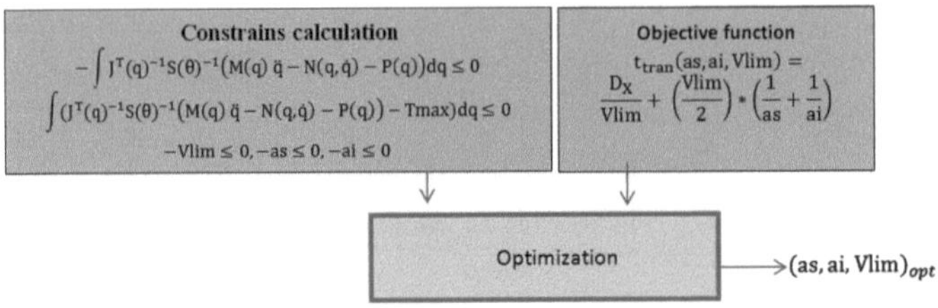

Fig. 6 Optimization process

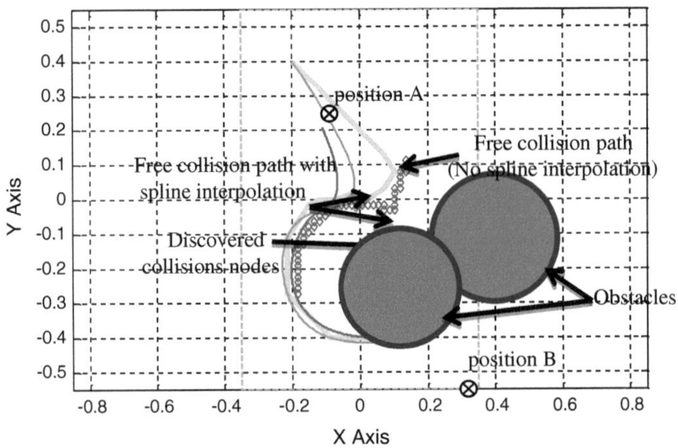

Fig. 7 Collision free trajectory

Figure 8 shows the obtained results, where cables are in red, the passive serial part is in blue and the trajectory is in green.

Figure 9 shows position tracking given by the controller. In blue the desired position and in red the real position of the robot.

Figure 10 shows that the cable tensions are within the desired interval.

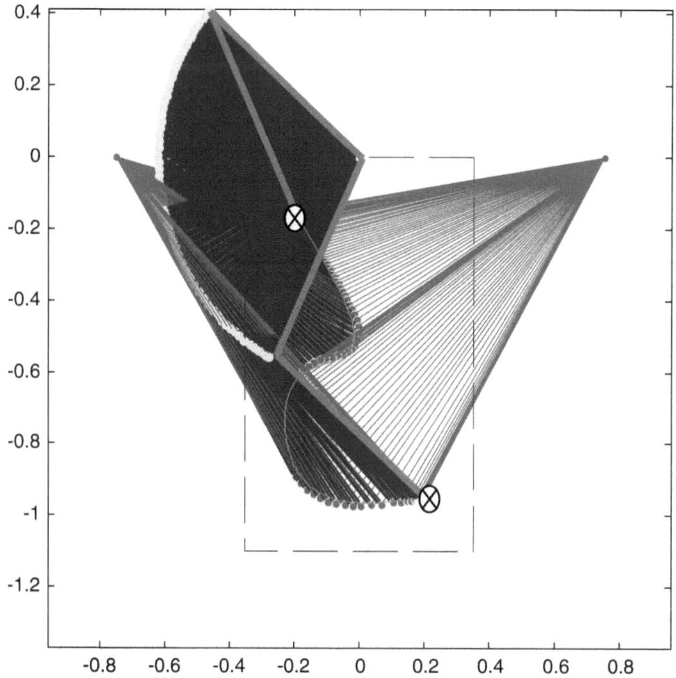

Fig. 8 Results with different positions of the robot

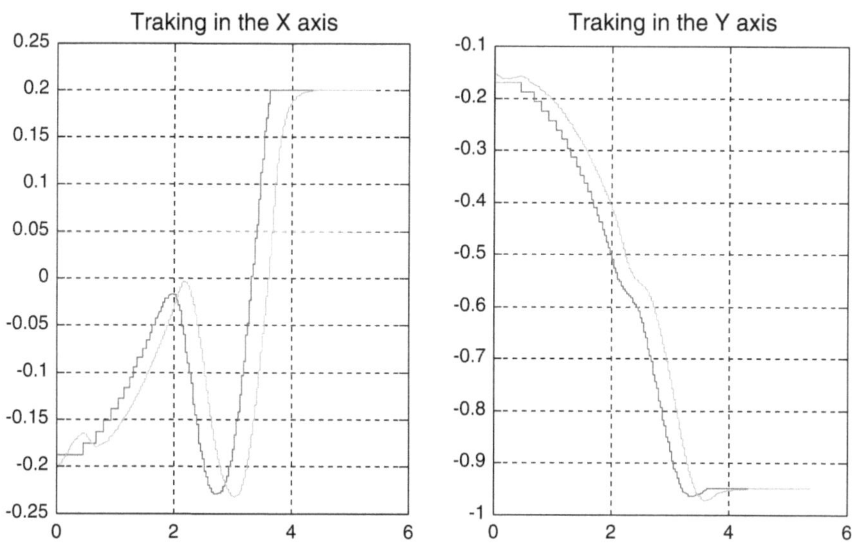

Fig. 9 Controller results

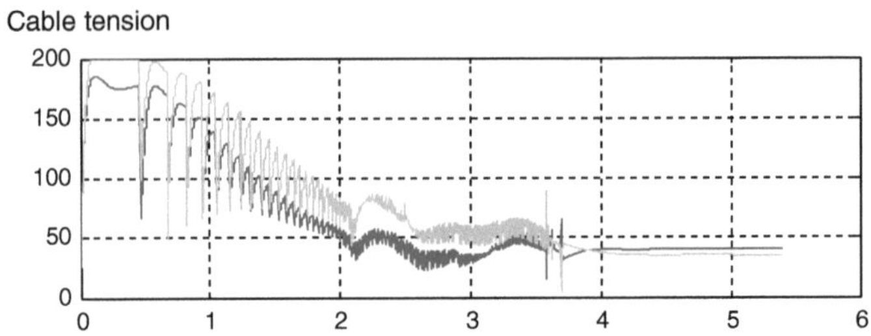

Fig. 10 Cable tensions

4 Conclusion

In this work, the problem of dynamic path planning for a hybrid cable serial robot has been addressed. The proposed approach starts by generating the shortest geometric obstacle-free path, followed by a dynamic path tracker. During this phase, the trajectory is tracked using the maximum velocity and acceleration, while keeping the tensions in the cables within an acceptable range. The objective is to reduce the total elapsed time from the initial to the final position. The obtained results show the efficiency of the method to find the shortest and the fastest path between two given positions.

References

1. Ming, A., Higuchi, T.: Study on multiple degree-of-freedom positioning mechanism using wires (part 1)—concept, design and control. Int. J. Jpn. Soc. Precis. Eng. **28**, 131–138 (1994)
2. Merlet, J.-P.: Parallel Robots. In: Solid Mechanics and Its Applications. Springer, New York (2006)
3. Albus, J., Bostelman, R., Dagalakis, N.: The NIST robocrane. J. Robotic Syst. **10**(5), 709–724 (1993)
4. Trevisani, A., Gallina, P., Robert, L.: Cable-direct-driven robot (CDDR) with passive SCARA support theory and simulation. J. Int. Robotic Syst. **46**(1), 73–94 (2006)
5. Lahouar, S., Ottaviano, E., Zeghoul, S., Romdhane, L., Ceccarelli, M.: Collision free path-planning for cable-driven parallel robots. Robotics Auton. Syst. **57**(11), 1083–1093 (2009)

An Integrated Software Package for Advanced Industrial Robot Applications

C. Liang, H. Yan, R. Li, I.-M. Chen, M.H. Ang Jr. and Z. Huang

Abstract This paper introduces an integrated robot software package for advanced industrial robot applications with the aims to make industrial robots easy-to-operation, easy-to-programming, much more flexible with advanced capabilities. This software package is developed based on ROS (Robot Operation System) middleware and ROS-Industrial open source packages by integrating different function modules. The module design implementation of a graphic user interface, visual programming, intermediate robot language and simulation environment are presented. Utility and flexibility of the software package are evaluated through two experiment demos: automated robot bin-picking and automated robot 3d object taping, which has a potential to be used for a wide of robot applications.

Keywords Industrial robot · Robot software development · Robot bin-picking · Robot taping

1 Introduction

For more than 30 years, industrial robots have been widely used for material handling, welding and dispensing in automobile industry, which are cost-effective for a high-volume and repetitive tasks [1]. However, low-volume, high-mixed parts

C. Liang (✉) · I.-M. Chen
School of Mechanical and Aerospace Engineering, Nanyang Technological University, Singapore, Singapore
e-mail: chliang@ntu.edu.sg

H. Yan · M.H. Ang Jr.
Department of Mechanical Engineering, National University of Singapore, Singapore, Singapore
e-mail: mpeyanh@nus.edu.sg

R. Li · Z. Huang
Institute for Infocomm Research, Agency for Science Technology and Research, Singapore, Singapore
e-mail: li-r@i2r.a-star.edu.sg

S. Zeghloul et al. (eds.), *Robotics and Mechatronics*,
Mechanisms and Machine Science 37, DOI 10.1007/978-3-319-22368-1_26

production are the characteristics for next-generation manufacturing. Thus, smarter, faster and more flexible industrial robots are required. Advanced capabilities like real-time collision avoidance, mobile manipulation, 3d perception enabled path planning and grasp planning are preferred for accomplishing particular tasks that have never done before. These requirements greatly challenge capabilities of tradition industrial robots in both hardware and software aspects [2].

To meet these challenges for future markets, robot manufactures are developing next-generation robots with more advanced capabilities for different application areas. There are several commercial products available in the market. The KUKA LBR lightweight robot is a torque controlled 7 d.o.f.s (degrees of freedom) robot manipulator with intrinsic safety, which has been widely used in academic laboratories for research purposes [3]. ABB Frida® is a dual-arm industry robot for small parts assembly tasks in 3C industries (Computers, Communications and Consumer electronics). This robot is capable of working side by side with human workers [4]. A Danish robot company Universal Robots provides UR5 and UR10 light weight robots, which have a wide application in both research fields and practical industry applications [5]. Additionally, Nextage-Open is dual-arm industrial humanoid robot developed by Kawada industry Co. Ltd in Japan. About 150 units are currently working in factories together with human workers [6].

There are also many efforts on the development of robot software [7]. A comparison of robotic middleware is presented in [8]. ROS (Robot Operating System) is being used extensively by the research community for service robotics [9]. In particular, ROS has been applied to industrial robotics through ROS-Industrial, which is an open-source project that extends the advanced capabilities of ROS software to manufacturing, which contains libraries, tools and drivers for industrial hardware [10]. However, this software package is still for experienced developers. It takes a long time to build a robot application. A Japanese startup company Mujin released the Mujin controller, which claims it is the world's first commercial industrial robot motion controller for real-time applications like bin-picking and multi-robot manipulation planning. Mujin is capable of connecting with several different industrial robots to enhance their capabilities [11].

In this paper, a software package is introduced for advanced industrial robot applications. This software package is developed as a part of RADOE (Robot Application and Operating Environment) work package in the Singapore A*STAR industrial robotics research program. The rest of this paper is organized as follows: In Sect. 2, an architecture design of the software package is proposed. Functions of each module are briefly described. In Sect. 3, software modules design implementation issues are addressed in terms of simulations, graphic interface and visual programming, intermediate robot language, program manager and system integration. In Sect. 4, an automated robot bin-picking application and an automated robot 3d object taping application are demonstrated to evaluate practical operation performances of the software package.

2 Architecture Design

Figure 1 shows a scheme of the integrated software package architecture for advanced industrial robot applications, which consists of six modules. Functions of each module can be described in the following:

(1) Simulation module is used to test developed control algorithms and for operation performances evaluation purposes. Simulation saves a lot of testing time for a complex system before the software deploying on a real robot. In particular, robot and devices can be conveniently enabled or disabled by configure macro parameters for a hardware in the loop simulation.
(2) Intermediate Robot Language (IRL) module is to describe a task that can be implemented on different robotics platform. IRL releases people from managing different robot programs in different languages that depend on manufactures'. It also reduces the effort required to plan, apply and maintain core robotic programming techniques, saving time and money.
(3) Advanced capabilities module is a cluster of open source software packages available from robotics research community. This module takes advantages of other people's work and advanced capabilities for industrial robot applications thus can be easily integrated.
(4) Interface module consists of two sub-modules: comparing with tedious manual creates ROS components and coding, visual programming tool is to specify programs by manipulating function modules. It allows access to software libraries for rapid robot application development with appropriate APIs (Application Programming Interfaces). Visual user interface sub-module is used to configure the programming and operation environment, setting environment variables, as well as for parameters tuning purposes.

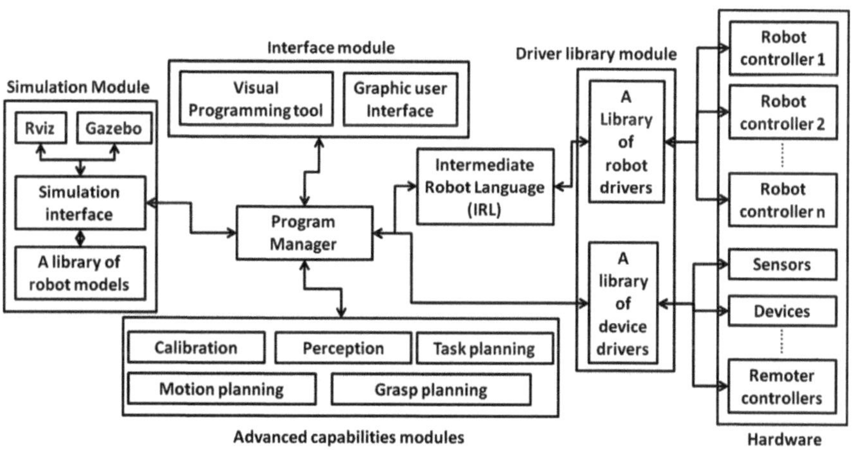

Fig. 1 A scheme illustrates an architecture design of the integrated software package

(5) Driver library module contains two sub-libraries. One library includes device drivers for communicating and control of a variety of robots. In general, robot manufactures provide these ROS compatible robot hardware drivers. Another library is the device drivers for peripherals (such as conveyors, mobile bases, rotary tables, etc.), actuators and sensors.
(6) Program manager module is the core of whole software package, which is responsible for setting up the whole program through parameters configuration files. It monitors operation sequence of each sub-module and plays a role of error recovering. In addition, program manager integrates other modules during the software package implementations.

3 Module Implementation

3.1 Simulation Module

Simulation module is developed based on Rviz and Gazebo in ROS environment. Rviz (ROS visualization) is used for displaying a virtual model of the industrial robot and live representations of sensor values coming over ROS Topics including camera data, laser scanner point cloud data and etc. Gazebo is a dynamic simulation environment, which can communicate with simulation interface via Gazebo-ROS API plugging. A library of robot models is included in the simulation module. Robot model are described in URDF (Unified Robot definition format) files. URDF files can be converted to SDF files via the use of XACRO for Gazebo simulations.

3.2 Intermediate Robot Language

Intermediate Robot Language (IRL) module enables users to conveniently program and design robotic applications in a generic and scalable way. The proposed IRL is completely independent to any robot manufactures' language. Development of the IRL commence in conjunction with the simulation platform. An IRL program consists of three components: .dat file-contains robot set up and target data;.prog file-contains individual robot programs; .task file-contains task level commands (for example call robot programs, timing, robot I/O etc.) These files can be individually categorised within an xml file as shown in Fig. 2.

3.3 Graphical Interface and Visual Programming

A user-friendly graphical interface and visual programming tool are created in C++ using QT creator. Visual programming tool allows a user to drag and drop different

```
<IRL_Program id="bk101">
 <data>
     ABB_2400 = {kin_model..}; rob_targ1 = {0,1,3,51,1...}; .......
 </data>
 <Programs>
   <prog id="Rob1_1"> <Robot> ABB_2400 </Robot>
     <motions>
      MoveJ{rob_targ1,100,100} MoveL{rob_targ2,100,100} .......
     </motions>
   </prog>                                          Program 1

   <prog id="Rob2_1"> <Robot> ABB_4400_60 </Robot>
      ......
   </prog>                                          Program 2
    ......
 </programs>
 <tasks>
   Call Program "Rob1_1" {run_time="15" Max_Vel="1.1" ... ...}
   Wait SIG[1] then Wait 10 and Sig[2]="true"
   Start Concurrent :"Rob1_2" "Rob2_1"
   End Concurrent
   ......
 </tasks>
</IRL_Prog>
```

Fig. 2 An example xml file illustrates the intermediate robot language file structure

components onto canvas from a library. Each component is a ROS launch file and the connections between components represent ROS messages being passed through ROS topics. The output of the visual programming tool is automatically created ROS launch files with parameters, "remap"-ing and dependencies. This module has following functions:

(1) Creating, saving, loading project files on a graphical interface
(2) Adding components such as motion planning and grasping planning
(3) Run simulations through corresponding launch files
(4) Ability to drill down into a module to see the various sub-components and rewire them as necessary
(5) Auto-populating all launch files available in a build ROS package

3.4 *Program Manager and Integration*

Program manager module bridges communications between different sub-modules by using ROS topics, services and actions. For each sub-module, different message types are created on the base of existing ROS messages format. These messages communication are base on the publishing and subscribing mechanism, which can be defined in ROS msg files. For services and actions, message types are defined in .srv and .action files, respectively.

In particular, two messages 'ErrorHandle.msg' and 'ModuleState.msg' are defined. When an error in one sub-module occurs, error message is published through 'ErrorHandle.msg' for error recovering purposes. As for 'ModuleState.msg', it is published repeatedly in every 1 s in order to monitor the running status of each sub module. In addition, configuration parameters for each sub-module are stored in.xml files, which are included in launch files. Each sub-module runs as a node in ROS environment.

4 Example Demos

Two experiment demos have been built in the laboratory to evaluate operation performances of the developed software package. One demo is an automated robot bin-picking system with aims to pick different shaped items from a stationed shelf according to a wish list. This demo demonstrates an advanced capability of an industrial robot by integrating 3d object recognition, real-time obstacles avoidance motion planning, grasping and task planning. The second demo is an automated robot 3d object taping system. This demo aims to show flexibility of the software package, which is capable of generating and executing a taping motion trajectory after receiving point clouds data of a scanned 3d object.

4.1 Demo I: Automated Robot Bin-Picking

Picking and placing is one of the most common tasks for traditional industrial robots. In general, the robot is programmed to perform repetitive picking for a high-volume object with similar shapes in a simple constructed environment. Automated picking of different shaped objects in a cluttered environment is still very challenge for traditional industrial robots. Figure 3a shows an experiment setup of the automated robot items picking system, which is composed of a UR5 lightweight robot manipulator, a versatile robot gripper, a Kinect 3d sensor and a shelf with different shaped grocery products. One challenge is automatic recognition of the target object, thus a library including RGB and depth images, 3d point clouds model of all candidate objects are built. In the experiment, the robot is capable of generating a collision avoidance path and deploying a suitable grasping strategy to pick up the object.

Figure 3b shows a snapshot of the simulation. Collision avoidance motion planning algorithm and tasking planning strategy can be tested with real streaming data from Kinect sensor, which saves a lot of time for testing. Figure 4 shows snapshots of experiment tests for a bin-picking movement. The robot manipulator successfully picks a yellow-duck-bath-toy for one shelf bin and places it to an order bin.

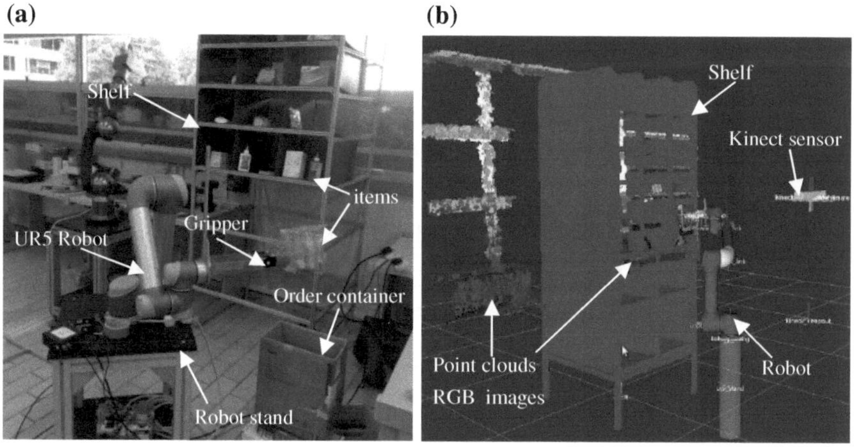

Fig. 3 Automated robot picking demo: **a** an experiment setup; **b** a snapshot of the simulation

Fig. 4 Snapshots of experiment test for picking an object from the shelf

4.2 Demo II: Automated Robot 3d Object Taping

Another experiment demo is the automated robot 3d object taping as shown in Fig. 5a. The automated taping system is composed of a UR10 lightweight robot manipulator, a taping mechanism mounted on the robot end-effector, a Kinect 3d sensor and a rotation platform. A 3d model of the tapped object can be obtained by receiving scanned data from Kinect sensor. Figure 5b shows an example 3d model of the scanned object. A motion planning algorithm is used to generate a motion trajectory on the surface of the tapped object. In particular, the 3d model scanning algorithm has been built in the software package so that it is capable of taping different objects in a very short of setup time. Furthermore, taping movement can be simulated in a simulation environment to evaluated prescribed taping velocity and dimension of overlap between each layer.

Figure 6 shows snapshots of the robot taping experimental test. It can be noted that tape is pressed on the object surface. The compliance of the taping mechanism maintains certain of taping force. In the taping procedure, the robot moves up-down

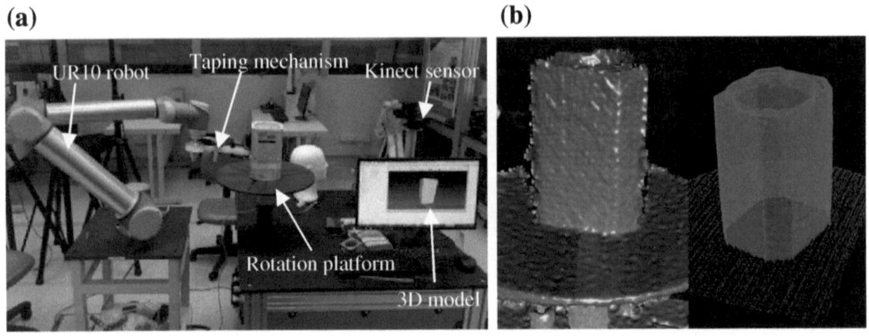

Fig. 5 Automated robot taping demo: **a** an experiment setup; **b** 3d models of the taped object

Fig. 6 Snapshots of experimental test for an automated robot taping

and keeps taping mechanism head normal to the object surface. The motion of rotation platform is synchronized with robot to ensure a good quality taping. Tapes are taped on the object surface for with prescribed velocity and overlap.

5 Conclusions

This paper introduces an integrated software package for advanced industrial robot applications. The open architecture of this software package makes it is very easy and flexible to integrate high-level robotic software modules and high value-added robot software library for diversified industrial robotic applications. Advantages of deploying this software package in industrial robots can be summarized as easy-to-use, easy-to-programming, and flexible with advanced capabilities. From two experiment demos, utility and flexibility of the developed software package are evaluated. This integrated software package has the potential for a wide of industrial robot applications.

Acknowledgments This work has been partially supported by the Singapore Agency for Science, Technology and Research (A*STAR) under SERC industrial robotics research program Project No. 1225100008.

References

1. Hägele, M., Nilsson, K., Pires, J.N.: Industry Robotics, (Springer Handbook of Robotics), pp. 963–986. Springer, New York (2008)
2. BMBF: Industry 4.0. http://www.bmbf.de/en/19955.php (2015)
3. Rainer, B., Kurth, J., Schreiber, G., Koeppe, R., Albu-Schäffer, A., Beyer, A., Eiberger, O.: The KUKA-DLR Lightweight Robot arm-a new reference platform for robotics research and manufacturing. In: Proceedings 2010 6th German conference on robotics (ROBOTIK), pp. 1–8. VDE (2010)
4. Jingguo, G., Zhaofu, C., Qingwei, L.: Small part assembly with dual arm robot and smart camera. In: Proceedings of International Symposium on Robotics, ISR/Robotik 2014, pp. 1–6. VDE (2014)
5. Ostergaard, E.H.: Light weight robot for everybody, IEEE Robot. Autom. Mag. **19**(4) (2012)
6. Christian, S., Karayiannidis, Y., Nalpantidis, L., Gratal, X., Qi, P., Dimarogonas, V., Kragic, D.: Dual arm manipulation—a survey. Robot. Auton. Syst. **10**, 1340–1353 (2012)
7. Nader, M., Al-Jaroodi, J., Jawhar, I.: Middleware for robotics: a survey, robotics, In: Proceedings of th 2008 IEEE Conference on Automation and Mechatronics AIM 2008, pp. 736–742 (2008)
8. Gergely, M., Sinčák, P., Krizsán, Z.: Comparison Study of Robotic Middleware for Robotic Applications, Emergent Trends in Robotics and Intelligent Systems, pp. 121–128. Springer International Publishing, New York (2015)
9. Quigley, M., Conley, K., Gerkey, B., Faust, J., Foote, T., Leibs, J., Wheeler, R., Andrew, YNg.: ROS: An open-source robot operating system. In: ICRA workshop on open source software, vol. 3.2, p. 5 (2009)
10. Edwards, S., Chris, L.: ROS-industrial–applying the robot operating system (ROS) to industrial applications. In: IEEE International Conference on Robotics and Automation, ECHORD Workshop (2012)
11. Diankov, R., James, K.: Openrave: A Planning Architecture for Autonomous Robotics, Robotics Institute Technical Report, CMU-RI-TR-08-34 79 (2008)

Part VII
Advanced Robotics

A Method for the Approximation of the Multiple IK Solutions of Regular Manipulators Based on the Uniqueness Domains and Using MLP

Vassilis C. Moulianitis, Evgenios M. Kokkinopoulos and Nikos A. Aspragathos

Abstract In this paper, a method for dividing the training data as well as training MLP systems to obtain the multiple solutions of IKP of regular manipulators is presented. The sets of training data for each system are strictly defined using the concept of uniqueness domains. The training data are obtained by the forward kinematics and the sign of the determinant of the manipulator Jacobian is used for the determination of the uniqueness domains. An illustrative example with a 3 dof robot with known IK solutions is presented for the verification of the proposed approach.

Keywords Inverse kinematics · Multiple solutions · MLP · Regular manipulators

1 Introduction

The kinematics of an open-chain manipulator are modeled by a set of non-linear and transcendental equations, expressing the map of the configuration space to Cartesian space. Generally, the inverse kinematic problem (IKP) has multiple

V.C. Moulianitis (✉)
Department of Product and Systems Design Engineering, University of the Aegean, 84100, Ermoupolis, Syros, Greece
e-mail: moulianitis@syros.aegean.gr

E.M. Kokkinopoulos · N.A. Aspragathos
Mechanical Engineering and Aeronautics Department, University of Patras, 26500 Rio, Patras, Greece
e-mail: kokkinopoulos.evgenios@gmail.com

N.A. Aspragathos
e-mail: asprag@mech.upatras.gr

S. Zeghloul et al. (eds.), *Robotics and Mechatronics*,
Mechanisms and Machine Science 37, DOI 10.1007/978-3-319-22368-1_27

solution; it is a difficult problem and its solvability depends on the anatomic characteristics of the robotic manipulator.

A variety of methods including analytic, numerical or based on soft computing have been developed to solve the IKP. In [6], the constraints of the structural parameters, concerning the distance and orientation between consecutive joint axes of the manipulator, that give closed-form solutions of IKP were investigated.

Among the numerical methods, the polynomial continuation, the Groebner Bases technique as well as the ones based on the methods were presented [7]. However, these methods are vunerable to local optimums, and time consuming. An evolutionary approach together with two niching strategies for finding all the IKP solutions [4], and another with adaptive niching and clustering [10] were presented. Both methods are based on the direct kinematics of the robot, however these methods can be used only for off-line processing.

Neuro-fuzzy techniques [5, 9] solved simple manipulators and provided only one solution to the IKP. An algorithm [8] which used radial basis function networks and two fuzzy logic methods to determine all the solutions of the IKP was presented. The training data were distinguished manually based on prior knowledge of the multiple configurations corresponding to a given pose of the robot. However, the manipulator anatomies used made the division of training data easier, since the robot used had known inverse kinematics solution and the simple anatomy pf the manipulators made easy to distinguish the multiplicity of configurations (e.g. above-below, left-right).

A method proposed [1] for speeding up the learning of robot kinematics by decomposition of the manipulator into two or more virtual robot arms and by the reduction of the number of training samples.

To our knowledge, there is no presented method or technique based on soft computing that could be used to find all the solutions of the IKP and to be on-line applicable.

In this paper, a method for approximating all the solutions for the IKP using Multilayer Perceptron systems is proposed. This method is based on the forward kinematic to obtain the training data so it could be used to resolve the IKP of any regular robot. The training data set is divided into subsets called uniqueness domains, corresponding to each solution. Since each solution of the IKP is located in a uniqueness domain and they are separated by the singularity locus then a change of posture implies passing through a singularity. This division is derived by using the singularity locus determined by the sign change of the manipulator Jacobian determinant.

Current robots are designed in a limited number of structure types in order to obtain analytical solutions. A generic approach for the accurate approximation of all the IK solutions opens a wide search space for new robot structures either modular metamorphic or fixed towards reconfigurable manufacturing systems.

2 Problem Definition

In this paper, MLP systems will be used to approximate all the solutions of the IKP only for regular manipulators. A systematic and generic approach is proposed for the structure of the system approximating all the solutions. The most important part for the structure of the proposed approximator of the IK solutions, is the appropriate division of the training data to subsets corresponding to each IK solution. The proposed division is based on the knowledge considering the division of the configuration space in uniqueness domains [11], which is equal to the number of IK solutions. Therefore the proposed structure of the IK approximator includes the $m \times n$ number of MLP systems where n ($n \leq 6$) is the number of DoF and m the number of uniqueness domains. For a general 6 DoF manipulator there are 96 MLP systems at maximum that must be trained since the maximum number of solutions are sixteen [6].

In the following a technique is proposed for the determination of the uniqueness domains with low computational time, in order to determine the training data for each group of MLP corresponding to each IK. Then the system of $m \times n$ approximators of the multiple solutions for each joint variable is structured. An approximator based on MLP is trained and tested. For verification purposes the simulated experiments were implemented using a regular manipulator with known IK solutions.

2.1 The Configuration Subspaces Determination

In this section, the determination of the uniqueness domains and their correspondance to the IK solutions are presented. This is critical for the structure of the system for the approximation of all the IK and for the training each of them. Generally, the IKP problem is represented by the following:

$$\underline{\theta} = \underline{f}^{-1}(\underline{p}) \tag{1}$$

where, $\underline{\theta}$ are the joint variables, $\underline{p}$ represents the position and orientation of the end-effector and $\underline{f}$ the forward kinematics function. The inverse kinematics solution exist only when the determinant of the Jacobian matrix is not singular.

Therefore, the sets will be separated by the singularity locus provided by $det(\mathbf{J}) = 0$. The open sets in the configuration space that are limited by the singularity locus has been presented in [2, 11]. These open sets are named as c-sheets or aspects or uniqueness domains. In regular manipulators, each of these open sets contain only one solution of the IKP, something that is not true for cuspidal manipulators [11]. In Fig. 1, the singularity locus of a three DoF regular manipulator is presented by red and green dotted lines and as it can be seen separates the

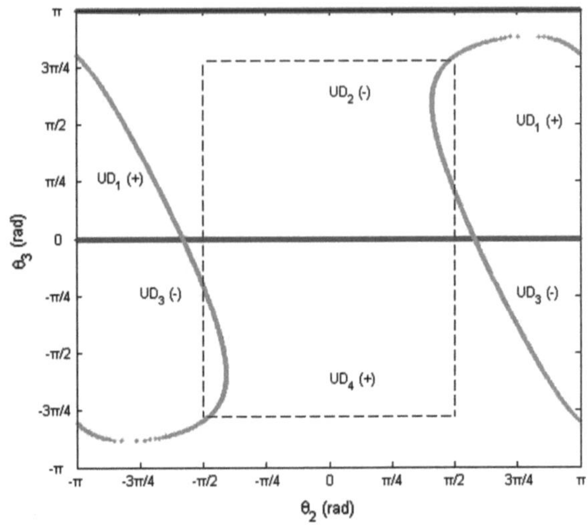

Fig. 1 Singularity locus and the inverse kinematics solutions of a 3 Dof manipulator with anatomical parameters presented in the beginning of Sect. 3

(θ_2, θ_3) domain in four areas. Each uniqueness domain is named as $UD_j, j = 1, \ldots, 4$.

Another interesting finding is that the sign of the determinant is preserved in every open set UD_j as it is shown in Fig. 1.

In addition, the Jacobian's determinant is θ_1 invariant [3] the uniqueness domains depend only on θ_2, θ_3. In manipulators with low degrees of freedom, $det(\mathbf{J}) = 0$ is easily solved, however, for 6 DoF the solution is much more complicated. In order to avoid complex computations it is easier to create a normal grid in the configuration space in order to determine the connected components by the change of the sign of it's Jacobian's determinant.

2.2 The Proposed Structure of the Approximator

Taking into account that the number of uniqueness domains is equal to the number of the IK solutions, and with the sign preservation technique it is possible to create the training data sets for each IK solution, by regularly griding the configuration space and calculating the sign of the Jacobian determinant. The following method is proposed to approximate all the solutions of the IKP:

1. Regularly grid the configuration space to obtain the set $\Theta = \{\underline{\theta}^l, l = 1, \ldots, k\}$ where $\underline{\theta}^l$ represents the joint coordinates at the lth point of the grid.
2. Calculate the Jacobian's determinant $detJ(\underline{\theta}^l), l = 1, \ldots, k$.
3. Determine the elements of each subset of connected components defined by the sign of the Jacobian's determinant of the tuple $(\underline{\theta}^l, sgn(detJ(\underline{\theta}^l)))$. The number of the subsets $UD_j, j = 1, 2, \ldots m$ is equal to the maximum number m of the IK

Fig. 2 Proposed training method of an MLP

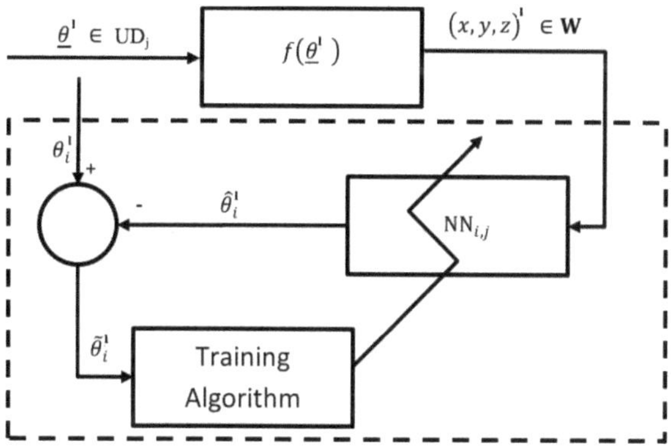

solutions. The union of these subsets is $\cup_{j=1}^{m} UD_j = \Theta - \{\text{Singular Locus}\}$, or equal to Θ apart from $\underline{\theta}^l$ belonging to the singularity locus.

4. For every $\underline{\theta}^l \in UD_j$ determine the TCP position, to obtain the following tuple. $(\underline{\theta}^l, x, y, z)_j$.
5. Form the set of tuples $(x, y, z, \theta_i^l)_j, i = 1, 2, 3, j = 1, 2, \ldots, m$.
6. Use $(x, y, z, \theta_i^l)_j$ to train a system that approximates θ_i. An MLP system is trained for every θ_i, as it is shown in Fig. 2. The position variables otained by the FK are used as inputs to the MLP system which approximates one configuration variable $\hat{\theta}_i$ and the error between the desired and approximated values $\tilde{\theta}_i = \theta_i - \hat{\theta}_i$ is calculated. The error is fed to the training algorithm that tunes the system. When the training is completed, the $NN_{i,j}$ system is used to approximate θ_i by the position data.
7. Repeat for all uniqueness domains and all joint variables θ_i.

In this paper, multilayer perceptron systems are used to approxiamte the IKP. The Levenberg-Marquardt backpropagation method for 100 epochs is used to train a set of MLP systems for every θ_i and every IK solution. The architecture of the MLP system is found by trial and error techniques by training various systems with one or two hidden layers with a total number of neurons between [10, 20]. In the case of one hidden layer the MLP system have the topology has the form $3 - HL_1 - 1$ where $HL_1 \in [10, 20]$. In the case of two hidden layers the MLP system has the topology of the form $3 - HL_1 - HL_2 - 1$ where $10 \le HL_1 - HL_2 \le 20$.

Since the manipulator tasks are located in high dexterity areas of the workspace the training and testing is limited to those areas to demonstrate the practical significance of the method. For a given joint error, the position error is lowest in areas of the workspace where the inverse condition number is close to 1. Consequently a small approximation error in joint space is amplified to a small error in Cartesian space.

3 Simulated Experiments and Results

A 3 dof regular manipulator has up to four solutions in the inverse kinematics problem. The D-H parameters of the typical puma-like manipulator used in the simulated experiments are the following: $\alpha_1 = \alpha_3 = 0^o$, $\alpha_2 = 90^o$, $a_1 = 0$, $a_2 = 0.3$, $a_3 = 0.707$ and $d_1 = d_2 = d_3 = 0$. The TCP is positioned by 0.5 m along the X axis relative to the 3rd frame. The limits of the configuration variables are $\theta_1 \in [-100^o, 100^o]$, $\theta_2 \in [-90^o, 90^o]$ and $\theta_3 \in [-140^o, 140^o]$.

The position Jacobian's determinant of this robot is the following:

$$det(\mathbf{J_p}) = -0.707/2sin(\theta_3)(cos(\theta_2 + \theta_3)/2 + 0.707cos(\theta_2) + 0.3) \quad (2)$$

The first factor of $det(\mathbf{J_p}) = 0$ is shown in Fig. 1 by the red lines while the second one is shown by the green lines. According to Eq. (2) the singularity locus of the robot is θ_1 invariant.

Following the method a uniform grid is defined in the proposed configuration space with a step of 2.5° taking into account the joint variable limits. The limited universe of discourse of (θ_2, θ_3) is shown by the black dashed line in Fig. 1. There are areas in the joint space where the KCI is equal or approximately equal to 1. In order to minimise the approximation error the following thresholds are selected: (a) $KCI > 0.7$ for the UD_1, UD_3. (b) $KCI > 0.1$ for the UD_2, UD_4.

Since there are four uniqueness domains, the following twelve sets are formed:

$$(x, y, z, \theta_i)_j, i = 1, 2, 3, j = 1, \ldots, 4 \quad (3)$$

An MLP system is trained for every set using the method referred in the previous section. The best systems for every set is shown in Table 1. Some systems converged to the specified MSE error (10^{-7}—1st stop training criterion) very early such as the first system which converged in the 13th epoch. Other systems did not reach the specified MSE after 100 epochs of training (2nd stop training criterion), however the MSE is close to the specified one. The worst case appeared in the 2nd system that presents an MSE equal to 3.4510^{-7}. Overall, the values of the MSE presented in tab. 1 are the lowest using the lowest number of neurons.

The computational time for training the system varies greatly, from a couple of minutes to several (10–20 min) according to the number of training data, the topology as well as the termination epoch.

The absolute errors in the training data for the 3rd uniqueness domain are shown in Fig. 3. The high majority of the test data show very low error comparable to the current industrial robots (<0.1 mm). A few data points present high error in the order of mm. The maximum joint error is presented in the system that approximates θ_2 and it is about 4×10^{-3} rad whole the maximum error in the Cartesian space is less than 2 mm. In Fig. 4, the position error in the workspace is shown for the 3rd set. Comparing the results with those found in the literature, the joint error is better than the one presented in [9] (in some cases is 100 times better). The mean error

Table 1 MLP system for every set and training results

Set	Joint		Neurons/layer	Epochs	MSE	Training data
1	1	θ_1^1	[11 9]	13	6.36×10^{-8}	11,826
1	2	θ_2^1	[11 9]	100	3.45×10^{-7}	1826
1	3	θ_3^1	[9 10]	100	2.06×10^{-7}	11,826
2	1	θ_1^2	[11 9]	66	9.95×10^{-8}	7695
2	2	θ_2^2	[10 9]	64	8.99×10^{-8}	7695
2	3	θ_3^2	[12 5]	74	9.73×10^{-8}	7695
3	1	θ_1^3	[9 10]	15	3.71×10^{-8}	11,826
3	2	θ_2^3	[10 8]	100	2.91×10^{-7}	11,826
3	3	θ_3^3	[15 5]	100	1.26×10^{-7}	11,826
4	1	θ_1^4	[11 9]	77	9.9×10^{-8}	7695
4	2	θ_2^4	[13 7]	14	7.78×10^{-8}	7695
4	3	θ_3^4	[13 7]	98	9.8×10^{-8}	7695

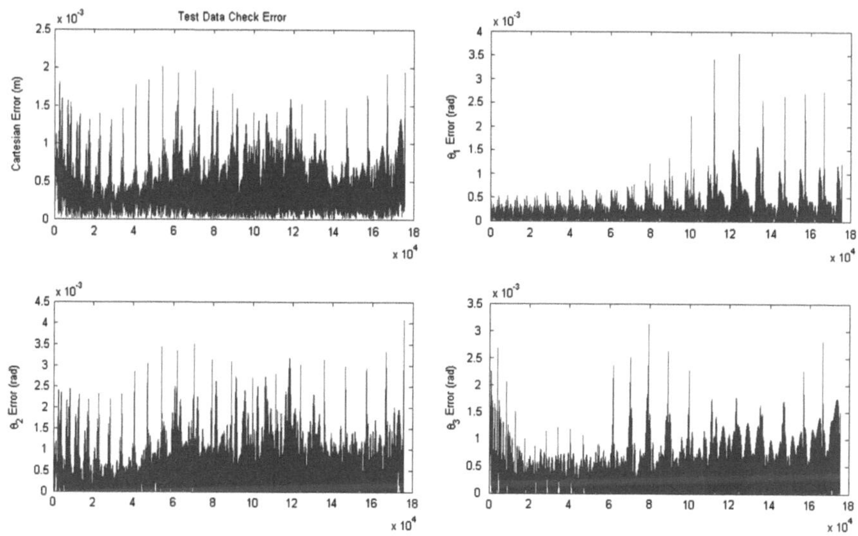

Fig. 3 Test error for the 3rd uniqueness domain

presented in [1] is 0.9 mm, with standard deviation of 2.9 mm implying that are points in the workspace that the error is higher than the one presented in this paper. Finally, the RMS error presented in [8] was less than 2 % which is much greater than the one presented in this paper.

The maximum position error is shown in the extremes of the workspace defined by the training data. However, it was expected that the maximum error in the 3rd set should be lower than the one presented in the 4th set due to the higher KCI, because

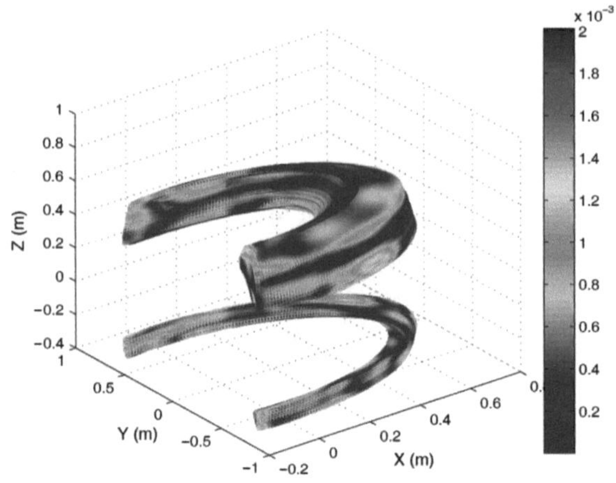

Fig. 4 Cartesian error in the workspace for the 3rd uniqueness domain

the training MSE error is higher in the case of the 4th set than the one presented in the 3rd set. The joint error can be lowered if the system is trained for more than 100 epochs, or if more than 20 neurons are used in the two hidden layers of the MLP. The Cartesian error can be improved either by training the system in areas where the KCI is relatively high and/or by using more training data.

4 Conclusions

In this paper, a method that could be used to find all the solutions of the inverse kinematics problem is presented based on training MLP systems. For the first time the structure of the approximator is based on the knowledge of the distribution of the IK solutions and the corresponding training data obtained by forward kinematics. The training data further elaborated to define the areas of high dexterity to reduce the approximation error in useful areas. The accuracy achieved is considerably higher than in previous methods based on soft computing while the presented method determined all of the solutions that was not the case in most of them. Future work is targeted to the application of the method for 6 DoF robots without analytical solutions.

References

1. de Angulo, V., Torras, C.: Learning inverse kinematics: reduced sampling through decomposition into virtual robots. IEEE Trans. Syst. Man Cybern. Part B: Cybern. **38**(6), 1571–1577 (2008). doi:10.1109/TSMCB.2008.928232
2. Burdick, J.W.: A classification of 3R regional manipulator singularities and geometries. Mech. Mach. Theory **30**(1), 71–89 (1995). doi:10.1016/0094-114X(94)00043-K

3. Gotlih, K., Troch, I.: Base invariance of the manipulability index. Robotica 22(4), 455–462 (2004). doi:10.1017/S0263574704000220
4. Kalra, P., Mahapatra, P., Aggarwal, D.: An evolutionary approach for solving the multimodal inverse kinematics problem of industrial robots. Mech. Mach. Theory **41**(10), 1213–1229 (2006). doi:http://dx.doi.org/10.1016/j.mechmachtheory.2005.11.005. http://www.sciencedirect.com/science/article/pii/S0094114X05002053
5. Lazarevska, E.: A neuro-fuzzy model of the inverse kinematics of a 4 dof robotic arm. In: Proceedings of 14th International Conference on Computer Modelling and Simulation (UKSim), 2012 UKSim, pp. 306–311 (2012). doi:10.1109/UKSim.2012.51
6. Mavroidis, C., Roth, B.: Structural parameters which reduce the number of manipulator configurations. J. Mech. Des. **116**(1), 3–10 (1994). doi:http://dx.doi.org/10.1115/1.2919373
7. Nielsen, J., Roth, B.: On the kinematic analysis of robotic mechanisms. Int. J. Robot. Res. **18** (12), 1147–1160 (1999). doi:10.1177/02783649922067771. http://ijr.sagepub.com/content/18/12/1147.abstract
8. Raptis, S.N., Tzafestas, E.S.: Robot inverse kinematics via neural and neurofuzzy networks: architectural and computational aspects for improved performance. J. Inf. Optim. Sci. **28**(6), 905–933 (2007). doi:10.1080/02522667.2007.10699780. http://dx.doi.org/10.1080/02522667.2007.10699780
9. Srinivasan, A., Nigam, M.: Neuro-fuzzy based approach for inverse kinematics solution of industrial robot manipulators. Int. J. Comput. Commun. Control **3**(3), 224–234 (2008)
10. Tabandeh, S., Clark, C., Melek, W.: A genetic algorithm approach to solve for multiple solutions of inverse kinematics using adaptive niching and clustering. In: Proceedings of the IEEE Congress on Evolutionary Computation, 2006. CEC 2006, pp. 1815–1822 (2006). doi:10.1109/CEC.2006.1688527
11. Wenger, P.: Uniqueness domains and regions of feasible paths for cuspidal manipulators. IEEE Trans. Robot. **20**(4), 745–750 (2004). doi:10.1109/TRO.2004.829467

An Approach to Symbolical Formulation of Forward Kinematics of Serial Robots

S. Krutikov

Abstract In this paper, an alternative representation of the forward kinematic model is proposed. Its form is a linear combination of the predefined functions of joint coordinates. Its coefficients referred to as *generalized kinematic parameters* determine the model completely. Derivation of forward instantaneous kinematics based on the suggested approach is presented. An example demonstrating its usage is also included.

Keywords Forward · Kinematic · Model · Symbolic · Jacobian

1 Introduction

The forward kinematic model determines the position and the orientation of the robot links with respect to its base depending on their relative positions. It is required in many applications such as visualization of the links movement, calculation of the feedback used in robot control algorithms and procedures for solving the inverse kinematics problem using optimization techniques [7]. The general approach to obtain this model is to attach a frame to every link of the robot starting from the base and to define Euclidean transformations which make each frame to coincide with the next one. The composition of these transformations determines the model. It depends on a number of constant geometrical parameters and variable joint coordinates. Partial derivatives of the resulting transformation with respect to these coordinates and parameters determine forward instantaneous kinematics and the identification model, respectively.

Many of the applications mentioned above are time-critical so it is important to maximize the computational efficiency of the algorithm being used. One of the

S. Krutikov (✉)
Department of Robotics and Mechatronics, Bauman Moscow State Technical University, Moscow, Russia
e-mail: sergey.krutikov@bmstu.ru

S. Zeghloul et al. (eds.), *Robotics and Mechatronics*,
Mechanisms and Machine Science 37, DOI 10.1007/978-3-319-22368-1_28

common ways to achieve this is to generate the underlying model symbolically. Today, there are a lot of computer programs for multibody systems which implement this optimization [2]. They use general-purpose computer algebra systems or special techniques of symbolic computations [6]. However, they do not take into account some features of the analytical representation of rigid body kinematics. In contrast to them this paper does.

Here, an alternative representation of the forward kinematic model for an extensive class of serial mechanisms is proposed. It is assumed that all the components of this model are elements of a linear space of functions with a predefined structure. Their coordinates referred to as *generalized kinematic parameters* are some combinations of geometrical parameters. They determine the model completely and uniquely. A recursive procedure is designed for calculating these parameters. Formulation of forward instantaneous kinematics using the developed approach is also presented. The approach is demonstrated for a general 3-DoF robot manipulator with all revolute joints.

2 Classical Formulation

Consider a serial robot which consists of $N + 1$ rigid bodies (links) connected with joints. Only the joints which are kinematically equivalent to a chain of the helical ones are allowed. Link 0 (base) is supposed to be fixed and link N is the end-effector. Suppose a frame i is attached to a link i of the robot ($i = 0, \ldots, N$). Let ${}^{i-1}\mathbf{T}_i$ be a homogenous transformation matrix which represents the position and the orientation of the frame i relative to the frame $i-1$. Then, the forward kinematic model is determined by the set of matrices ${}^0\mathbf{T}_i$ [4]. According to the rule of homogenous transformation composition, we have:

$$ {}^0\mathbf{T}_i = {}^0\mathbf{T}_1 \ldots {}^{i-1}\mathbf{T}_i. \tag{1} $$

Since every joint is thought to be a compound one, the resulting transformation ${}^{i-1}\mathbf{T}_i$ can be calculated as a composition of r_i elementary screw transformations ${}^{i-1}\mathbf{T}_{i,k_i}$:

$$ {}^{i-1}\mathbf{T}_i = {}^{i-1}\mathbf{T}_{i,1} \ldots {}^{i-1}\mathbf{T}_{i,r_i}. \tag{2} $$

Here “elementary” means that the screw axis is one of the coordinate axes of a moving frame. Substitution of Eq. 2 into Eq. 1 gives us

$$ {}^0\mathbf{T}_i = \mathbf{A}_1 \ldots \mathbf{A}_{m_i}, \tag{3} $$

where $m_i = \sum_{l=1}^{i} r_l$ and $\mathbf{A}_j = {}^{i-1}\mathbf{T}_{i,k_i}$ iff $j = m_{i-1} + k_i$. To determine each elementary transformation $\mathbf{A}_j$, a pitch h_j must be specified [7]. We also need an angle of rotation ϕ_j, if the pitch is finite, or a linear displacement p_j, otherwise. Then, if the screw axis is X, the transformation matrix can be constructed as follows:

$$\mathbf{A}_j = \begin{pmatrix} 1 & 0 & 0 & h_j\phi_j \\ 0 & \cos\phi_j & -\sin\phi_j & 0 \\ 0 & \sin\phi_j & \cos\phi_j & 0 \\ 0 & 0 & 0 & 1 \end{pmatrix}. \tag{4}$$

For Y and Z axes this matrix has a very similar form. Pitch values for all transformations are supposed to be constant due to a fixed kinematic structure of the robot. So each elementary transformation is completely determined either by ϕ_j or p_j, which are referred to as *screw coordinates*. Thus, the total amount of geometrical parameters is $(m + v)$ where $m = m_N$ and v is the number of nonzero finite pitches.

Now assume that only n out of m screw transformations are variable and their indices j_k are ordered as follows: $j_{k-1} < j_k$ $(k = 2, \ldots, n)$. Then, once all $(m + v - n)$ constant geometrical parameters are given, the set of n variable screw coordinates determines the model uniquely. Therefore, this set can be chosen as a set of generalized coordinates of the robot. Let us divide the set $M = \{1, \ldots, m\}$ of integers into n subsets $M_k = \{b_k, \ldots, e_k\}$ so that $j_k \in M_k$, $b_k = e_{k-1} + 1$, $e_0 = 0$ and $e_n = m$. We also suppose that there exist integers $n_i \leq n$ $(i = 1, \ldots, N)$ which satisfy $e_{n_i} = m_i$. Then, if we introduce matrices

$$\mathbf{T}_k = \mathbf{A}_{b_k} \ldots \mathbf{A}_{e_k} \tag{5}$$

and take into account Eq. 3, we have

$${}^0\mathbf{T}_i = \mathbf{T}_1 \ldots \mathbf{T}_{n_i}. \tag{6}$$

Each matrix $\mathbf{T}_k$ depends on only one generalized coordinate q_k which is equal to ϕ_{j_k} if h_{j_k} is finite and to p_{j_k}, otherwise.

For example, for robot manipulators we have $r_i = 4$, $m_i = 4i$, $n = N$, $e_i = m_i$ and $n_i = i$. This makes ${}^{i-1}\mathbf{T}_i = \mathbf{T}_i$. If the classical Denavit-Hartenberg notation is used, each transformation $\mathbf{T}_k$ consists of two elementary rotations (zero pitch) and two elementary translations (infinite pitch):

$$\mathbf{T}_k = \mathbf{Rot}(Z, \theta_k)\mathbf{Trans}(Z, d_k)\mathbf{Trans}(X, a_k)\mathbf{Rot}(X, \alpha_k).$$

If joint k is revolute, $j_k = b_k$ and $q_k = \theta_k$. If it is prismatic, $j_k = b_k + 1$ and $q_k = d_k$.

3 Alternative Representation

Consider the elementary screw transformation $\mathbf{A}_{j_k}$. Relative to Eq. 4 it can be represented as a linear combination of some predefined functions of generalized coordinates:

$$\mathbf{A}_{j_k}(q_k) = \sum_{l=1}^{4} \mathbf{A}_{j_k}^{l} f_k^l(q_k),$$

where $\mathbf{A}_{j_k}^{l}$ are constant matrices, $f_k^l(q_k) \in F_k$ and $F_k = \{1(q_k), \cos q_k, \sin q_k, q_k\}$. Taking into account Eq. 5, we have

$$\mathbf{T}_k(q_k) = \sum_{l=1}^{4} \mathbf{T}_k^l f_k^l(q_k),$$

where $\mathbf{T}_k^l = \mathbf{A}_{b_k} \ldots \mathbf{A}_{j_k}^l \ldots \mathbf{A}_{e_k}$. Substitution into Eq. 6 gives us

$$^0\mathbf{T}_i(\mathbf{q}) = \sum_{l_1=1}^{4} \cdots \sum_{l_{n_i}=1}^{4} \left[\mathbf{T}_1^{l_1} \ldots \mathbf{T}_{n_i}^{l_{n_i}}\right] f_1^{l_1}(q_1) \ldots f_{n_i}^{l_{n_i}}(q_{n_i}), \tag{7}$$

where $\mathbf{q}$ is the column vector of generalized coordinates.

Consider a linear space $\mathbb{F}_k = \text{span}\,(F_k)$. It is isomorphic to $\mathbb{R}^4$ since all elements of the set F_k are linearly independent as continuous functions of q_k. Let us define a linear space $\mathbb{B}_n$ as a tensor product of $\mathbb{F}_k$:

$$\mathbb{B}_n = \mathbb{F}_1 \otimes \ldots \otimes \mathbb{F}_n.$$

If the operation $\otimes$ is a regular scalar multiplication, the space $\mathbb{B}_n$ is a finite dimensional linear subspace of all continuous functions of $\mathbf{q}$. Let us define a set of tensor indices I^k as a kth cartesian power of the set $I = \{1, \ldots, 4\}$ and denote the product $f_1^{l_1} \ldots f_n^{l_n}$ by b_n^l, where $l = (l_1, \ldots, l_n) \in I^n$. It can be proved that the functions $b_n^l(\mathbf{q})$ are linearly independent and constitute the natural basis of $\mathbb{B}_n$. Then, Eq. 7 can be rewritten as follows:

$$^0\mathbf{T}_i(\mathbf{q}) = \sum_{l \in I^n} \mathbf{T}_i^l b_n^l(\mathbf{q}), \tag{8}$$

where $\mathbf{T}_i^l = \mathbf{T}_1^{l_1} \ldots \mathbf{T}_{n_i}^{l_{n_i}}$ if $l = (l_1, \ldots, l_{n_i}, 1, \ldots, 1)$ and $\mathbf{T}_i^l = 0$, otherwise. This equation means that the components of each matrix $^0\mathbf{T}_i$ are the elements of the space $\mathbb{B}_n$ and the components of the matrices $\mathbf{T}_i^l$ are their coordinates in the natural basis of $\mathbb{B}_n$. Let us define the whole set of these coordinates as the set of *generalized kinematic parameters* of the link i. It determines the position and the orientation of this link completely and uniquely according to the properties of bases.

Generalized parameters are some generally nonlinear combinations of classical parameters, e.g., DH parameters for robot manipulators. Technically, the total amount of generalized parameters of the link i is $12 \cdot 4^{n_i}$ as the last row of the transformation matrix is meaningless. However, a lot of these parameters are actually zero since the matrices $\mathbf{A}_j$ and $\mathbf{A}_{j_k}^l$ are sparse. Also, there is a dependency between the generalized parameters of adjacent links due to recursive relations

$$^0\mathbf{T}_i = ^0\mathbf{T}_{i-1}(\mathbf{T}_{n_{i-1}+1} \ldots \mathbf{T}_{n_i}), \quad k = 1, .., N, \, ^0\mathbf{T}_0 = \mathbf{I}_4.$$

Taking this into account, these parameters can be calculated using a recursive procedure. Let us define the matrices $\mathbf{T}^{i_k}$ as follows:

$$\mathbf{T}^{i_k} = \mathbf{T}_1^{l_1} \ldots \mathbf{T}_k^{l_k},$$

where $i_k = (l_1, \ldots, l_k) \in I^k$. This leads to a natural recursion

$$\mathbf{T}^{i_k} = \mathbf{T}^{i_{k-1}} \mathbf{T}_k^{l_k}, \quad k = 1, \ldots, n, \; \mathbf{T}^{i_0} = \mathbf{I}_4.$$

Then, if k is equal to one of n_i, we assign the value of the current matrix $\mathbf{T}^{i_k}$ to the matrix $\mathbf{T}_i^l$, where $l = (i_k, 1, \ldots, 1)$. The computational efficiency of this recursive algorithm is significantly increased if a special implementation of sparse matrices is used. Another optimization is to detect matrices $\mathbf{T}^{i_k}$ which are zero, stop the current recursion branch and switch to another tensor index i_k.

4 Forward Instantaneous Kinematics

The forward instantaneous kinematic model determines linear and angular velocities of the robot links with respect to the base frame if joint positions and velocities are given. It is typically represented using the Jacobian matrices $\mathbf{J}_i$ for each link [4]

$$\mathbf{s}_i = \mathbf{J}_i(\mathbf{q})\dot{\mathbf{q}},$$

where $\mathbf{s}_i$ is a spatial velocity vector of the link i. The part of the Jacobian corresponding to the angular velocity can be obtained from the geometrical model. Let U_i be the set of all indices j_k $(k \leq n_i)$ such that the respective screw transformation $\mathbf{A}_{j_k}$ has a finite pitch. The absolute angular velocity ω_i of the link i is the sum of the relative ones [1]:

$$\omega_i = \sum_{k \in U_i} \mathbf{u}_k \dot{q}_k,$$

where $\mathbf{u}_k$ is a unit vector of the coordinate axis which is a screw axis of the elementary transformation $\mathbf{A}_{j_k}$. Its nonzero coordinates $\mathbf{u}_k^l$ with respect to the natural basis of $\mathbb{B}_n$ can be calculated as follows:

$$\mathbf{u}_k^l = (\mathbf{I}_3\ \mathbf{0})\mathbf{T}^{i_{k-1}}\mathbf{A}_{b_k} \ldots \mathbf{A}_{j_k-1}\mathbf{e}_r, \quad \forall i_{k-1} \in I^{k-1},$$

where $l = (i_{k-1}, 1, \ldots, 1) \in I^n$ and $\mathbf{e}_r$ is a unit vector whose rth component is 1. If the screw axis is X then $r = 1$, if it is Y then $r = 2$, and if it is Z then $r = 3$.

The other part of the Jacobian corresponds to the linear velocity of the link frame origin. In symbolical computations it is usually obtained by calculating the time derivative of the respective position vector $\mathbf{p}_i$ [4]. Its coordinates with respect to the base frame are the first three components of the last column of the transformation matrix ${}^0\mathbf{T}_i$. Let $\mathbf{p}_i^l$ be a column vector of generalized kinematic parameters corresponding to the position part of the matrix $\mathbf{T}_i^l$. Then, Eq. 8 yields:

$$\mathbf{p}_i = \sum_{l \in I^n} \mathbf{p}_i^l b_n^l.$$

Differentiating the latter equation with respect to time and denoting $\dot{\mathbf{p}}_i$ by $\mathbf{v}_i$, we have

$$\mathbf{v}_i = \sum_{l \in I^n} \mathbf{p}_i^l \left(\sum_{k=1}^{n} \frac{\partial b_n^l}{\partial q_k} \dot{q}_k \right). \tag{9}$$

Due to the structure of space $\mathbb{B}_n$, we obviously obtain

$$\frac{\partial b_n^l}{\partial q_k} = f_1^{l_1} \ldots \frac{\partial f_k^{l_k}}{\partial q_k} \ldots f_n^{l_n}.$$

It can be easily checked that each space $\mathbb{F}_k$ is closed under differentiation with respect to q_k, i.e.,

$$\frac{\partial f_k^l}{\partial q_k} = c(l) f_k^{h(l)}, \quad \forall l \in I.$$

The functions $c(l)$ and $h(l)$ map the set I to the set $C = \{-1, 0, 1\}$ of integers and to itself, respectively. They are defined in Table 1. This yields the space $\mathbb{B}_n$ to be closed under differentiation as well. Given a tensor index $l = (l_1, \ldots, l_n) \in I^n$, let us introduce the set of functions $c_k(l)$ and $h_k(l)$ $(k = 1, .., n)$ which map the set I^n to the set C and to itself, respectively. They are defined as follows: $c_k(l) = c(l_k)$ and $h_k(l) = (l_1, \ldots, h(l_k), \ldots, l_n)$. Then, we have

Table 1 The definition of the functions $c(l)$ and $h(l)$

l	$c(l)$	$h(l)$
1	0	4
2	−1	3
3	1	2
4	1	1

$$\frac{\partial b_n^l}{\partial q_k} = c_k(l) b_n^{h_k(l)}, \quad \forall l \in I^n.$$

Substitution of the last equation into Eq. 9 gives us

$$\mathbf{v}_i = \sum_{k=1}^{n} \left(\sum_{l \in I^n} c_k(l) \mathbf{p}_i^l b_n^{h_k(l)} \right) \dot{q}_k. \tag{10}$$

The expression in parentheses is convinient for calculation of the linear velocity Jacobian, since it is equal to a partial derivative $\partial \mathbf{v}_i / \partial \dot{q}_k$. Given all nonzero generalized parameters $\mathbf{p}_i^l$, all nonzero projections $\mathbf{v}_{ik}^l$ of the kth Jacobian column on the space $\mathbb{B}_n$ can be directly calculated.

5 Example

The algorithm of calculating the generalized kinematic parameters described above has been implemented in C++ language using GiNaC library [3] for symbolic manipulations. The results of its application to a general 3-DoF robot manipulator with all revolute joints are presented in Table 2. Only the end-effector parameters corresponding to its position and to axis Z of its frame are shown. The classical Denavit-Hartenberg notation has been used to construct the matrices. Then, the

Table 2 The generalized kinematic parameters of the end-effector

$l \in I^3$	$\mathbf{p}_3^l$ (transposed)	$\mathbf{z}_3^l$ (transposed)
$(1,1,1)$	$(0,\ 0,\ d_1 + d_2 \cos\alpha_1 + d_3 \cos\alpha_1 \cos\alpha_2)$	$(0,\ 0,\ \cos\alpha_1 \cos\alpha_2 \cos\alpha_3)$
$(1,2,1)$	$(0,\ 0,\ -d_3 \sin\alpha_1 \sin\alpha_2)$	$(0,\ 0,\ -\sin\alpha_1 \sin\alpha_2 \cos\alpha_3)$
$(1,1,2)$	$(0,\ 0,\ 0)$	$(0,\ 0,\ -\cos\alpha_1 \sin\alpha_2 \sin\alpha_3)$
$(1,2,2)$	$(0,\ 0,\ 0)$	$(0,\ 0,\ -\sin\alpha_1 \cos\alpha_2 \sin\alpha_3)$
$(1,3,3)$	$(0,\ 0,\ 0)$	$(0,\ 0,\ \sin\alpha_1 \sin\alpha_3)$
$(1,3,1)$	$(0,\ 0,\ a_2 \sin\alpha_1)$	$(0,\ 0,\ 0)$
$(1,1,3)$	$(0,\ 0,\ a_3 \cos\alpha_1 \sin\alpha_2)$	$(0,\ 0,\ 0)$
$(1,2,3)$	$(0,\ 0,\ a_3 \sin\alpha_1 \cos\alpha_2)$	$(0,\ 0,\ 0)$

(continued)

Table 2 (continued)

$l \in I^3$	$\mathbf{p}_3^l$ (transposed)	$\mathbf{z}_3^l$ (transposed)
$(1,3,2)$	$(0,\ 0,\ a_3 \sin\alpha_1)$	$(0,\ 0,\ 0)$
$(2,1,1)$	$(a_1,\ -d_2 \sin\alpha_1 - d_3 \sin\alpha_1 \cos\alpha_2,\ 0)$	$(0,\ -\sin\alpha_1 \cos\alpha_2 \cos\alpha_3,\ 0)$
$(2,1,2)$	$(0,\ 0,\ 0)$	$(0,\ \sin\alpha_1 \sin\alpha_2 \sin\alpha_3,\ 0)$
$(3,1,1)$	$(d_2 \sin\alpha_1 + d_3 \sin\alpha_1 \cos\alpha_2,\ a_1,\ 0)$	$(\sin\alpha_1 \cos\alpha_2 \cos\alpha_3,\ 0,\ 0)$
$(2,2,1)$	$(a_2,\ -d_3 \cos\alpha_1 \sin\alpha_2,\ 0)$	$(0,\ -\cos\alpha_1 \sin\alpha_2 \cos\alpha_3,\ 0)$
$(2,3,1)$	$(d_3 \sin\alpha_2,\ a_2 \cos\alpha_1,\ 0)$	$(\sin\alpha_2 \cos\alpha_3,\ 0,\ 0)$
$(3,2,1)$	$(d_3 \cos\alpha_1 \sin\alpha_2,\ a_2,\ 0)$	$(\cos\alpha_1 \sin\alpha_2 \cos\alpha_3,\ 0,\ 0)$
$(3,3,1)$	$(-a_2 \cos\alpha_1,\ d_3 \sin\alpha_2,\ 0)$	$(0,\ \sin\alpha_2 \cos\alpha_3,\ 0)$
$(2,1,3)$	$(0,\ -a_3 \sin\alpha_1 \sin\alpha_2,\ 0)$	$(0,\ 0,\ 0)$
$(2,2,3)$	$(0,\ a_3 \cos\alpha_1 \cos\alpha_2,\ 0)$	$(\sin\alpha_3,\ 0,\ 0)$
$(2,3,2)$	$(0,\ a_3 \cos\alpha_1,\ 0)$	$(\cos\alpha_2 \sin\alpha_3,\ 0,\ 0)$
$(3,2,2)$	$(0,\ a_3,\ 0)$	$(\cos\alpha_1 \cos\alpha_2 \sin\alpha_3,\ 0,\ 0)$
$(3,3,3)$	$(0,\ -a_3 \cos\alpha_2,\ 0)$	$(-\cos\alpha_1 \sin\alpha_3,\ 0,\ 0)$
$(2,2,2)$	$(a_3,\ 0,\ 0)$	$(0,\ -\cos\alpha_1 \cos\alpha_2 \sin\alpha_3,\ 0)$
$(2,3,3)$	$(-a_3 \cos\alpha_2,\ 0,\ 0)$	$(0,\ \cos\alpha_1 \sin\alpha_3,\ 0)$
$(3,1,3)$	$(a_3 \sin\alpha_1 \sin\alpha_2,\ 0,\ 0)$	$(0,\ 0,\ 0)$
$(3,2,3)$	$(-a_3 \cos\alpha_1 \cos\alpha_2,\ 0,\ 0)$	$(0,\ \sin\alpha_3,\ 0)$
$(3,3,2)$	$(-a_3 \cos\alpha_1,\ 0,\ 0)$	$(0,\ \cos\alpha_2 \sin\alpha_3,\ 0)$
$(3,1,2)$	$(0,\ 0,\ 0)$	$(-\sin\alpha_1 \sin\alpha_2 \sin\alpha_3,\ 0,\ 0)$

Table 3 The parameters of the linear velocity Jacobian of the end-effector

$l \in I^3$	$\mathbf{v}_{31}^l$	$l \in I^3$	$\mathbf{v}_{32}^l$	$l \in I^3$	$\mathbf{v}_{33}^l$
$(2,1,1)$	$\mathbf{p}_3^{(3,1,1)}$	$(1,2,1)$	$\mathbf{p}_3^{(1,3,1)}$	$(1,1,2)$	$\mathbf{p}_3^{(1,1,3)}$
$(3,1,1)$	$-\mathbf{p}_3^{(2,1,1)}$	$(1,3,1)$	$-\mathbf{p}_3^{(1,2,1)}$	$(1,2,2)$	$\mathbf{p}_3^{(1,2,3)}$
$(3,2,1)$	$-\mathbf{p}_3^{(2,2,1)}$	$(1,3,3)$	$-\mathbf{p}_3^{(1,2,3)}$	$(1,3,3)$	$-\mathbf{p}_3^{(1,3,2)}$
$(3,3,1)$	$-\mathbf{p}_3^{(2,3,1)}$	$(1,2,2)$	$\mathbf{p}_3^{(1,3,2)}$	$(2,1,2)$	$\mathbf{p}_3^{(2,1,3)}$
$(2,3,1)$	$\mathbf{p}_3^{(3,3,1)}$	$(2,3,1)$	$-\mathbf{p}_3^{(2,2,1)}$	$(2,2,2)$	$\mathbf{p}_3^{(2,2,3)}$
$(2,2,1)$	$\mathbf{p}_3^{(3,2,1)}$	$(2,2,1)$	$\mathbf{p}_3^{(2,3,1)}$	$(2,3,3)$	$-\mathbf{p}_3^{(2,3,2)}$
$(3,1,3)$	$-\mathbf{p}_3^{(2,1,3)}$	$(3,3,1)$	$-\mathbf{p}_3^{(3,2,1)}$	$(3,2,3)$	$-\mathbf{p}_3^{(3,2,2)}$
$(3,2,3)$	$-\mathbf{p}_3^{(2,2,3)}$	$(3,2,1)$	$\mathbf{p}_3^{(3,3,1)}$	$(3,3,2)$	$\mathbf{p}_3^{(3,3,3)}$
$(3,3,2)$	$-\mathbf{p}_3^{(2,3,2)}$	$(2,3,3)$	$-\mathbf{p}_3^{(2,2,3)}$	$(2,2,3)$	$-\mathbf{p}_3^{(2,2,2)}$
$(2,2,2)$	$\mathbf{p}_3^{(3,2,2)}$	$(2,2,2)$	$\mathbf{p}_3^{(2,3,2)}$	$(2,3,2)$	$\mathbf{p}_3^{(2,3,3)}$
$(2,3,3)$	$\mathbf{p}_3^{(3,3,3)}$	$(3,3,2)$	$-\mathbf{p}_3^{(3,2,2)}$	$(3,1,2)$	$\mathbf{p}_3^{(3,1,3)}$
$(3,3,3)$	$-\mathbf{p}_3^{(2,3,3)}$	$(3,2,3)$	$\mathbf{p}_3^{(3,3,3)}$	$(3,2,2)$	$\mathbf{p}_3^{(3,2,3)}$
$(3,2,2)$	$-\mathbf{p}_3^{(2,2,2)}$	$(2,3,2)$	$-\mathbf{p}_3^{(2,2,2)}$	$(3,3,3)$	$-\mathbf{p}_3^{(3,3,2)}$

nonzero parameters of the linear velocity Jacobian of the end-effector have been calculated using Eq. 10. Some of them are presented in Table 3. It illustrates that in software terms the velocity parameters are just pointers to the position ones up to the opposite sign. This fact can significantly reduce the computational cost of symbolic formulation of instantaneous kinematics.

6 Conclusions

In this paper, the components of both finite and differential kinematic models of serial robots are represented as elements of a linear space. Due to the predefined structure of this space, only generalized parameters must be calculated to obtain these models. A recursive relations for calculation of these parameters have been obtained and implemented in software. So the computational complexity of explicit formulation of forward kinematics is reduced using the presented approach. Also it can be useful for the optimization of the automatically generated source code which implements an algorithm based on forward kinematics. However, such a procedure is still under development. A comparison with an existing software package (e.g. OpenSymoro [5]) will be performed after it has been finished. Other interesting applicatons of the proposed approach are derivation of the robot calibration model and analysis of the parameter identifiability.

References

1. Amirouche, F.M.L.: Fundamentals of Multibody Dynamics: Theory and Applications. Birkhuser, Boston (2006)
2. Brown, I.C., Larcombe, P.J.: A survey of customised computer algebra programs for multibody dynamic modelling. In: Munro, N. (ed.) The Use of Symbolic Methods in Control System Analysis and Design, chap. 3, pp. 53–77. The Institute of Engineering and Technology, London (1999)
3. Johannes Gutenberg University Mainz: GiNaC 1.6.3. An open framework for symbolic computation within the C++ programming language. http://www.ginac.de/tutorial.pdf
4. Khalil, W., Dombre, E.: Modelling, Identification and Control of Robots. Butterworth-Heinemann, Oxford (2004)
5. Khalil, W., Vijayalingam, A., Khomutenko, B., Mukhanov, I., Lemoine, P.: Opensymoro: an open-source software package for symbolic modelling of robots. In: IEEE/ASME International Conference on Advanced Intelligent Mechatronics, , pp. 1206–1211. Besancon, France, July 2014 (2014)
6. Samin, J.C., Fisette, P.: Symbolic Modeling of Multibody Systems. Kluwer Academic Publishers, Dordrecht (2003)
7. Siciliano, B., Khatib, O. (eds.): Springer Handbook of Robotics. Springer, Heidelberg (2008)

Grasp Database Generator for Anthropomorphic Robotic Hands

H. Mnyusiwalla, P. Vulliez, J.P. Gazeau and S. Zeghloul

Abstract Grasp databases can be useful for data-driven grasp synthesis algorithm or benchmarking grasp quality metrics. This paper presents an algorithm to produce a grasp database for anthropomorphic robotic hands. The method can generate different types of grasps from the human grasp taxonomy, it relies on the notion of hand preshapes. For each type of grasp there is a hand preshape defined, containing information on how the hand should approach the object and close its fingers. The proposed approach is validated on the anthropomorphic RoBioSS hand.

Keywords Grasp database · Robotic hand · Grasp synthesis · Hand preshapes

1 Introduction

Stable grasp synthesis has been researched for several decades and has led to numerous grasping algorithms as detailed in [12]. Most of these algorithms rely on different metrics [11] to evaluate the stability of a grasp. To analyze grasp quality measures like in the papers [5, 7] we need to have a big grasp database with a variety of grasps. One notable related work is the Columbia Grasp Database [6]. This publicly available database consists of grasp sets of a human hand model and of the Barrett robotic hand. However this grasping database is more focused

H. Mnyusiwalla (✉) · P. Vulliez
Department GMSC, Pprime Institute CNRS - University of Poitiers - ENSMA, UPR 3346 Poitiers, France
e-mail: hussein.mnyusiwalla@univ-poitiers.fr

P. Vulliez
e-mail: philippe.vulliez@univ-poitiers.fr

J.P. Gazeau · S. Zeghloul
Institut PPRIME, UPR 3346, University of Poitiers, Poitiers, France
e-mail: jean.pierre.gazeau@univ-poitiers.fr

S. Zeghloul
e-mail: said.zeghloul@univ-poitiers.fr

S. Zeghloul et al. (eds.), *Robotics and Mechatronics*,
Mechanisms and Machine Science 37, DOI 10.1007/978-3-319-22368-1_29

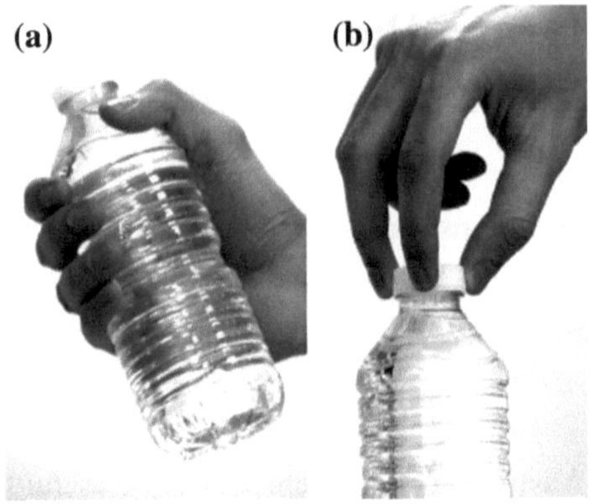

Fig. 1 Grasp types. **a** Power-grasp. **b** Precision-grasp

towards grasps with a large contact area between the hand and the objects called power-grasps instead of grasps with a contact zone limited to the fingertips known as precision-grasps [10]; an example of both grasps is shown on Fig. 1.

Analyzing quality measures with the Columbia Grasp Database wouldn't inform us on the effectiveness of these metrics with precision-grasps. However, precision-grasping for in-hand manipulation is the main target for robotic hands as power-grasping can be achieve easily with a gripper.

There are several existing methods to generate grasps [12], however these approaches usually generate a specific type of grasp and are difficult to adapt to other objects or to other hands. The method proposed in this paper is based on the notion of hand preshapes [14] and can generate datasets of different grasps from the grasp taxonomy [4]. This method is easily adaptable to different hands and works with all objects modeled by triangular meshes.

This paper is structured as follows. Section 2 details the targeted grasps for our algorithm and why they were chosen. Section 3 presents the notion of hand preshapes and Sect. 4 explains the generation process. Section 5 discusses the obtained results. Finally Sect. 6 concludes the paper and outlines the future work.

2 Target Grasps

To design robotic hands that have the same dexterity as the human hand, researchers have been studying human grasping behavior. This research led to several taxonomies classifying human diverse set of grasps. The most recent taxonomy is the one established by Feix et al. [4]. The authors divide human grasping behavior into 33 types of grasp arranged in three categories: power, intermediate and precision grasps.

It's useful to have the same variety of grasps for anthropomorphic hands, therefore we will base our grasp database on this taxonomy. We will limit our study in this paper to the grasps presented on Fig. 2. This set of grasps was chosen based

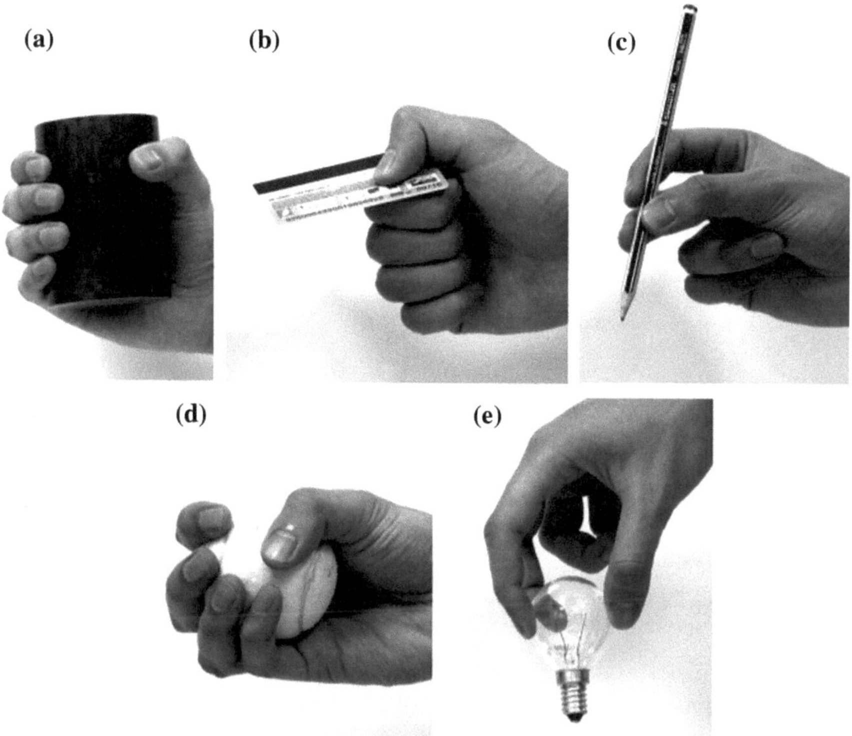

Fig. 2 Target grasps. **a** Medium wrap. **b** Lateral pinch. **c** Thumb-2-finger. **d** Power sphere. **e** Tripod

on the study in [3]. The authors analyzed nearly 10,000 grasp instances performed by two housekeepers and two machinists in their daily work. They found that these five grasp types cover almost 70 % of the performed grasps.

The Medium Wrap and the Power Sphere are power-grasps. The Thumb-2-finger and tripod are precision-grasps. The Lateral Pinch grasp is an intermediate grasp in the Feix's taxonomy. However, for simplification, it is referred to as a power-grasp further in this paper because it has a similar generation process as the other power-grasps in the proposed method, as detailed in Sect. 4.

3 Grasp Preshapes

The grasp generator is based on hand preshapes [14]. For each type of grasp from the taxonomy, the user has to define the corresponding preshape for the robotic hand. This concept of preshape expands the notion of pregrasps defined in [9] as it also includes information on how the hand should approach the object and how

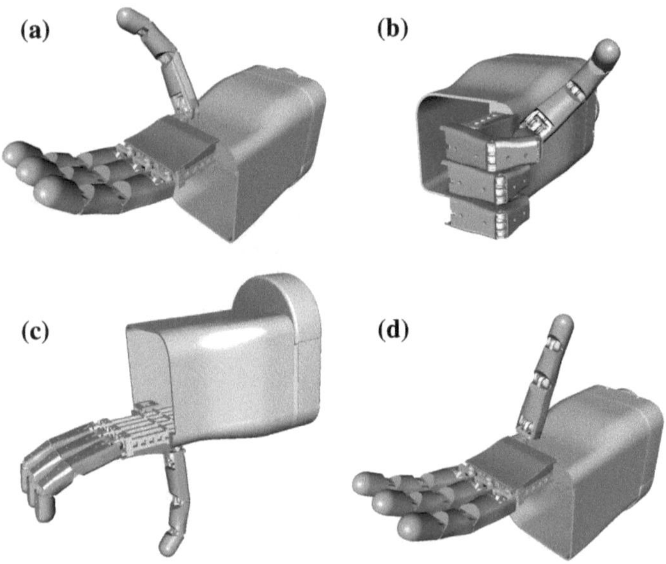

Fig. 3 Preshapes for the target grasps. **a** Medium wrap. **b** Lateral pinch. **c** Thumb-2-finger and tripod. **d** Power sphere

fingers should move. A preshape consists of initial joint values, hand orientation for the approach phase and details on moving and fixed fingers for the hand closing phase.

Similar to the notion of eigengrasps [1], a preshape is based on joint coupling. The joint coupling for the eigengrasps are built on the Principal Component Analysis of human grasping movement done in [13]. However this analysis leads to coupling even between fingers, therefore the determination of which eigengrasps to use for a specific type of grasp is non-trivial. Consequently the notion of eigengrasp is hard to adapt to reproduce the taxonomy.

The joint coupling in our preshapes is used mostly for the flexion movement of the fingers which leads to more natural grasps with anthropomorphic hands. Figure 3 shows the different preshapes for the target grasps. The Tripod grasp and the Thumb-2-finger have the same initial joint values. The difference is during the finger closing phase. For the Tripod grasp, abduction-adduction joints move while they remain still for the Thumb-2-finger grasp.

4 Grasps Generation

The pregrasp moves around the object and for each position the fingers are closed and the grasp is evaluated. For each grasp we check if they are force closure by verifying if the convex hull of the grasp wrench space includes the wrench space

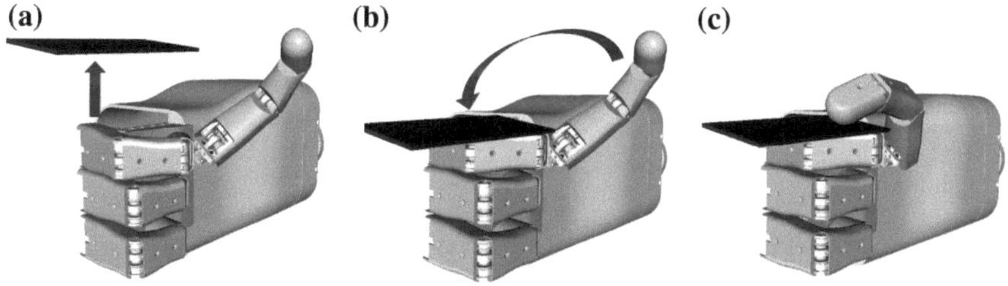

Fig. 4 Power-grasp generation. **a** Approaching phase. **b** Finger-closing phase. **c** Grasp validation

origin [8]. For precision-grasps we also check if the number of contact points is accurate for the grasp. For example, for the tripod grasp we verify if there is only three contact points.

The robotic hand is positioned around the object using a spherical coordinate frame associated to the object. The center of the spherical frame is sampled along the object longest dimension. This process is repeated with varying hand orientations. We adjust the sampling range by exploiting the object symmetries to reduce the computing time.

Power-grasps are sampled with a single radius value, the preshape moves closer to the object until the hand collides with the object, and then the fingers are closed. The process is illustrated on Fig. 4. Once the grasp is valid, the hand position and orientation in the object frame, the joint-values and the contact points with their normals are saved.

5 Results

The robotic hand used to generate the database is a new human size anthropomorphic hand developed in the RoBioSS laboratory. The hand kinematic model is presented on Fig. 5. The hand has four fingers and sixteen tendon actuated degrees of freedom; four degrees of freedom for each finger. All the fingers have almost the same kinematics, the only difference is that the thumb has slightly different link

Fig. 5 Kinematic model of the RoBioSS hand

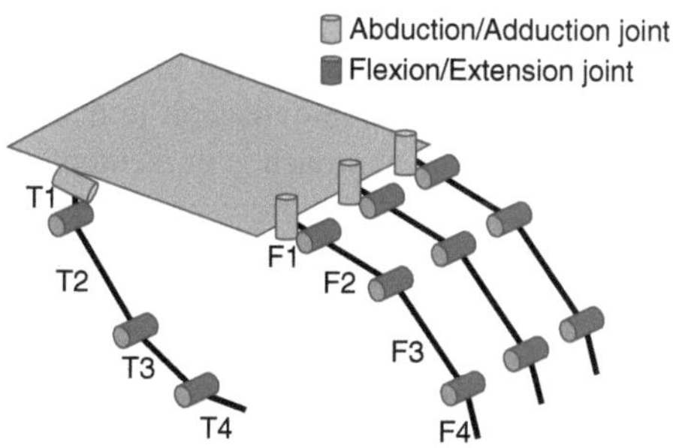

Table 1 Link lengths

Thumb	T1	T2	T3	T4
Length in (mm)	15.2	53	40	21
Fingers	F1	F2	F3	F4
Length in (mm)	15.2	45	35	22

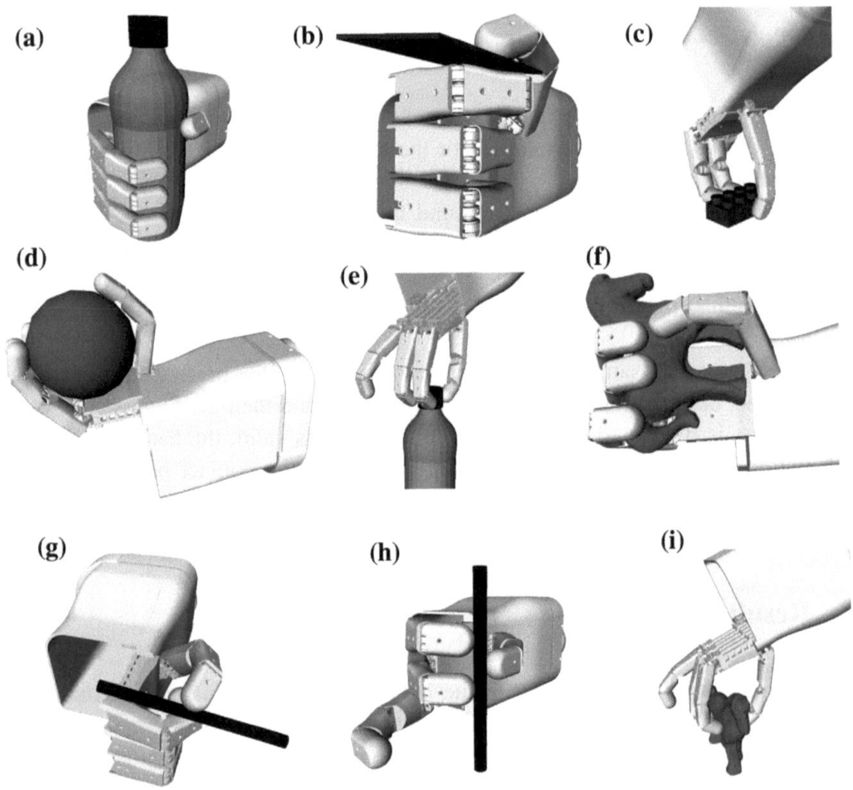

Fig. 6 Generated grasps. **a** Medium wrap. **b** Lateral pinch. **c** Thumb-2-finger. **d** Power sphere. **e** Tripod. **f** Medium wrap. **g** Lateral pinch. **h** Thumb-2-finger. **i** Tripod

lengths. The lengths are given in Table 1. Each finger has an abduction-adduction joint and three flexion-extension joints.

The proposed approach is implemented in the open-source software OpenRAVE [2]. To have a good variety of grasps, different objects with varying geometries were selected to construct the database. Figure 6 gives examples of the obtained grasps.

The force closure criterion is quite selective therefore the generated power-grasps don't need any post-processing. Precision-grasps however require more user intervention because they produce many unstable grasps. To filter out those grasps automatically we need to improve the grasp validation process by finding suitable quality measures for precision-grasps.

6 Conclusion

We presented an approach which can produce a large variety of grasps. This method works with any object modeled by a triangle mesh and is easily adaptable to any anthropomorphic hands. For each desired type of grasp, we defined a hand preshape for the robotic hand. This preshape was then used on several objects to generate the database. The generation of power-grasps works quite well and doesn't require too much user intervention as opposed to the generation of precision-grasps.

Precision-grasp data-sets needed much more post-processing to filter unstable grasps. To automate the database generation furthermore we need to introduce new quality measures for the validation process. This will be done in our future work as our next step will be to use this grasp database to evaluate grasp quality measures. Once the process will be fully automated we will add more grasp types and objects to create a larger database.

Acknowledgements This work has been sponsored by the French government research program Investissements d'Avenir through the Robotex Equipment of Excellence (ANR-10-EQPX-44).

References

1. Ciocarlie, M., Goldfeder, C., Allen, P.: Dexterous grasping via eigengrasps: a low-dimensional approach to a high-complexity problem. In: Proceedings of the Robotics: Science and Systems 2007 Workshop-Sensing and Adapting to the Real World, Electronically published (2007)
2. Diankov, R., Kuffner, J.: OpenRAVE: A Planning Architecture for Autonomous Robotics. Technical Report (CMU-RI-TR-08-34) (2008)
3. Feix, T., Bullock, I.M., Dollar, A.M.: Analysis of human grasping behavior: object characteristics and grasp type. IEEE Trans. Haptics (in press) **7**(3), 311–323 (2014). doi:10.1109/TOH.2014.2326871
4. Feix, T., Pawlik, R., Romero, J., Kragic, D.: The generation of a comprehensive grasp taxonomy. Archives of Physical Medicine
5. Goins, A.K., Carpenter, R., Wong, W.k., Balasubramanian, R.: Evaluating the efficacy of grasp metrics for utilization in a Gaussian process-Based grasp predictor. In: International Conference on Intelligent Robots and Systems (IROS 2014), 2014 IEEE/RSJ, pp. 3353–3360 (2014). doi:10.1109/IROS.2014.6943029
6. Goldfeder, C., Ciocarlie, M., Dang, H., Allen, P.K.: The columbia grasp database. In: Proceedings—IEEE International Conference on Robotics and Automation, pp. 1710–1716 (2009). doi:10.1109/ROBOT.2009.5152709
7. León, B., Rubert, C., Morales, A.: Characterization of grasp quality measures for evaluating robotic hands prehension. In: IEEE International Conference on Robotics and Automation, pp. 3688–3693 (2014). doi:10.1109/ICRA.2014.6907393

8. Miller, A., Allen, P.: Examples of 3D grasp quality computations. In: Proceedings 1999 IEEE International Conference on Robotics and Automation (Cat. No.99CH36288C) vol. 2, pp. 1240–1246 (1999). doi:10.1109/ROBOT.1999.772531
9. Miller, A., Knoop, S., Christensen, H., Allen, P.: Automatic grasp planning using shape primitives. In: 2003 IEEE International Conference on Robotics and Automation (Cat. No.03CH37422), vol. 2, pp. 1824–1829 (2003). doi:10.1109/ROBOT.2003.1241860
10. The prehensile movements of the human hand: Napier. J. Bone Joint Surg. **38**, 902–913 (1956)
11. Roa, M.A., Suárez, R.: Grasp quality measures: review and performance (2014). doi:10.1007/s10514-014-9402-3
12. Sahbani, A., El-Khoury, S., Bidaud, P.: An overview of 3D object grasp synthesis algorithms. Rob. Auton. Syst. **60**(3), 326–336 (2012). doi:10.1016/j.robot.2011.07.016
13. Santello, M., Flanders, M., Soechting, J.F.: Postural hand synergies for tool use. J. Neurosci. **18**(23), 10105–10115 (1998) (citeulike-article-id:423192)
14. Wren, D., Fisher, R.: Dextrous hand grasping strategies using preshapes and digit trajectories. In: 1995 IEEE International Conference on Systems, Man and Cybernetics. Intelligent Systems for the 21st Century, vol. 1, (1995). doi:10.1109/ICSMC.1995.537883

From Human Motion Capture to Industrial Robot Imitation

P. Laguillaumie, M.A. Laribi, P. Seguin, P. Vulliez, A. Decatoire and S. Zeghloul

Abstract Making a robot duplicate human operator gestures is a difficult task. To achieve this goal, we have defined a methodology and have started designing software tools to simplify and automate this process. The first step is the motion acquiring of the tool or of the object manipulated by the human subject. The recorded motion is then processed to be transformed as the trajectory of the Tool Center Point (TCP) of a robotic arm-hand set whose joint architecture and kinematics are different from the human arm ones. All robot axis coordinates are computed by solving the inverse kinematics model. We demonstrated the feasibility of our approach by experimentally having a robot duplicate the movement of a tool motioned by an operator. The 3D motion capture system used is based on passive marker technology which does not perturb the human gesture. After being processed, the tool trajectory is used to generate a list cyclic positions controls synchronously applied by a real-time controller to the robot motor drives. The robotic set-up consists of a newly designed 4-finger hand embedded on an industrial 6-axis serial robot provided with an innovative open robotic drive.

Keywords Biomimetics · Motion capture · Human gesture · Duplicate movement · Kinematic model · Industrial robot

1 Introduction

In industrial production or in service robotics, mechanical systems are increasingly used to perform tasks originally entrusted to human workers or helpers. One way to assign the task to a robot is to use biomimetics in the sense of teaching a robot to

P. Laguillaumie (✉) · P. Seguin · P. Vulliez · A. Decatoire
Department GMSC, Pprime Institute CNRS - University of Poitiers - ENSMA, UPR 3346 Poitiers, France
e-mail: pierre.laguillaumie@univ-poitiers.fr

M.A. Laribi · S. Zeghloul
Institut PPRIME, UPR 3346, University of Poitiers, Poitiers, France

S. Zeghloul et al. (eds.), *Robotics and Mechatronics*,
Mechanisms and Machine Science 37, DOI 10.1007/978-3-319-22368-1_30

use a tool as a human operator would do. The RoBioSS team of the Pprime Institute originally investigated biomimetics as a way to control robots in the contexts of defining strategies for reaching and grasping in prehension [1] and improving the gait of walking humanoids robots [2].

In this paper, we present a complete methodology for reproducing a human gesture by an industrial robot equipped with a suitable gripper. The process starts with motion capture of the movement of a tool manipulated by a human and ends with the generation of a similar movement on a robotic hand-arm platform.

This method was first put into practice for a French TV scientific show[1] whose topic was the interactions between humans and robots for entertainment purpose.

2 Experimental Set-up

The experimental set-up uses two scientific platforms of Pprime Institute funded by the French Robotex research funding project: a 3D-motion capture system and a robotic cell designed for cobotics (cooperation between robots and between a robot and human).

The motion capture system is equipped with a Vicon system which consists of up to 20 ccd video cameras and an acquisition device able to record the three dimensional motions of passive reflective markers. In addition, the scientific platform offers the opportunity to measure ground forces reactions with seven 6-axis force plates. The video and force plate measurements are synchronized and recorded by the Vicon acquisition device so that they can be post-processed with Nexus software and MathWorks Matlab software.

The set-up is routinely used for human gait clinical analysis, athlete gesture optimization and ergonomics design of workstation.

For our application, we used the Vicon motion analysis system with ten cameras to record at 100 Hz frequency the motions of the tool and the operator arm during the task that we aimed to reproduce.

The captured movement is to be reproduced by a robotic arm-hand set designed for precise manipulation. The robotic handler is a 4-finger, 12-actuator, tendon-driven anthropomorphic hand designed by the RoBioSS team in 2014 [3]. It has human-size geometry and each finger has three flexion-extension and an abduction-adduction motions. The hand is mounted on a TX60 6-axis Stäubli serial robot (see Fig. 1) powered by a uniVAL drive industrial robotic controller, which offers the innovative feature of a possible integration which an external controller [4].

All 18 axes are put into motion by a B&R Automation industrial PC [5]. Communication between the robot controller, the DC motor power stages and the PC runs through a high speed real-time Ethernet Powerlink fieldbus which guaranties that all motions are synchronized. Low level control loops for the hand

[1]http://www.france5.fr/emissions/on-n-est-pas-que-des-cobayes.

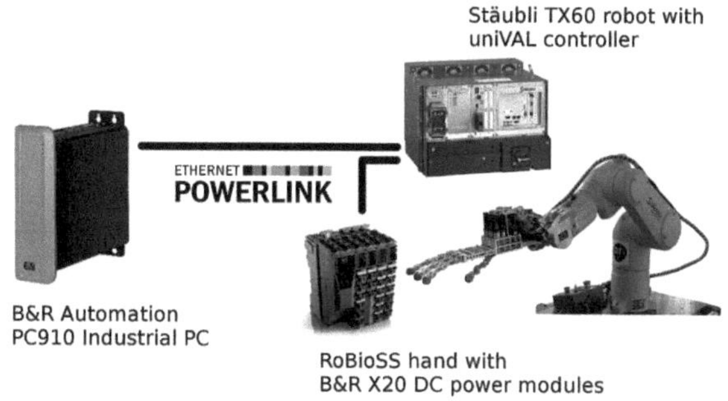

Fig. 1 Experimental set-up

motors also run on the industrial PLC (see Fig. 1). Robot-level software implements PLCopen Motion standard. In our application, all axes are given 4 ms periodic position commands.

3 Biomimetics-Based Approach

3.1 Motion Adaptation

The aim of this work is to imitate the captured human motion with a serial robot. The approach principle could be to control the robot as a marionette, imposing the human joint angles to the serial robot. This approach will not create a feasible motion since the robot has not necessarily the same geometry. The motion adaptation transforms the human motion to a feasible motion for a chosen robot. Two main constraints must be respected by the adapted motion.

First, all coordinates of each marker on the object must be inside the serial robot workspace. Secondly, the Inverse Kinematic Model (IKM) of the serial robot must be respected at any time. The synopsis for the «tool movement duplication» process is presented in Fig. 2 and may be described as follows.

At first, the motion of the tool is recorded by a 3D motion capture set-up. The trajectory of the point inside the hand and the orientation are processed with a Matlab routine in order to optimally place the complete tool motion inside the robot workspace, taking into account the different arm configurations and avoiding singularities. The tool is virtually placed into the robotic hand. The positions of the joint variables of the Stäubli robot are computed using the inverse kinematic model. The new trajectories are then smoothed to limit the speed and acceleration discontinuities originating from the noise and measurement errors. At this point, we have computed the temporal series of joint variables.

Before being applied to the robot, the sets of joint commands are evaluated in a 3D simulation environment in order to check that they respect the limits of each joint range as well as the possible environmental constraints of the robotic scene. If all safety and technological conditions are satisfied, the motions can be played on the arm-hand set-up by our real-time robotic control-command software.

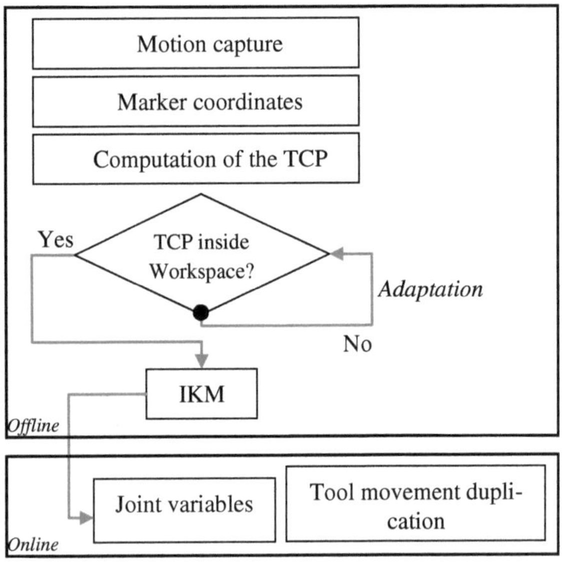

Fig. 2 Process overview

3.2 *Motion Capture*

To collect kinematic data in order to record the trajectories of the markers, a Vicon motion system is used with ten cameras located around the studied area. The observed volume is about 1 m long, 1.8 m high, and 1 m wide.

Seven reflective markers are placed on the subject skin and six on the tool (Fig. 3). To position the markers on the human arm, we made the following choice: three markers for the shoulder (clavicle, scapular acromion and scapula), two for the elbow (humeral medial and lateral epicondyles) and two for the wrist (styloid processes of the radius and of the ulna).

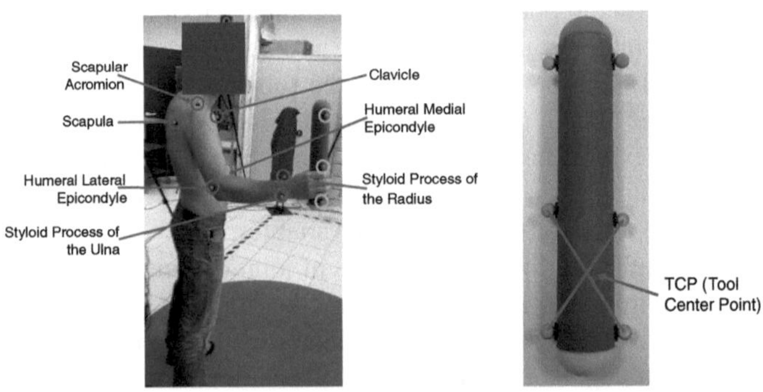

Fig. 3 Markers set (*left*) and Tool Center Point (*right*)

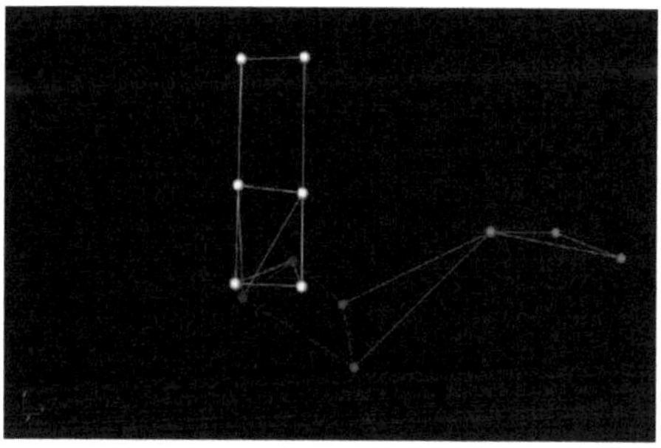

Fig. 4 Model of the human's arm + tool under Vicon Nexus software

The second step to motion capture is to define the intersegment model in Vicon Nexus video processing software.

In our case study, the number of segments chosen is four (Fig. 4): shoulder (green), arm (red), forearm (blue) and tool (yellow). The segment axes are chosen according to the International Society of Biomechanics' (ISB) recommendations.

3.3 Manipulator Kinematics

The description of the structure of the Stäubli TX-60 robot is shown in Fig. 5. The transformation between the joint space and the Cartesian space of the robot is very important since robots are controlled in the joint space, whereas tasks are defined and objects are manipulated in the Cartesian space. The kinematic problem deals with the analytical study of the relation between these two spaces. The kinematic problem consists of two subproblems: the forward kinematics defined as the transformation from the joint space to the Cartesian space and the inverse kinematics defined as the transformation from the Cartesian space to the joint space.

The Stäubli TX-60 robot is a RRR anthropomorphic type and the wrist consists of three intersecting revolute axes, equivalent to a spherical joint. Specified link frames and modified Denavit-Hartenberg parameters [6] are depicted in Fig. 5 and Table 1.

1. *Forward kinematics*

The forward kinematic problem aims at determining the position and orientation of the end effector given the values of the joint variables of the robot. Now to access the kinematic equations, the transformation matrices are computed. The following relation specifies the end effector condition with respect to the base.

$$\mathbf{T}^{06} = \mathbf{T}^{01}\mathbf{T}^{12}\mathbf{T}^{23}\mathbf{T}^{34}\mathbf{T}^{45}\mathbf{T}^{56}$$

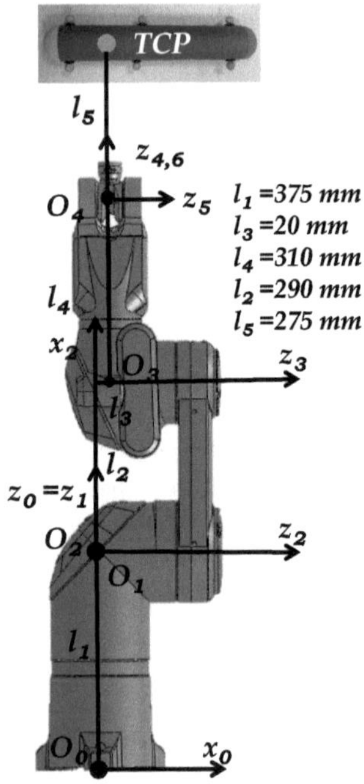

Fig. 5 Denavit-Hartenberg parameters of the Stäubli TX-60 robot

Table 1 Denavit-Hartenberg parameters of the Stäubli TX-60 robot

Link	Joint offset	Joint angle	Link length	Twist angle
1	l_1	q_1	0	0
2	0	q_2	l_2	$-\frac{\pi}{2}$
3	l_3	q_3	0	0
4	l_4	q_4	0	$\frac{\pi}{2}$
5	0	q_5	0	$-\frac{\pi}{2}$
6	0	q_6	0	$\frac{\pi}{2}$

In order to simplify the problem, we consider here only the first three joints. The operational coordinates of point TCP (X_P, Y_P, Z_P) are defined as follow:

$$\begin{cases} X_P = \sin q_1 [l_4 \sin(q_2 + q_3) + l_2 \sin q_2] + l_3 \cos q_1 \\ Y_P = \cos q_1 [l_4 \sin(q_2 + q_3) + l_2 \sin q_2] - l_3 \sin q_1 \\ Z_p = l_4 \cos(q_2 + q_3) + l_2 \cos q_2 + l_1 + l_5 \end{cases} \quad (1)$$

2. *Inverse kinematics*

The inverse problem consists of calculating the joint variables corresponding to a given location of the end-effector. When it exists, the explicit form which gives all possible solutions (there is rarely uniqueness of solution) constitutes what we call the inverse kinematic model. All possible solutions obtained for the inverse kinematic model given by the equation system (1) are given as follow.

$$\text{With}: \begin{cases} \text{if } q_1 \neq \frac{\pi}{2}[\pi], \bar{X} = \frac{X_P + l_2 \sin q_1}{\cos q_1} \text{ else } \bar{X} = \frac{Y_P - l_2 \cos q_1}{\sin q_1} \text{ and } \bar{Z} = Z_P - l_1 - l_5 \\ A = l_4 \cos q_3 + l_3 \text{ and } B = l_4 \sin q_3 \end{cases}$$

3. *Workspace and robot placement*

In order to duplicate the tool movement, the coordinates of the TCP must be inside the industrial robot workspace. This section is dedicated to the analytic formulation of the robot workspace and the robot placement.

The robot workspace is depicted in Fig. 6. The orthogonal projections of the workspace on the planes (o, **x**, **y**) and (o, **x**, **z**) allow to draw a simple geometric representation based on elementary curves.

Let X_P, Y_P, Z_P be the coordinates of the TCP in the Cartesian space. The TCP is inside the workspace if and only if the following conditions are fulfilled.

$$\begin{aligned} Eq_1 &= X_P^2 + Y_P^2 - l_m^2 > 0 \\ Eq_2 &= X_P^2 + Y_P^2 - l_M^2 < 0 \\ Eq_3 &= (X_P - X_{O_2})^2 + (Z_P - Z_{O_2})^2 - (l_2 + l_4 + l_5)^2 < 0 \\ Eq_4 &= (X_P - X_{O_2})^2 + (Z_P - Z_{O_2})^2 - l_2^2 > 0 \\ Eq_5 &= (Z_P - Z_{O_2}) \tan q_{2\min} - (X_P - X_{O_2}) > 0 \end{aligned}$$

where $l_m = l_2$, $l_M = l_1 + l_2 + l_4 + l_5$, $q_{2\min} = -135°$.

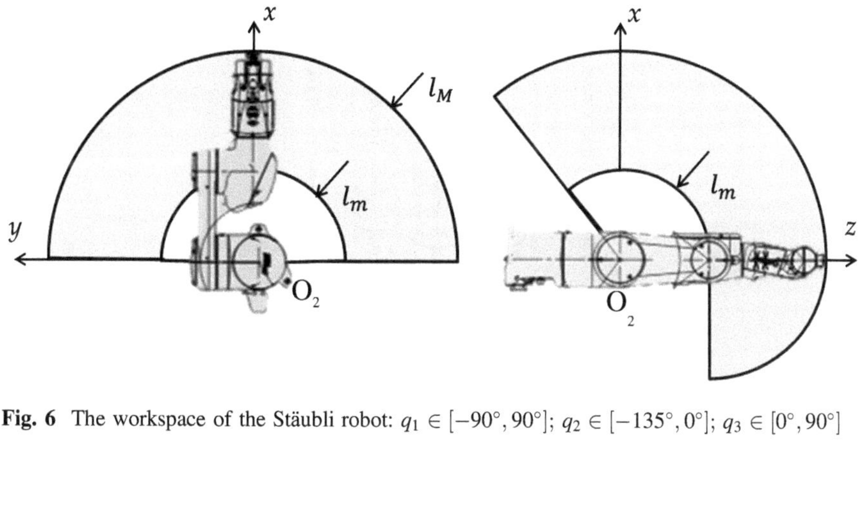

Fig. 6 The workspace of the Stäubli robot: $q_1 \in [-90°, 90°]$; $q_2 \in [-135°, 0°]$; $q_3 \in [0°, 90°]$

Table 2 Solutions obtained for the inverse kinematic model

Solution	q_1	q_2	q_3
1	$\mathrm{acos}\left(\frac{l_2}{\sqrt{X_P^2+Y_P^2}}\right) - ATAN2(X_P, Y_P)$	$ATAN2(A\bar{X} - B\bar{Z}, B\bar{X} + A\bar{Z})$	$\mathrm{acos}\left(\frac{\bar{X}^2+\bar{Z}^2-l_3^2-l_4^2}{2l_4l_3}\right)$
2	$\mathrm{acos}\left(\frac{l_2}{\sqrt{X_P^2+Y_P^2}}\right) - ATAN2(X_P, Y_P)$	$ATAN2(A\bar{X} - B\bar{Z}, B\bar{X} + A\bar{Z})$	$-\mathrm{acos}\left(\frac{\bar{X}^2+\bar{Z}^2-l_3^2-l_4^2}{2l_4l_3}\right)$
3	$-\mathrm{acos}\left(\frac{l_2}{\sqrt{X_P^2+Y_P^2}}\right) - ATAN2(X_P, Y_P)$	$ATAN2(A\bar{X} - B\bar{Z}, B\bar{X} + A\bar{Z})$	$\mathrm{acos}\left(\frac{\bar{X}^2+\bar{Z}^2-l_3^2-l_4^2}{2l_4l_3}\right)$
4	$-\mathrm{acos}\left(\frac{l_2}{\sqrt{X_P^2+Y_P^2}}\right) - ATAN2(X_P, Y_P)$	$ATAN2(A\bar{X} - B\bar{Z}, B\bar{X} + A\bar{Z})$	$-\mathrm{acos}\left(\frac{\bar{X}^2+\bar{Z}^2-l_3^2-l_4^2}{2l_4l_3}\right)$

In case of violation of these conditions, the adaptation algorithm is involved in the aim to compute the displacement vector $(\Delta x, \Delta y, \Delta z)$, needed to satisfy the workspace constraints.

4 Results

This section deals with an example of duplication of a human gesture. The joint variables of the Stäubli robot are computed using the inverse kinematic model detailed in Table 2. The configuration with elbow-down of the articulated arm is considered here. Each curve of the joint variable is smoothed in order to limit the speed and acceleration discontinuities.

Figures 7 and 8 present the corresponding TCP positions acquired using motion capture and the obtained joint variables for the first three links of the Stäubli robot. Figure 9 presents a screenshot for the experimental duplication of the human motion by the robotic arm-hand set using the obtained joint variables (Fig. 8).

The first tests of human gesture duplication have shown good performance and robustness of our open robotic command software: several real-speed human gestures have been reproduced.

The only constraint imposed to the operator was to limit the wrist motion relatively to the forearm since it proved difficult to integrate arbitrary tool orientation in our control scheme. This restriction will be removed in future developments.

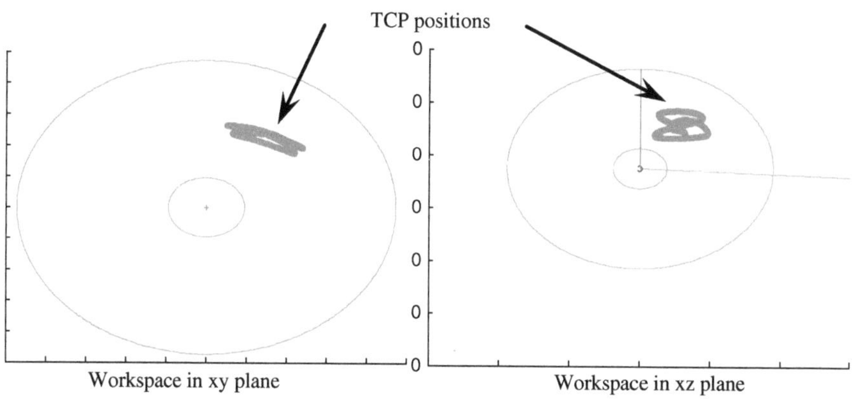

Fig. 7 TCP positions

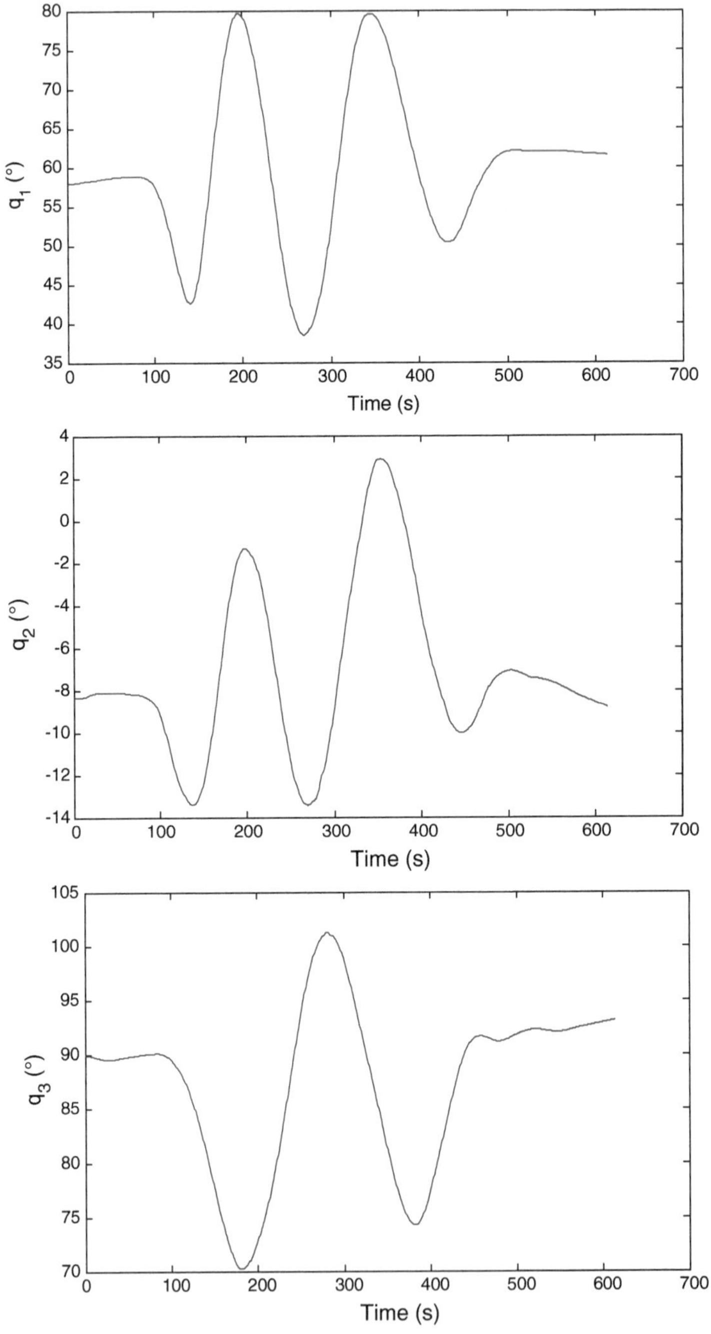

Fig. 8 Joint variables q_1, q_2 and q_3 corresponding to TCP positions

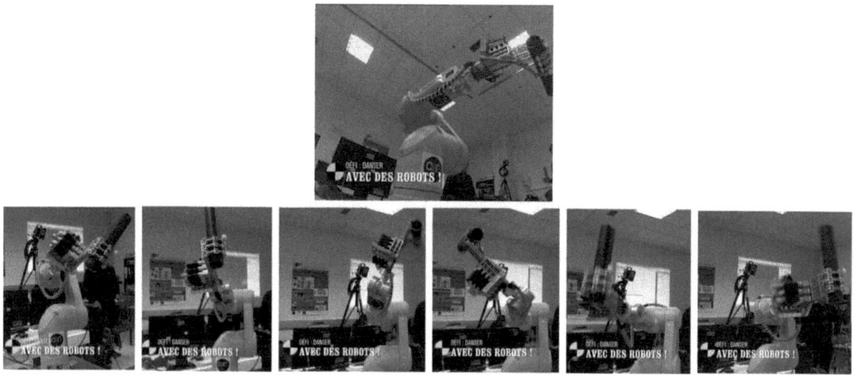

Fig. 9 Experimental validation by the robotic arm-hand set

5 Conclusions

This paper presented an integrated approach to the problem of reproducing a human hand or tool motion task with a robot. The methodology introduced here has four steps: record the tool motion with a motion capture system, extract the tool center point trajectory, adapt the motion to the robot workspace and compute the temporal series of joint coodinates for the robotic arm + hand system. The motion can then be played on a robot with an open controller by imposing cyclic joint position commands. The whole process has been successfully implemented on RoBioSS Pprime scientific platforms: a Vicon Nexus motion capture system and a newly designed arm-hand robot.

Future work on this subject will consist in improving the robustness of the method in terms of controlling the tool orientation in space and of being able to reproduce tasks performed by a larger variety of human operators. We are also planning to introduce a validation step dedicated to the compliance of the motion to be reproduced and ensure that can be supported by the mechanical limits of the robot.

Acknowledgments This research is supported by the Poitou-Charentes region 2007–2013 (program project 10 Images and interactivities), in partnership with the European Union (FEDER/ERDF, European Regional Development Fund). This research is supported by ROBOTEX, the French national network of robotics platforms (N° ANR-10- EQPX-44-01).

References

1. Touvet, F., Daoud, N., Gazeau, J.P., Zeghloul, S., Maier, M.A., Eskiizmirliler, S.: A biomimetic reach and grasp approach for mechanical hands. Robot. Auton. Syst. **60**(3), 473–486 (2012)
2. Boutin, L., Eon, A., Zeghloul, S., Lacouture, P.: From human motion capture to humanoid locomotion imitation application to the robots HRP-2 and HOAP-3. Robotica **29**, 325–334 (2011)

3. Gazeau, J.P., Zeghloul, S., Arsicault, M., Rivière, T., Vulliez, P.: "DOIGT ROBOTIQUE MODULAIRE POUR LA PREHENSION ET LA MANIPULATION DEXTRE", demande de brevet d'invention français n°1459956 du 16 octobre 2014, Titulaire: CNRS
4. Stäubli Unival Drive Technology: http://www.unival-drive.com/
5. B&R Automation: PC910 Industrial PC datasheet: http://www.br-automation.com/en/products/industrial-pcs/automation-pc-910/, X20MM2436 DC motor drive datasheet: http://www.br-automation.com/en/products/control-systems/x20-system/motor-controllers/x20mm2436/
6. Khalil W., Kleinginger J.F.: A new geometric notation for open and closed loop robots. In: Proceedings of IEEE International Conference on Robotics and Automation, San Francisco, pp. 1174–1179 (1986)

Dynamic Decoupling of Adjustable Serial Manipulators Taking into Account the Changing Payload

J.L. Xu, V. Arakelian and J.-P. Le Baron

Abstract This paper deals with the problem of dynamic decoupling of adjustable serial manipulators with changing payload. Without payload, this kind of manipulator can accomplish the dynamic decoupling via opposite rotation of links and optimal redistribution of masses. However, the introduction of the changing payload leads to the perturbation of dynamic decoupling of the manipulator. In order to eliminate such a perturbation, the optimal control design is discussed. The proposed approach, which is a symbiosis of mechanical and control solutions, improves the known design concepts permitting the dynamic decoupling of serial manipulators taking into account the changing payload. The efficiency of the proposed technique is evaluated by using MATLAB software.

Keywords Mechatronic design · Serial manipulators · Decoupled dynamics · Adjustable mechanisms

1 Introduction

It is known that the dynamic model of a robot manipulator is usually described by a set of nonlinear, highly coupled differential equations. The complicated dynamics results from varying inertia, interactions between the different joints, and nonlinear forces such as Coriolis and centrifugal forces. Nonlinear forces cause errors in

J.L. Xu · V. Arakelian (✉) · J.-P. Le Baron
Department of M&CSE, INSA-Rennes, Rennes, France
e-mail: Vigen.Arakelyan@irccyn.ec-nantes.fr

J.L. Xu
e-mail: Jiali.Xu@insa-rennes.fr

J.-P. Le Baron
e-mail: Jean-Paul.Le-Baron@insa-rennes.fr

V. Arakelian
IRCCyN, UMR CNRS 6597, Nantes, France

S. Zeghloul et al. (eds.), *Robotics and Mechatronics*,
Mechanisms and Machine Science 37, DOI 10.1007/978-3-319-22368-1_31

position response at high speed, and have been shown to be significant even at slow speed [1]. Thus, the dynamic decoupling of robot mechanisms has attracted researchers' attention and different solutions have been proposed [2–4].

In recent study [5], the problem of dynamic decoupling of 2-DOF adjustable serial manipulators has been discussed. The cancellation of the coefficients of nonlinear terms in the manipulator's kinetic has been achieved by adding to the initial architecture with adjustable lengths of links a double Scott-Russell mechanism [6] combined with the identical and opposite rotations of manipulator's links, i.e. $\ddot{\theta}_1 = -\ddot{\theta}_2$ (Fig. 1).

When the second link of the adjustable serial manipulator is statically balanced, i.e. $l_{BS_2} = 0$ (Fig. 1) and the gravitational force is perpendicular to the motion plane xOy, the equations of motion can be written as:

$$\begin{bmatrix} \tau_1 \\ \tau_2 \end{bmatrix} = \begin{bmatrix} I_{S2} + I_{S1} + m_1 l_{AS1}^2 + m_2 l_1^2 & I_{S2} \\ I_{S2} & I_{S2} \end{bmatrix} \begin{bmatrix} \ddot{\theta}_1 \\ \ddot{\theta}_2 \end{bmatrix} = \begin{bmatrix} a+b & a \\ a & a \end{bmatrix} \begin{bmatrix} \ddot{\theta}_1 \\ \ddot{\theta}_2 \end{bmatrix} \tag{1}$$

where, $a = I_{S2}; b = I_{S1} + m_1 l_{AS1}^2 + m_2 l_1^2$; I_{S1} is the axial moment of inertia of link 1 relative to the centre of mass S_1 of link 1; I_{S2} is the axial moment of inertia of link 2 relative to the centre of mass S_2 of link 2; m_1, m_2 are the masses of links 1 and 2; l_{AS1} is the distance between the centre of mass S_1 of link 1 and joint centre A; l_{BS2} is the distance between the centre of mass S_2 of link 2 and joint centre B; $\ddot{\theta}_1$ is the angular acceleration of link 1 relative to the base; $\ddot{\theta}_2$ is the angular acceleration of link 2 relative to link 1;

From these equations it follows that

$$\text{if:} \quad \ddot{\theta}_1 = -\ddot{\theta}_2 \quad \begin{cases} \tau_1 = b\ddot{\theta}_1 \\ \tau_2 = 0 \end{cases} \tag{2}$$

i.e. the dynamic equations are decoupled and the second actuator torque is cancelled.

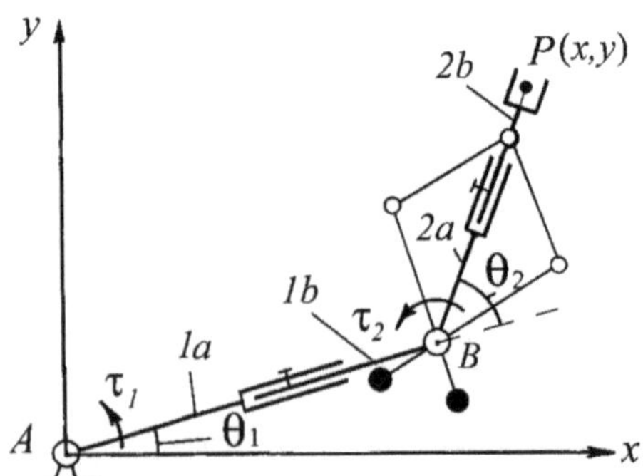

Fig. 1 The 2-DOF adjustable serial manipulator with decoupled dynamics

2 Dynamic Decoupling Taking into Account the Payload

The mentioned dynamic decoupling of the equations of motion corresponds to the manipulator without payload. However, it is obvious that the changing payload creates the variable loads on the actuators, which are also nonlinear.

$$\begin{cases} \tau_1 = \psi(\theta_2)\ddot{\theta}_1 + [\gamma_2 + \beta(\theta_2)]\ddot{\theta}_2 - 2\alpha(\theta_2)\dot{\theta}_1\dot{\theta}_2 - \alpha(\theta_2)\dot{\theta}_2^2 \\ \tau_2 = [\gamma_2 + \beta(\theta_2)]\ddot{\theta}_1 + \gamma_2\ddot{\theta}_2 + \alpha(\theta_2)\dot{\theta}_1^2 \end{cases} \quad (3)$$

where,

$$\alpha(\theta_2) = M_2 l_1 l_{BS2}\sin(\theta_2);\ \beta(\theta_2) = M_2 l_1 l_{BS2}\cos(\theta_2);$$
$$\psi(\theta_2) = \gamma_1 + \gamma_2 + M_2 l_1^2 + 2\beta(\theta_2);$$
$$\gamma_1 = \left[I_{S1} + m_1 l_{AS1}^2\right];\ \gamma_2 = \left[I_{S2} + M_2 l_{BS2}^2\right]$$

with $M_2 = m_2 + \Delta_m$ and $l_{BS2} = \frac{\Delta_m}{m_2+\Delta_m} l_2$, Δ_m is the mass of the payload.

Thus, the introduction of the payload leads to an unbalance of the second link of the manipulator. According to proposed solution mentioned above, the dynamic equations can be rewritten as:

$$\begin{cases} \tau_1 = \left[I_{S1} + m_1 l_{AS1}^2 + m_2 l_1^2\right]\ddot{\theta}_1 + \Delta_m l_1[l_1 + l_2\cos(\theta_2)]\ddot{\theta}_1 \\ \qquad - [\Delta_m l_1 l_2 \sin(\theta_2)]\dot{\theta}_1\dot{\theta}_2 \\ \tau_2 = \Delta_m l_1 l_2\left[\cos(\theta_2)\ddot{\theta}_1 + \sin(\theta_2)\dot{\theta}_1^2\right] \end{cases} \quad (4)$$

It is obvious that, the torques of actuators are great simplified, but they are still not completely decoupled. Compared with Eq. (2), the payload compensation can be obtained by:

$$\begin{cases} \Delta\tau_1 = \Delta_m l_1[l_1 + l_2\cos(\theta_2)]\ddot{\theta}_1 - \Delta_m l_1 l_2 \sin(\theta_2)\dot{\theta}_1\dot{\theta}_2 \\ \Delta\tau_2 = \Delta_m l_1 l_2\left[\cos(\theta_2)\ddot{\theta}_1 + \sin(\theta_2)\dot{\theta}_1^2\right] \end{cases} \quad (5)$$

3 Computer Simulation of Example Study

In order to evaluate the performance of the proposed technique, the simulations with payload compensation and without payload compensation are performed.

First, to demonstrate the influence of the payload compensation, the open-loop control system, which is a non-feedback system, is used.

3.1 Open-Loop Control System

The open-loop control law can be written as:

$$\begin{cases} \tau_1(t) =(a+b)\,\ddot{\theta}_{1R}(t) + a\ddot{\theta}_{2R}(t) + \Delta\tau_{1R} \\ \tau_2(t) =a\ddot{\theta}_{1R}(t) + a\ddot{\theta}_{2R}(t) + \Delta\tau_{2R} \end{cases} \quad (6)$$

where, the payload compensation is given by:

$$\begin{cases} \Delta\tau_{1R} = \Delta_m l_1 [\, l_1 + l_2 \cos(\theta_{2R})\,]\ddot{\theta}_{1R} - \Delta_m l_1 l_2 \sin(\theta_{2R})\,\dot{\theta}_{1R}\dot{\theta}_{2R} \\ \Delta\tau_{2R} = \Delta_m l_1 l_2 \left[\cos(\theta_{2R})\,\ddot{\theta}_{1R} + \sin(\theta_{2R})\dot{\theta}_{1R}^2\right] \end{cases} \quad (7)$$

with

$$\begin{cases} \theta_{1R}(t) =\theta_1^i + \left(\theta_1^f - \theta_1^i\right)\left(\frac{t}{T}\right)^3 \left[10 - 15\left(\frac{t}{T}\right) + 6\left(\frac{t}{T}\right)^2\right] \\ \theta_{2R}(t) = \theta_2^i + \left(\theta_2^f - \theta_2^i\right)\left(\frac{t}{T}\right)^3 \left[10 - 15\left(\frac{t}{T}\right) + 6\left(\frac{t}{T}\right)^2\right] \end{cases} \quad (8)$$

where, T is the amount of time taken to execute the trajectories, $0 \le t \le T$. And the $\dot{\theta}_{1R}, \dot{\theta}_{2R}, \ddot{\theta}_{1R}, \ddot{\theta}_{2R}$ can be obtained by continuous derivation.

The initial and final positions of the manipulator are: $\theta_1^i = 0.576$; $\theta_2^i = 0$ and $\theta_1^f = 1.92$; $\theta_2^f = -1.344$. The trajectories θ_{1R}(t) and $\theta_{2R}(t)$ are functions of time such that $[\theta_{1R}(0)\theta_{2R}(0)]^T = \left[\theta_1^i \quad \theta_2^i\right]^T$ and $[\theta_{1R}(\mathrm{T})\ \theta_{2R}(T)] = \left[\theta_1^f \quad \theta_2^f\right]^T$. Since the trajectories are parameterized by time, we can obtain velocities and accelerations along the trajectories by differentiation.

As $\theta_2^f = \theta_1^i + \theta_2^i - \theta_1^f \Rightarrow \ddot{\theta}_{2R}(t) = -\ddot{\theta}_{1R}(t)\, et\, \dot{\theta}_{2R}(t) = -\dot{\theta}_{1R}(t)$

For $T = 1$ s and $\Delta_m = 1$ kg, the responses with MATLAB software of the manipulator given for illustration in Sect. 3, are presented in Figs. 2 and 3.

The dashed curves show the torques and the angular displacements of the manipulator without payload compensation and the solid curves show the same parameters with payload compensation. It can thus be seen that there are errors between these two cases. With payload compensation, both links of the manipulator can rotate exactly to the target angles. However, without payload compensation, the errors of angular displacements of links 1 and 2 are respectively 39 % and 16.5 %. The effect of feed forward control taking into account the payload is verified. Thus, in the case without payload compensation, a feedback control is needed to reduce the errors.

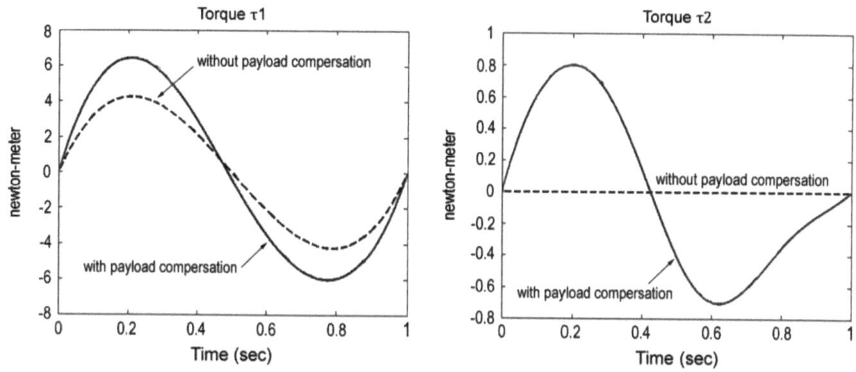

Fig. 2 Torques with payload compensation (*solid line*) and without it (*dashed line*)

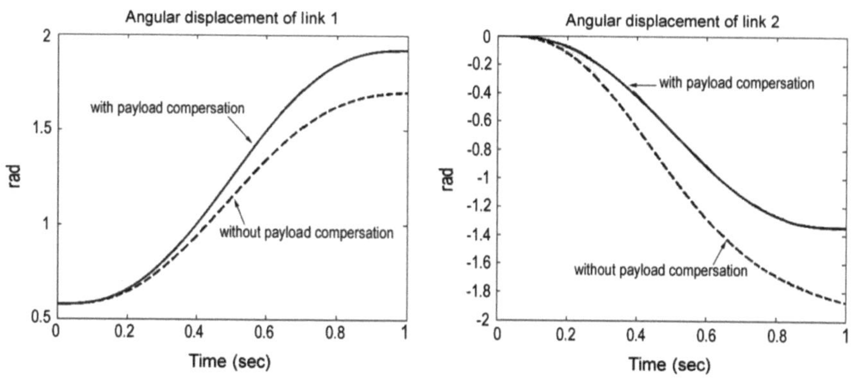

Fig. 3 Angular displacements of links with payload compensation (*solid line*) and without it (*dashed line*) for the open-loop system of equilibrium model

3.2 Closed-Loop Control System

The closed-loop control law can be written as:

$$\left\{\begin{aligned} \tau_1(t) =& (a+b)\,\ddot{\theta}_{1R}(t) + a\ddot{\theta}_{2R}(t) + \Delta\tau_{1R} \\ & - g_{11}\left[(a+b)\left[\dot{\theta}_1(t) - \dot{\theta}_{1R}(t)\right] + a\left[\dot{\theta}_2(t) - \dot{\theta}_{2R}(t)\right]\right] \\ & - g_{12}\left[(a+b)\left[\theta_1(t) - \theta_{1R}(t)\right] + a\left[\theta_2(t) - \theta_{2R}(t)\right]\right] \\ \tau_2(t) =& a\ddot{\theta}_{1R}(t) + a\ddot{\theta}_{2R}(t) + \Delta\tau_{2R} \\ & - g_{21}\left[a\left[\dot{\theta}_1(t) - \dot{\theta}_{1R}(t)\right] + a\left[\dot{\theta}_2(t) - \dot{\theta}_{2R}(t)\right]\right] \\ & - g_{22}\left[a\left[\theta_1(t) - \theta_{1R}(t)\right] + a\left[\theta_2(t) - \theta_{2R}(t)\right]\right] \end{aligned}\right. \quad (9)$$

where, $\Delta\tau_{1R}(\mathrm{t})$ *and* $\Delta\tau_{2R}(\mathrm{t})$ are given by the Eq. (7).

The constant gain elements g_{11}, g_{12}, g_{21} and g_{22} are obtained by an optimal pole-placement design through state feedback. The state space representation of the 2-DOF adjustable serial manipulator, which is statically balanced, will be written as:

$$\begin{bmatrix} \dot{\theta}_1 \\ \ddot{\theta}_1 \\ \dot{\theta}_2 \\ \ddot{\theta}_2 \end{bmatrix} = \begin{bmatrix} 0 & 1 & 0 & 0 \\ 0 & 0 & 0 & 0 \\ 0 & 0 & 0 & 1 \\ 0 & 0 & 0 & 0 \end{bmatrix} \begin{bmatrix} \theta_1 \\ \dot{\theta}_1 \\ \theta_2 \\ \dot{\theta}_2 \end{bmatrix} + \begin{bmatrix} 0 & 0 \\ \frac{1}{b} & -\frac{1}{b} \\ 0 & 0 \\ -\frac{1}{b} & \frac{1}{a}+\frac{1}{b} \end{bmatrix} \begin{bmatrix} \tau_1 \\ \tau_2 \end{bmatrix} \tag{10}$$

The controllable canonical form is given by two independent subsystems as follows:

$$\begin{cases} \begin{bmatrix} (a+b)\,\ddot{\theta}_1 + a\ddot{\theta}_2 \\ (a+b)\,\dot{\theta}_1 + a\dot{\theta}_2 \end{bmatrix} = \begin{bmatrix} 0 & 0 \\ 1 & 0 \end{bmatrix} \begin{bmatrix} (a+b)\,\dot{\theta}_1 + a\dot{\theta}_2 \\ (a+b)\,\theta_1 + a\theta_2 \end{bmatrix} + \begin{bmatrix} 1 \\ 0 \end{bmatrix} \tau_1 \\ \begin{bmatrix} a\ddot{\theta}_1 + a\ddot{\theta}_2 \\ a\dot{\theta}_1 + a\dot{\theta}_2 \end{bmatrix} = \begin{bmatrix} 0 & 0 \\ 1 & 0 \end{bmatrix} \begin{bmatrix} a\dot{\theta}_1 + a\dot{\theta}_2 \\ a\theta_1 + a\theta_2 \end{bmatrix} + \begin{bmatrix} 1 \\ 0 \end{bmatrix} \tau_2 \end{cases} \tag{11}$$

A double integrator $\underbrace{\begin{bmatrix} \ddot{\varphi}(t) \\ \dot{\varphi}(t) \end{bmatrix}}_{\mathrm{x}(\dot{\mathrm{t}})} = \underbrace{\begin{bmatrix} 0 & 0 \\ 1 & 0 \end{bmatrix}}_{A} \underbrace{\begin{bmatrix} \dot{\varphi}(t) \\ \varphi(t) \end{bmatrix}}_{\mathrm{x(t)}} + \underbrace{\begin{bmatrix} 1 \\ 0 \end{bmatrix}}_{B} u(t)$ is completely controllable.

We seek $u(t)$ that minimize the cost $J = \int\limits_0^\infty [\, x^T(t)Qx(t) + u^2(t)\,]\,dt$ where the matrix Q is based on the controllability grammien defined by:

$$G_c(0,T_p) = \int\limits_0^{T_p} \left[e^{At}BB^Te^{A^Tt} \right] dt$$

The matrix $Q = \left[T_pG_c(0,T_p) \right]^{-1}$ is symmetric and positive definite. The parameter $\mathrm{T_p}$ assume that poles of closed-loop system may be placed, in the S plane, at the left of the vertical straight with the abscissa $-\frac{1}{T_p}$.

The linear quadratic controller is unique, optimal $u(t) = -Gx(t)$, full state feedback control law with $G = B^T\Sigma$ that minimizes the cost J.

The matrix Σ is the unique, symmetric, positive definite solution to the algeric Riccati equation: $A^T\Sigma + \Sigma A - \Sigma BB^T\Sigma + Q = 0$. As a result,

$$g_{11} = g_{12} = \frac{2\sqrt{1+\sqrt{3}}}{T_P},\; g_{21} = g_{22} = \frac{2\sqrt{3}}{T_P^2}$$

Then the closed-loop characteristic polynomial is:

$$P(s) = s^2 + \frac{2\sqrt{1+\sqrt{3}}}{T_P}s + \frac{2\sqrt{3}}{T_P^2}$$

If $P(s) = s^2 + 2\zeta\omega_n s + \omega_n^2$, we have: $\omega_n = \frac{\sqrt{2\sqrt{3}}}{T_P}$ and $\zeta = \frac{\sqrt{1+\sqrt{3}}}{\sqrt{2\sqrt{3}}} = 0.89$

For $T = 1$ s, $T_p = 0.2$ s and $\Delta_m = 1$ kg, the responses with MATMAB software are presented at Figs. 4 and 5.

As in the previous case, the dashed curves show the torques and the angular displacements of the manipulator without payload compensation and the solid curves show the same parameters with compensation. Like in previous simulations, the payload compensation allows an exact reproduction of manipulator motions. In

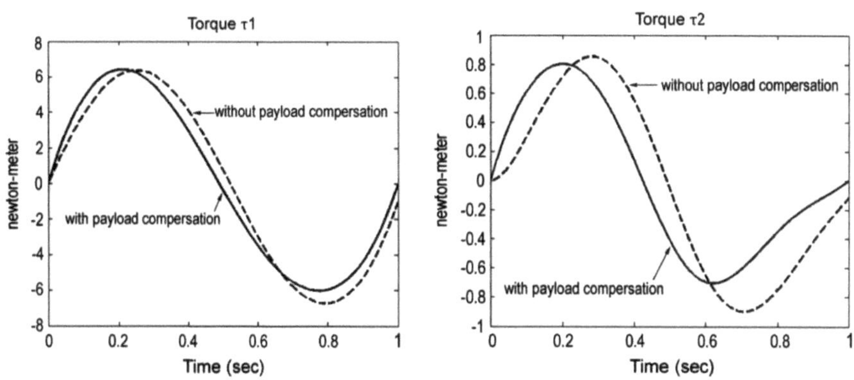

Fig. 4 Torques with payload compensation (*solid line*) and without it (*dashed line*) for the closed-loop system of equilibrium model

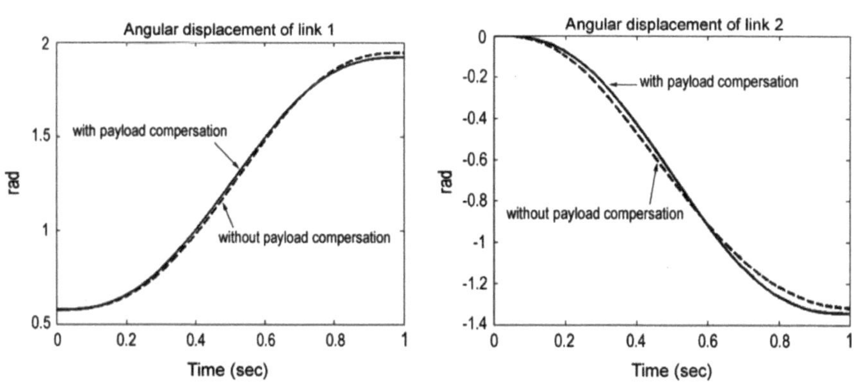

Fig. 5 Angular displacements of links with payload compensation (*solid line*) and without it (*dashed line*) for the closed-loop system of equilibrium model

comparison with these results, the errors of angular displacements of links 1 and 2 without payload compensation (Fig. 4 and 5) are not more than 2 %. Thus, the use of the feedback can further reduce the errors which are caused by eliminating the payload compensation.

4 Conclusions

The serial manipulator with adjustable lengths of links studied in previous work can accomplish the dynamic decoupling via opposite rotation of links and optimal redistribution of masses. However, it is obvious that the introduction of the changing payload leads to the perturbation of the dynamic decoupling of the manipulator. To ensure linearized and decoupled dynamics of the manipulator for changing payload, an optimal control technique is applied.

As a result, the complete dynamic decoupling of the manipulator taking into account the changing payload is accomplished. In order to evaluate the efficiency of the proposed technique, the simulations with payload compensation and without payload compensation are performed with open-loop control system which is a non-feedback system and closed-loop control system. The obtained results, carried out by using MATLAB software, have showed the efficiency of the proposed solution.

Finally, it should be noted that the proposed design concept relative to other solutions permits to create manipulators with decoupled dynamics by a relatively small increase in the total mass of the moving links and it takes into account the changing payload. Such a solution can find a wide application in the design of fixed-sequence manipulators for a precise reproduction of gripper motions.

References

1. Brady, M., Hollerbach, J., Johnson, T., Lozano-Perez, T.: Robot Motion, Planning and Control. MIT Press, Cambridge (1982)
2. Filaretov, V.F., Vukobratovic, M.K.: Static balancing and dynamic decoupling of the motion of manipulation robots. Mechatronics, **3**, 767–782 (1993)
3. Coelho, T., Yong, L., Alves, V.: Decoupling of dynamic equations by means of adaptive balancing of 2-dof open-loop mechanisms. J. Mech. Mach. Theory **39**, 871–881 (2004)
4. Arakelian, V., Sargsyan, S.: On the design of serial manipulators with decoupled dynamics. J. Mechatronics **22**, 904–909 (2012)
5. Arakelian, V., Le Baron, J.-P., Besnard, N., Lardeau, F.: On the design of reconfigurable serial manipulators with decoupled dynamics. In: Proceedings of the 5th International Mechanical Engineering Forum, Prague, pp. 106–114 (2012)
6. Rao, J.S., Dukkipati, R.V.: Mechanism and machine theory, 2nd edn. New Age International Ltd, New Delhi (1992)

Author Index

S. Zeghloul et al. (eds.), *Robotics and Mechatronics*,
Mechanisms and Machine Science 37, DOI 10.1007/978-3-319-22368-1

MIX
Papier aus verantwortungsvollen Quellen
Paper from responsible sources
FSC® C105338
FSC
www.fsc.org

If you have any concerns about our products,
you can contact us on
ProductSafety@springernature.com

In case Publisher is established outside the EU,
the EU authorized representative is:
Springer Nature Customer Service Center GmbH
Europaplatz 3, 69115 Heidelberg, Germany

Printed by Libri Plureos GmbH
in Hamburg, Germany